AF598012

Methods in Molecular Biology

Series Editor
John M. Walker
School of Life and Medical Sciences
University of Hertfordshire
Hatfield, Hertfordshire, AL10 9AB, UK

For further volumes:
http://www.springer.com/series/7651

Peptide Antibodies

Methods and Protocols

Edited by

Gunnar Houen

Statens Serum Institut, Copenhagen, Denmark

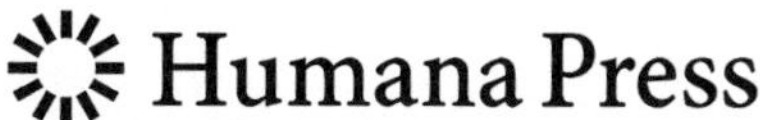

Editor
Gunnar Houen
Statens Serum Institut
Copenhagen, Denmark

ISSN 1064-3745 ISSN 1940-6029 (electronic)
Methods in Molecular Biology
ISBN 978-1-4939-2998-6 ISBN 978-1-4939-2999-3 (eBook)
DOI 10.1007/978-1-4939-2999-3

Library of Congress Control Number: 2015948252

Springer New York Heidelberg Dordrecht London

Cover Illustration: Image provided by Tim Vickers.

Printed on acid-free paper

Humana Press is a brand of Springer
Springer Science+Business Media LLC New York is part of Springer Science+Business Media (www.springer.com)

Preface

The ability to design and produce peptide antibodies has had a large impact on molecular cell biology and immunology, including all the different techniques involved (immunoassays, immunoprecipitation, immunoblotting, immunohistochemistry, etc).

This large impact is a result of several technical and scientific advances: solid phase peptide synthesis, peptide carrier conjugation and immunization, genomics, transcriptomics, proteomics and elucidation of the molecular basis of antigen presentation and recognition by dendritic cells, macrophages, B cells, and T cells.

Moreover, although peptide antibodies have been available for many years, they continue to be a field of active research and method development. For example, peptide antibodies which are dependent on specific posttranslational modifications are of great interest (phosphorylation, citrullination, etc.) and different forms of recombinant peptide antibodies are gaining interest (nanobodies, single chain antibodies, TCR-like antibodies, etc.).

This volume covers basic and advanced aspects of peptide antibody production, characterization, and uses.

I thank all contributors and editorial staff for their work, especially the series editor, John Walker. Also, I want to thank all my collaborators and students throughout the years.

Copenhagen, Denmark *Gunnar Houen*

Contents

Contributors

ROSITA ACCARDI • *Infections and Cancer Biology Group, International Agency for Research on Cancer-World Health Organization, Lyon, France*

NORAH A. ALTURKI • *Human Health Therapeutics Portfolio, National Research Council Canada, Ottawa, ON, Canada; College of Applied Medical Sciences, King Saud University, Riyadh, Saudi Arabia*

HEIDI GERTZ ANDERSEN • *Innate Immunology Group, Section for Immunology and Vaccinology, National Veterinary Institute, Technical University of Denmark, Frederiksberg, Denmark*

MEHDI ARBABI-GHAHROUDI • *Human Health Therapeutics Portfolio, National Research Council Canada, Ottawa, ON, Canada; School of Environmental Sciences, University of Guelph, Guelph, ON, Canada; Department of Biology, Carleton University, Ottawa, ON, Canada*

ANN-LOUISE BERGSTRÖM • *Innate Immunology Group, Section for Immunology and Vaccinology, National Veterinary Institute, Technical University of Denmark, Frederiksberg, Denmark; Department Neurodegeneration, H. Lundbeck A/S, Valby, Denmark*

YVONNE BEITER • *Natural and Medical Sciences Institute at the University of Tuebingen, Reutlingen, Germany*

ANGELO BOLCHI • *Biochemistry and Molecular Biology Unit, Department of Life Sciences, University of Parma, Parma, Italy*

SØREN BUUS • *Laboratory of Experimental Immunology, Faculty of Health Sciences, Department of Immunology and Microbiology, University of Copenhagen, Copenhagen, Denmark*

ELENA CANALI • *Biochemistry and Molecular Biology Unit, Department of Life Sciences, University of Parma, Parma, Italy*

ANNA CHAILYAN • *Department of Biochemistry and Molecular Biology, University of Southern Denmark, Odense, Denmark*

NAI-KONG V. CHEUNG • *Gerstner Sloan Kettering Graduate School of Biomedical Sciences and Department of Pediatrics, Memorial Sloan-Kettering Cancer Center, New York, NY, USA*

HUI NA CHUA • *Sunway University, Kuala Lumpur, Malaysia*

GIORGIO COLOMBO • *Department of Computational Biology, Institute for Molecular Recognition Chemistry (ICRM), Italian National Research Council, Milan, Italy*

HENRIETTE CORDES • *Innate Immunology Group, Section for Immunology and Vaccinology, National Veterinary Institute, Technical University of Denmark, Frederiksberg, Denmark; Novo Nordisk A/S, Bagsværd, Denmark*

DARIO CORRADA • *Institute for Molecular Recognition Chemistry (ICRM), Italian National Research Council, Milan, Italy; Department of Earth and Environmental Sciences, University of Milano-Bicocca, Milan, Italy*

XAVIER DAURA • *Institut de Biotecnologia i de Biomedicina (IBB), Universitat Autònoma de Barcelona (UAB), Barcelona, Spain; Catalan Institution for Research and Advanced Studies (ICREA), Barcelona, Spain*

ANGELA FILOMENA • *Natural and Medical Sciences Institute at the University of Tuebingen, Reutlingen, Germany*
ANDERS FOMSGAARD • *Virus R&D Laboratory, Department of Microbiology Diagnostics and Virology, Statens Serum Institut, Copenhagen, Denmark; Infectious Disease Research Unit, Clinical Institute, University of Southern Denmark, Odense, Denmark*
TINA FRIIS • *Department of Autoimmunology and Biomarkers, Statens Serum Institut, Copenhagen, Denmark*
ANKUR GAUTAM • *Bioinformatics Centre, CSIR-Institute of Microbial Technology, Chandigarh, India*
ALESSANDRO GORI • *Department of Computational Biology, Institute for Molecular Recognition Chemistry (ICRM), Italian National Research Council, Milan, Italy*
LARA GROLLO • *Swinburne University, Melbourne, Australia*
SUDHEER GUPTA • *Bioinformatics Centre, CSIR-Institute of Microbial technology, Chandigarh, India*
PAUL R. HANSEN • *Department of Drug Design and Pharmacology, Faculty of Health and Medical Sciences, University of Copenhagen, Copenhagen, Denmark*
PETER M.H. HEEGAARD • *Innate Immunology Group, Section for Immunology and Vaccinology, National Veterinary Institute, Technical University of Denmark, Frederiksberg, Denmark*
KEVIN A. HENRY • *Human Health Therapeutics Portfolio, National Research Council Canada, Ottawa, ON, Canada*
KARIN HJERNØ • *Department of Biochemistry and Molecular Biology, University of Southern Denmark, Odense, Denmark*
PETER HØJRUP • *Department of Biochemistry and Molecular Biology, University of Southern Denmark, Odense, Denmark*
GUNNAR HOUEN • *Department of Autoimmunology and Biomarkers, Statens Serum Institut, Copenhagen, Denmark*
DAVID HOUGAARD • *Department of Congenital Diseases, Statens Serum Institut, Copenhagen, Denmark*
KRISTIN E. ILLIGEN • *Department of Quality Control, Statens Serum Institut, Copenhagen, Denmark*
ALBERT JELTSCH • *Institute of Biochemistry, Faculty of Chemistry, University Stuttgart, Stuttgart, Germany*
THOMAS O. JOOS • *Natural and Medical Sciences Institute at the University of Tuebingen, Reutlingen, Germany*
MASATOSHI KATAOKA • *Biomarker Analysis Research Group, Health Research Institute, National Institute of Advanced Industrial Science and Technology (AIST), Takamatsu, Japan*
DONGBUM KIM • *Department of Microbiology, College of Medicine, Hallym University, Gangwon-do, Republic of Korea*
KATHERINE KIRK • *Swinburne University, Melbourne, Australia*
GORAN KUNGULOVSKI • *Institute of Biochemistry, Faculty of Chemistry, University Stuttgart, Stuttgart, Germany*
HYUNG-JOO KWON • *Department of Microbiology and Center for Medical Science Research, College of Medicine, Hallym University, Gangwon-do, Republic of Korea*
INA KYCIA • *The Jackson Laboratory for Genomic Medicine, Farmington, CT, USA*
YOUNGHEE LEE • *Department of Biochemistry, College of Natural Sciences, Chungbuk National University, Chungbuk, Republic of Korea*

HONG LIU • *Eureka Therapeutics, Emeryville, CA, USA*
CHENG LIU • *Eureka Therapeutics, Emeryville, CA, USA*
HUI MA • *School of Biotechnology and Biomedical Diagnostics Institute, Dublin City University, Dublin, Ireland*
C. ROGER MACKENZIE • *Human Health Therapeutics Portfolio, National Research Council Canada, Ottawa, ON, Canada; School of Environmental Sciences, University of Guelph, Guelph, ON, Canada*
PAOLO MARCATILI • *Department of Systems Biology, Center for Biological Sequence Analysis, Technical University of Denmark, Lyngby, Denmark*
TAISUKE MATSUO • *Institute for Genome Research and Faculty of Pharmaceutical Sciences, University of Tokushima, Tokushima, Japan*
REBEKKA MAUSER • *Institute of Biochemistry, Faculty of Chemistry, University Stuttgart, Stuttgart, Germany*
PAOLA MIGLIORINI • *Clinical Immunology and Allergy Unit, Department of Clinical and Experimental Medicine, University of Pisa, Pisa, Italy*
OSMAN MIRZA • *Department of Drug Design and Pharmacology, Faculty of Health and Medical Sciences, University of Copenhagen, Copenhagen, Denmark*
ANNE MORTENSEN • *Department of Autoimmunology and Biomarkers, Statens Serum Institut, Copenhagen, Denmark*
MARTIN MÜLLER • *German Cancer Research Center, Heidelberg, Germany*
MORTEN NIELSEN • *Department of Systems Biology, Center for Biological Sequence Analysis, Technical University of Denmark, Lyngby, Denmark; Instituto de Investigaciones Biotecnológicas, Universidad Nacional de San Martín, Buenos Aires, Argentina*
RICHARD O'KENNEDY • *School of Biotechnology and Biomedical Diagnostics Institute, Dublin City University, Dublin, Ireland*
ALBERTO ODDO • *Department of Drug Design and Pharmacology, Faculty of Health and Medical Sciences, University of Copenhagen, Copenhagen, Denmark*
KAZUTO OHKURA • *Faculty of Pharmaceutical Science, Suzuka University of Medical Science, Suzuka, Japan*
DORTHE T. OLSEN • *Department of Autoimmunology and Biomarkers, Statens Serum Institut, Copenhagen, Denmark*
ANJA OLSEN • *Department of Infectious Disease Immunology, Statens Serum Institut, Copenhagen, Denmark*
THOMAS ØSTERBYE • *Laboratory of Experimental Immunology, Faculty of Health Sciences, Department of Immunology and Microbiology, University of Copenhagen, Copenhagen, Denmark*
SIMONE OTTONELLO • *Biochemistry and Molecular Biology Unit, Department of Life Sciences, University of Parma, Parma, Italy; Dipartimento di Bioscienze, Università di Parma, Parma, Italy*
FILOMENA PANZA • *Clinical Immunology and Allergy Unit, Department of Clinical and Experimental Medicine, University of Pisa, Pisa, Italy*
KLAUS BOBERG PEDERSEN • *Department of Autoimmunology and Biomarkers, Statens Serum Institut, Copenhagen, Denmark*
CLAUDIO PERI • *Department of Computational Biology, Institute for Molecular Recognition Chemistry (ICRM), Italian National Research Council, Milan, Italy*
TINA H. PIHL • *Department of Large Animal Sciences, Medicine, and Surgery, Faculty of Health and Medical Sciences, University of Copenhagen, Copenhagen, Denmark*

OLIVER POETZ • *Natural and Medical Sciences Institute at the University of Tuebingen, Reutlingen, Germany*
CHIT LAA POH • *Sunway University, Kuala Lumpur, Malaysia*
BALA K. PRABHALA • *Department of Drug Design and Pharmacology, Faculty of Health and Medical Sciences, University of Copenhagen, Copenhagen, Denmark*
FEDERICO PRATESI • *Clinical Immunology and Allergy Unit, Department of Clinical and Experimental Medicine, University of Pisa, Pisa, Italy*
GAJENDRA P.S. RAGHAVA • *Bioinformatics Centre, CSIR-Institute of Microbial Technology, Chandigarh, India*
FELICIANA REAL-FERNÁNDEZ • *Laboratory of Peptide and Protein Chemistry and Biology, Division of Pharmaceutical Sciences and Nutraceutic, Department of NeuroFarBa, University of Florence, Sesto Fiorentino, Italy*
IDA ROSENKRANDS • *Department of Infectious Disease Immunology, Statens Serum Institut, Copenhagen, Denmark*
GIADA ROSSI • *Laboratory of Peptide and Protein Chemistry and Biology, Division of Pharmaceutical Sciences and Nutraceutic, Department of NeuroFarBa, University of Florence, Sesto Fiorentino, Italy*
PAOLO ROVERO • *Laboratory of Peptide and Protein Chemistry and Biology, Division of Pharmaceutical Sciences and Nutraceutic, Department of NeuroFarBa, University of Florence, Sesto Fiorentino, Italy*
BRIAN H. SANTICH • *Gerstner Sloan Kettering Graduate School of Biomedical Sciences, Memorial Sloan-Kettering Cancer Center, New York, NY, USA*
ANDREA SANTONI • *Biochemistry and Molecular Biology Unit, Department of Life Sciences, University of Parma, Parma, Italy*
ANNETTE SCHIOLBORG • *Department of Autoimmunology and Biomarkers, Statens Serum Institut, Copenhagen, Denmark*
NICOLE SCHNEIDERHAN-MARRA • *Natural and Medical Sciences Institute at the University of Tuebingen, Reutlingen, Germany*
YASUO SHINOHARA • *Institute for Genome Research and Faculty of Pharmaceutical Sciences, University of Tokushima, Tokushima, Japan*
HARINDER SINGH • *Bioinformatics Centre, CSIR-Institute of Microbial Technology, Chandigarh, India*
OSCAR C. SOLÉ • *Institut de Biotecnologia i de Biomedicina (IBB), Universitat Autònoma de Barcelona (UAB), Barcelona, Spain*
GLORIA SPAGNOLI • *Biochemistry and Molecular Biology Unit, Department of Life Sciences, University of Parma, Parma, Italy*
MARKUS F. TEMPLIN • *Natural and Medical Sciences Institute at the University of Tuebingen, Reutlingen, Germany*
MASSIMO TOMMASINO • *Infections and Cancer Biology Group, International Agency for Research on Cancer-World Health Organization, Lyon, France*
NICOLE HARTWIG TRIER • *Department of Autoimmunology and Biomarkers, Statens Serum Institut, Copenhagen, Denmark*
DANIELE VIARISIO • *German Cancer Research Center, Heidelberg, Germany*
RYOHEI YAMAGOSHI • *Institute for Genome Research and Faculty of Pharmaceutical Sciences, University of Tokushima, Tokushima, Japan*
ATSUSHI YAMAMOTO • *Faculty of Pharmaceutical Science, Suzuka University of Medical Science, Suzuka, Japan*
TAKENORI YAMAMOTO • *Institute for Genome Research and Faculty of Pharmaceutical Sciences, University of Tokushima, Tokushima, Japan*

Chapter 1

Peptide Antibodies: Past, Present, and Future

Gunnar Houen

Abstract

Peptide antibodies recognize epitopes with amino acid residues adjacent in sequence ("linear" epitopes). Such antibodies can be made to virtually any sequence and have been immensely important in all areas of molecular biology and diagnostics due to their versatility and to the rapid growth in protein sequence information. Today, peptide antibodies can be routinely and rapidly made to large numbers of peptides, including peptides with posttranslationally modified residues, and are used for immunoblotting, immunocytochemistry, immunohistochemistry, and immunoassays. In the future, peptide antibodies will continue to be immensely important for molecular biology, TCR- and MHC-like peptide antibodies may be produced routinely, peptide antibodies with predetermined conformational specificities may be designed, and peptide-based vaccines may become part of vaccination programs.

Key words Peptides, Antibodies, Epitopes, Three-dimensional, Linear, Continuous, Contact residues, Recombinant, Single chain, TCR, MHC

1 The Development of Peptide Antibodies

Early work on protein structure and epitopes for antibodies (Abs) revealed that most epitopes were three dimensional (*see* **Notes 1** and **2**) and that a small percentage of Abs reacted with linear (continuous epitopes) [1–7]. Peptide Abs (Fig. 1) were described in 1980 and the use of synthetic peptides (coupled to a carrier protein) to induce specific Abs was developed in the following decades together with methods for epitope mapping and a general understanding of immunogenicity and antigenicity (Tables 1 and 2) (*see* **Notes 3–5**). This development was facilitated by the introduction of solid-phase peptide synthesis [41–43], the understanding of immunological T cell help for efficient stimulation of B cells to produce Abs [44–48], and the rapid growth in DNA and protein sequence information (Table 2).

Gunnar Houen (ed.), *Peptide Antibodies: Methods and Protocols*, Methods in Molecular Biology, vol. 1348, DOI 10.1007/978-1-4939-2999-3_1, © Springer Science+Business Media New York 2015

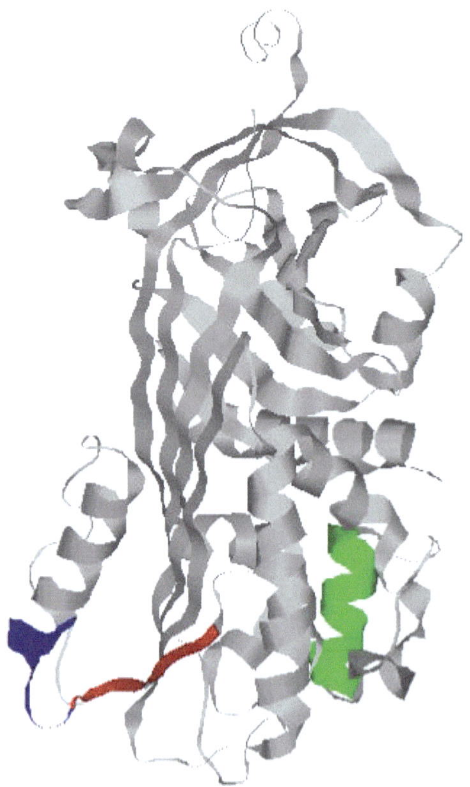

Fig. 1 Examples of peptide epitopes in ovalbumin [52]. Three different linear/continuous epitopes for monoclonal epitopes are marked in *blue*, *red*, and *green* respectively

2 Current Status of Peptide Antibodies

Currently, peptide synthesis, conjugation, and immunization protocols have been optimized and the applications of peptide Abs have expanded to include a variety of immunoassays (e.g., sandwich assays), immunoprecipitation, immunoblotting, immunocytochemistry, and immunohistochemistry (Table 1). Moreover, posttranslational modification-specific Abs (e.g., phosphorylation and citrullination), cleavage site-specific Abs (e.g., amyloid beta 1–40/1–42), tag-specific Abs (e.g., hexa-histidine, FLAG, myc), and conformation-dependent Abs (Tables 1 and 2, [49–51]) are available for the different applications. Methods for epitope prediction have been refined but must always be verified by experimental results and compared with available structural data (*see* **Note 6**).

Table 1
History, status, and future developments of peptide antibodies

A. History of peptide antibodies (selected publications)

Target[a]	Immunogen (residues (n))	References
MMLV putative protein	C-terminal pentadecapeptide	[8]
SV40 large T	N-terminal heptapeptide, C-terminal undecapeptide	[9]
HBV sAg	Several peptides (5–34)	[10]
FMDV VP1	Several peptides (15–40)	[11]
FMDV VP1	Hexadecapeptide	[12]
TCR	Branched lysine constructs	[13, 14]
RB	C-terminal decapeptide synthesised on carrier protein	[15]
CSFV E2	Dendrimeric peptide construct	[16]

B. Current applications of peptide antibodies (i.e., methods used for detection of proteins and studies of protein modification and processing)

Application	References
Immunoblotting	[17, 18]
Immunoassays (direct, sandwich, etc.)	[19, 20]
Immunocytochemistry and histochemistry	[21, 22]
Flow cytometry	[23, 24]
Immunoprecipitation	[25, 26]

C. Future developments of peptide antibodies

Application	References
TCR-like Abs	[27, 28]
Therapeutic peptide Abs/vaccines	[29, 30]
Predesigned, conformation-specific peptide Abs	[31, 32]
MHC-like Abs	?

[a] *CSFV* Classical swine fever virus, *FMDV* Foot and mouth disease virus, *HBV* Hepatitis B virus, *MMLV* Moloney murine leukemia virus, *RB* Retinoblastoma protein, *SV* Simian virus, *TCR* T cell receptor

Table 2
Peptide antibody reviews and resources

A. Review and handbooks

Subject	References
Peptide Abs	[33]
Peptide vaccines	[34]
FMDV vaccines and peptide Abs	[35]
Peptide Ab immunoassays	[36]
Peptide Ab laboratory techniques	[37]
Peptide antigenicity and immunogenicity	[38]
Peptide-based autoimmune serology	[39]
Posttranslational modification-specific peptide Abs	[40]

B. Websites

Epitope database	www.iedb.org
Epitope prediction	www.cbs.dtu.dk
Human protein atlas	www.proteinatlas.org
NCBI	www.ncbi.nlm.nih.gov
Uniprot/Swissprot	www.expasy.org
Protein database	www.pdb.org

3 Future Developments of Peptide Antibodies

Despite the achievements described above, the potential for peptide Abs has not been exhausted and many new uses have been recently established, are under development, or have been suggested (Table 1) including recombinant peptide Abs, single-chain peptide Abs, TCR-like Abs, predesigned conformation-dependent peptide Abs, and therapeutic peptide Abs. One of the original goals of peptide Abs, the development of clinical useful peptide vaccines, is getting closer to realization but still has to make it into clinical everyday use. MHC-like Abs, i.e., Abs, where the antibody-peptide complex mimics an MHC molecule, would be a desirable, although challenging goal.

4 Notes

1. All epitopes are three dimensional, but this term is here restricted to epitopes containing parts of a polypeptide chain not directly continuous in sequence. Thus, three-dimensional epitopes depend on a folded, native structure of the antigen (Ag). Three-dimensional epitopes may also be denoted "composite" epitopes.
2. Peptide epitopes are usually denoted "linear" or "continuous" epitopes but may also be denoted "simple" epitopes and are smaller than 20 residues continuous in sequence. The difference between peptides and polypeptides is not well defined but lies somewhere between 20 and 30 residues. All proteins are polypeptides, but this term is usually confined to polypeptides larger than 100 residues.
3. Epitope mapping: Mapping of amino acid residues with direct influence on Ab binding (e.g., by peptide scanning, X-ray crystallography, or NMR spectroscopy). Epitope residues may be contact residues or structural (conformational residues) or may contribute through backbone amide bonds.
4. Peptides may be antigenic (i.e., react with Ab (defined by Kd)) but not immunogenic (i.e., incapable of inducing an immune response (i.e., specific Abs and/or T cells)) [38]. The Ab response is quantified by the titers of a serum (defined by midpoint or endpoint titration) and by the average antigenicity of the Abs (Kd).
5. Contact residues: Amino acid residues directly interacting with the epitope or paratope (site on Ab interacting with epitope) as determined by X-ray crystallography and/or NMR spectroscopy of Ag-Ab complexes. The contact may take place between side chains or through backbone amide bonds.
6. See relevant chapters in this volume or see [53–55] for recent reviews.

References

1. Sela M, Schechter B, Schechter I, Borek F (1967) Antibodies to sequential and conformational determinants. Cold Spring Harbor Symp Quant Biol 32:537–545
2. Amit AG, Mariuzza RA, Phillips SE, Poljak RJ (1986) Three-dimensional structure of an antigen-antibody complex at 2.8 A resolution. Science 233:747–753
3. Colman PM, Laver WG, Varghese JN, Baker AT, Tulloch PA, Air GM, Webster RG (1987) Three-dimensional structure of a complex of antibody with influenza virus neuraminidase. Nature 326:358–363
4. Sheriff S, Silverton EW, Padlan EA, Cohen GH, Smith-Gill SJ, Finzel BC, Davies DR (1987) Three-dimensional structure of an antibody-antigen complex. Proc Natl Acad Sci U S A 84:8075–8079
5. Mariuzza RA, Phillips SE, Poljak RJ (1987) The structural basis of antigen-antibody recognition. Annu Rev Biophys Biophys Chem 16:139–159
6. Colman PM, Tulip WR, Varghese JN, Tulloch PA, Baker AT, Laver WG, Air GM, Webster RG (1989) Three-dimensional structures of influenza virus neuraminidase-antibody complexes. Philos Trans R Soc Lond B Biol Sci 323:511–518
7. Scherf T, Hiller R, Naider F, Levitt M, Anglister J (1992) Induced peptide conformations in different antibody complexes: molecular modeling of the three-dimensional structure of peptide-antibody complexes using NMR-derived distance restraints. Biochemistry 31:6884–6897
8. Sutcliffe JG, Shinnick TM, Green N, Liu FT, Niman HL, Lerner RA (1980) Chemical synthesis of a polypeptide predicted from nucleotide sequence allows detection of a new retroviral gene product. Nature 287: 801–805
9. Walter G, Scheidtmann KH, Carbone A, Laudano AP, Doolittle RF (1980) Antibodies specific for the carboxy- and amino-terminal regions of simian virus 40 large tumor antigen. Proc Natl Acad Sci U S A 77:5197–5200
10. Lerner RA, Green N, Alexander H, Liu FT, Sutcliffe JG, Shinnick TM (1981) Chemically synthesized peptides predicted from the nucleotide sequence of the hepatitis B virus genome elicit antibodies reactive with the native envelope protein of Dane particles. Proc Natl Acad Sci U S A 78:3403–3407
11. Bittle JL, Houghten RA, Alexander H, Shinnick TM, Sutcliffe JG, Lerner RA, Rowlands DJ, Brown F (1982) Protection against foot-and-mouth disease by immunization with a chemically synthesized peptide predicted from the viral nucleotide sequence. Nature 298:30–33
12. Pfaff E, Mussgay M, Böhm HO, Schulz GE, Schaller H (1982) Antibodies against a preselected peptide recognize and neutralize foot and mouth disease virus. EMBO J 1:869–874
13. Posnett DN, McGrath H, Tam JP (1988) A novel method for producing anti-peptide antibodies. Production of site-specific antibodies to the T cell antigen receptor beta-chain. J Biol Chem 263:1719–1725
14. Tam JP (1988) Synthetic peptide vaccine design: synthesis and properties of a high-density multiple antigenic peptide system. Proc Natl Acad Sci U S A 85:5409–5413
15. Hansen PR, Holm A, Houen G (1993) Solid-phase peptide synthesis on proteins. Int J Pept Protein Res 41:237–245
16. Li GX, Zhou YJ, Yu H, Li L, Wang YX, Tong W, Hou JW, Xu YZ, Zhu JP, Xu AT, Tong GZ (2012) A novel dendrimeric peptide induces high level neutralizing antibodies against classical swine fever virus in rabbits. Vet Microbiol 156:200–204
17. Petrasovits LA (2014) Protein blotting protocol for beginners. Methods Mol Biol 1099:189–199
18. Kurien BT, Dorri Y, Dillon S, Dsouza A, Scofield RH (2011) An overview of Western blotting for determining antibody specificities for immunohistochemistry. Methods Mol Biol 717:55–67
19. Wheeler MJ (2013) Immunoassay techniques. Methods Mol Biol 1065:7–25
20. Wild D (ed) (2013) The immunoassay handbook. Elsevier, Oxford
21. Brooks SA (2012) Basic immunocytochemistry for light microscopy. Methods Mol Biol 878:1–30
22. Ramos-Vara JA (2011) Principles and methods of immunohistochemistry. Methods Mol Biol 691:83–96
23. Davies D (2012) Cell separations by flow cytometry. Methods Mol Biol 878:185–199
24. Givan AL (2011) Flow cytometry: an introduction. Methods Mol Biol 699:1–29
25. Isono E, Schwechheimer C (2010) Co-immunoprecipitation and protein blots. Methods Mol Biol 655:377–387
26. Uljon SN, Mazzarelli L, Chait BT, Wang R (2000) Analysis of proteins and peptides directly from biological fluids by immunoprecipitation/mass spectrometry. Methods Mol Biol 146:439–452

27. Dahan R, Reiter Y (2012) T-cell-receptor-like antibodies—generation, function and applications. Expert Rev Mol Med. doi:10.1017/erm.2012.2
28. Neumann F, Sturm C, Hülsmeyer M, Dauth N, Guillaume P, Luescher IF, Pfreundschuh M, Held G (2009) Fab antibodies capable of blocking T cells by competitive binding have the identical specificity but a higher affinity to the MHC-peptide-complex than the T cell receptor. Immunol Lett 125:86–92
29. Naz RK, Dabir P (2007) Peptide vaccines against cancer, infectious diseases, and conception. Front Biosci 12:1833–1844
30. Yamada A, Sasada T, Noguchi M, Itoh K (2013) Next-generation peptide vaccines for advanced cancer. Cancer Sci 104:15–21
31. Paduch M, Koide A, Uysal S, Rizk SS, Koide S, Kossiakoff AA (2013) Generating conformation-specific synthetic antibodies to trap proteins in selected functional states. Methods 60:3–14
32. Lu SM, Hodges RS (2002) A de novo designed template for generating conformation-specific antibodies that recognize alpha-helices in proteins. J Biol Chem 277:23515–23524
33. Sutcliffe JG, Shinnick TM, Green N, Lerner RA (1983) Antibodies that react with predetermined sites on proteins. Science 219: 660–666
34. Shinnick TM, Sutcliffe JG, Green N, Lerner RA (1983) Synthetic peptide immunogens as vaccines. Annu Rev Microbiol 37:425–446
35. Brown F (1988) Use of peptides for immunization against foot-and-mouth disease. Vaccine 6:180–182
36. Van Regenmortel MH (1993) Synthetic peptides versus natural antigens in immunoassays. Ann Biol Clin (Paris) 51:39–41
37. Van Regenmortel MH, Briand JP, Muller S, Plaue S (Eds) (1988) Synthetic polypeptides as antigens. Laboratory techniques in biochemistry and molecular biology vol 19. Elsevier: Amsterdam
38. Van Regenmortel MH (2001) Antigenicity and immunogenicity of synthetic peptides. Biologicals 29:209–213
39. Fournel S, Muller S (2003) Synthetic peptides in the diagnosis of systemic autoimmune diseases. Curr Protein Pept Sci 4:261–274
40. Papini AM (2009) The use of post-translationally modified peptides for detection of biomarkers of immune-mediated diseases. J Pept Sci 15:621–628
41. Merrifield RB (1963) Solid phase peptide synthesis. I The synthesis of a tetrapeptide. J Am Chem Soc 85:2149–2154
42. Merrifield RB (1969) Solid-phase peptide synthesis. Adv Enzymol Relat Areas Mol Biol 32:221–296
43. Atherton E, Sheppard RC (1989) Solid Phase peptide synthesis: a practical approach. IRL Press, Oxford, England. ISBN 0-19-963067-4
44. Braciale TJ, Morrison LA, Sweetser MT, Sambrook J, Gething MJ, Braciale VL (1987) Antigen presentation pathways to class I and class II MHC-restricted T lymphocytes. Immunol Rev 98:95–114
45. Parker DC (1993) T cell-dependent B cell activation. Annu Rev Immunol 11:331–360
46. Fairchild PJ (1998) Presentation of antigenic peptides by products of the major histocompatibility complex. J Pept Sci 4:182–194
47. Appella E, Padlan EA, Hunt DF (1995) Analysis of the structure of naturally processed peptides bound by class I and class II major histocompatibility complex molecules. EXS 73:105–119
48. Maffei A, Harris PE (1998) Peptides bound to major histocompatibility complex molecules. Peptides 19:179–198
49. Blaydes JP, Vojtesek B, Bloomberg GB, Hupp TR (2000) The development and use of phospho-specific antibodies to study protein phosphorylation. Methods Mol Biol 99: 177–189
50. Miller DL, Potempska A, Wegiel J, Mehta PD (2011) High-affinity rabbit monoclonal antibodies specific for amyloid peptides amyloid-β40 and amyloid-β42. J Alzheimers Dis 23: 293–305
51. Terpe K (2003) Overview of tag protein fusions: from molecular and biochemical fundamentals to commercial systems. Appl Microbiol Biotechnol 60:523–533
52. Holm BE, Bergmann AC, Hansen PR, Koch C, Houen G, Trier NH (2014) Antibodies with specificity for native and denatured forms of ovalbumin differ in reactivity between enzyme-linked immunosorbent assays. APMIS. doi:10.1111/apm.12329
53. Soria-Guerra RE, Nieto-Gomez R, Govea-Alonso DO, Rosales-Mendoza S (2014) An overview of bioinformatics tools for epitope prediction: implications on vaccine development. J Biomed Inform S1532–0464(14): 00233. doi:10.1016/j.jbi.2014.11.003
54. Ansari HR, Raghava GP (2013) In silico models for B-cell epitope recognition and signaling. Methods Mol Biol 993:129–138
55. Ponomarenko JV, van Regenmortel MHV (2009) B-cell epitope prediction. In: Bourne PE, Gu J (eds) Structural bioinformatics. Wiley, New York, NY, pp 849–879

Chapter 2

The Structure of Natural and Recombinant Antibodies

Hui Ma and Richard O'Kennedy

Abstract

Immunoglobulins (Ig) isotypes A, D, E, G, and M are glycoproteins which are mainly composed of a "Y"-shaped Ig monomer (~150 kDa), consisting of two light and two heavy chains. Both light and heavy chains contain variable (N-terminal) and constant regions (C-terminal). Each light chain consists of one variable domain and one constant domain, whereas each heavy chain has one variable domain and three constant domains. However, heavy-chain antibodies consisting of only heavy chains and lacking the light chains are found in camelids and cartilaginous fishes. Unlike other immunoglobulins, the heavy chain of avian antibody IgY (~180 kDa) consists of four constant domains. The single-chain variable fragment (scFv; ~25 kDa) of an antibody contains variable regions of antibody heavy and light chains. The fragment antigen-binding (Fab; ~50 kDa) region has the full antibody light chain but the heavy chain is composed of a variable region and one constant domain.

Key words IgG, IgY, ScFv, Fab

1 Structure of Immunoglobulins

Antibodies, also known as immunoglobulins (Igs), enable antigen recognition in the serum. They are produced by B-cell-derived plasma cells. Antibodies are mainly located in blood, spleen, bone marrow, egg yolk for birds, as well as interstitial fluids and exocrine secretions. Antibodies can be effectively used by the immune system to identify, kill, or neutralize invading bacteria, parasites, toxins, and viruses and to destroy other foreign compounds [1].

Mammalian immunoglobulins are classified into five isotypes, namely IgM, IgD, IgG, IgE, and IgA. The synonymous "Y" shape associated with a basic immunoglobulin unit (Ig) monomer (or subunit) (~150 kDa) consists of two light and two heavy chains, which are connected by disulfide bonds (*see* Fig. 1a) [2]. Each light chain has two regions composed of one variable region (V_L) and one constant region (C_L), whereas each heavy chain contains one variable domain (V_H) and three constant domains (C_H1–3). All of the antibodies perform specific binding to defined antigens through

Gunnar Houen (ed.), *Peptide Antibodies: Methods and Protocols*, Methods in Molecular Biology, vol. 1348,
DOI 10.1007/978-1-4939-2999-3_2, © Springer Science+Business Media New York 2015

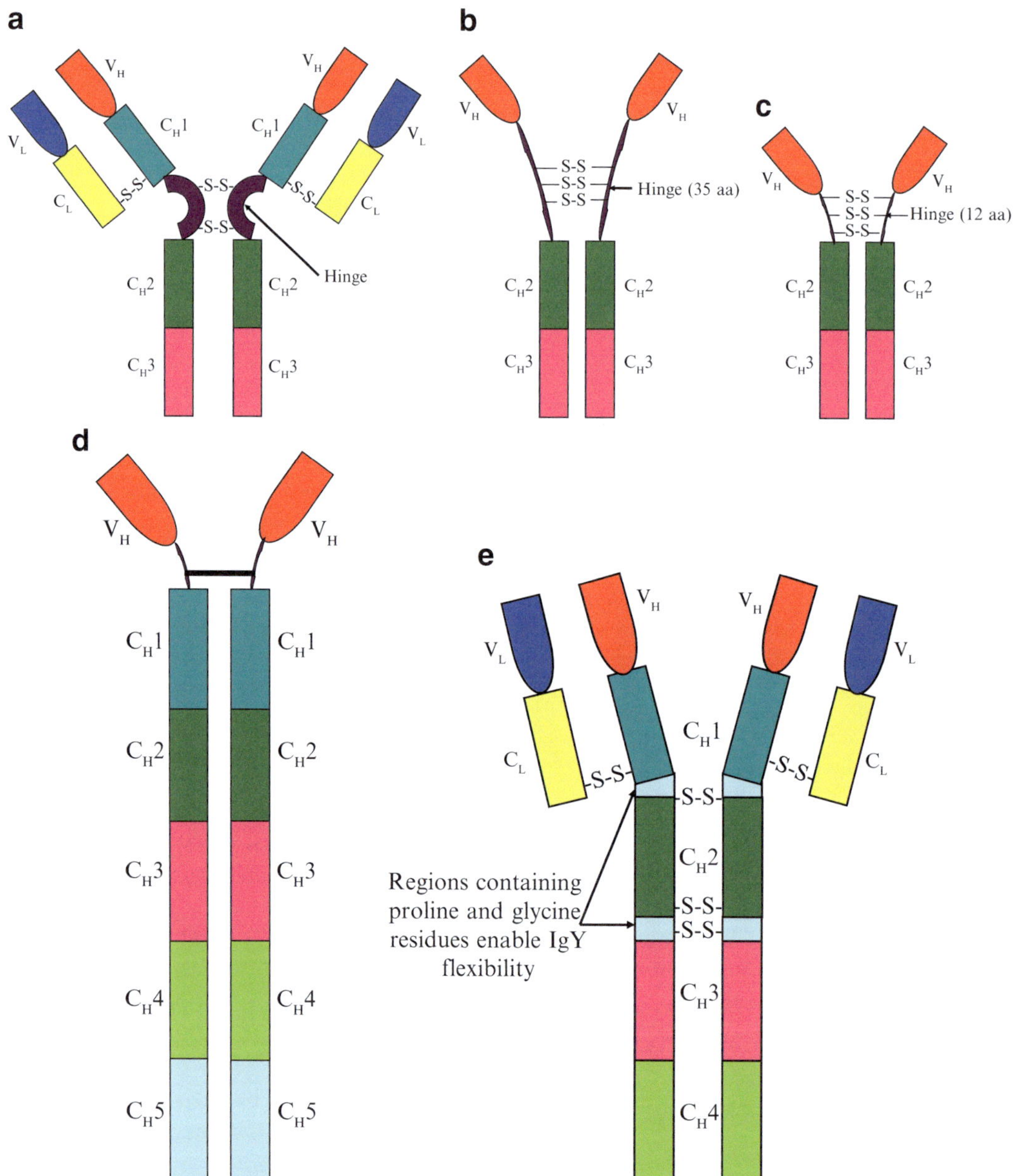

Fig. 1 Structure of basic immunoglobulins. (**a**) Shows structure of a basic immunoglobulin monomer. (**b–d**) Show structure of Camelidae IgG2 (**b**), Camelidae IgG3 (**c**), and avian IgY antibody (**d**). NH_2 = amino group; COOH = carboxylic acid group; V_H = variable region of antibody heavy chain; V_L = variable region of antibody light chain; C_L = constant region of antibody light chain; C_H1,2,3,4,5 = constant domain one, two, three, four, and five of antibody heavy chain; S–S = disulfide bond; and aa = amino acid

the variable regions. Almost all five isotypes, IgA (dimer), IgD (monomers), IgE (monomer), IgG (monomer), and IgM (pentamer), are composed of the same basic immunoglobulin unit with some modifications [3]. However, all the members of *Camelidae* family have a heavy-chain antibody, which consists of two heavy chains (one variable region and two constant regions per chain), and lacks the two light chains. There are three subclasses of IgG in camels and llamas, i.e., the conventional IgG1 (~160 kDa, with full-length light and heavy chains), IgG2 (~92 kDa; with a long hinge; *see* Fig. 1b) and IgG3 (~86 kDa; with a short hinge; *see* Fig. 1c), which lack both light chains and C_H1 [4]. Heavy-chain antibodies are also found in cartilaginous fish. They are called immunoglobulin new antigen receptors (IgNARs; ~175 kDa; *see* Fig. 1d). An IgNAR contains only two heavy chains and each chain has one variable region and five constant regions (C_H1–5) [5].

Avian immunoglobulins are of three principal classes, IgA, IgM, and IgY (the 180 kDa homologue of mammal IgG). Unlike the heavy chain of mammalian immunoglobulins, the heavy chain of IgY consists of four constant Ig domains (*see* Fig. 1e). Female chickens (hens) are favored for producing large amounts of IgY, as this can be harvested from egg yolk. The process is more convenient than isolation of antibodies from blood and other organs (e.g., spleen and bone marrow). IgG contains regions between C_H1 and C_H2, while in IgY two regions (one between C_H1 and C_H2 and the other between C_H2 and C_H3), containing proline and glycine residues, enable limited flexibility [6].

2 Structure of Recombinant Antibodies

A recombinant antibody does not exist naturally but is assembled from DNA by combining antibody heavy-chain and light-chain gene sequences. The single-chain variable fragment (scFv) and fragment antigen-binding (Fab) region are the most popular recombinant antibody formats used due to their short generation time and high antigen affinity and structural stability [7].

A scFv consists of variable (binding) regions of the antibody heavy (V_H) and light (V_L) chains, with a flexible linker [e.g., $(GGGGS)_3$ linker] joining the terminal ends of either the V_H to V_L (or V_L to V_H) (*see* Fig. 2a). It is popular and effective to use a disease-specific scFv for targeted therapy through fusing to therapeutic proteins or genes.

The Fab fragment, which is double the size of the scFv, is formed by one variable and one constant domain of both light and heavy chains, and is linked by a disulfide bridge (*see* Fig. 2b) [8]. There are many medicines, which have been derived from Fabs that are now approved by the Food and Drug Administration (FDA).

Moreover, various kinds of scFv and/or Fab-derived antibodies have been generated for clinical applications. Bivalent or trivalent scFvs consist of two or three scFvs linked with a short amino acid

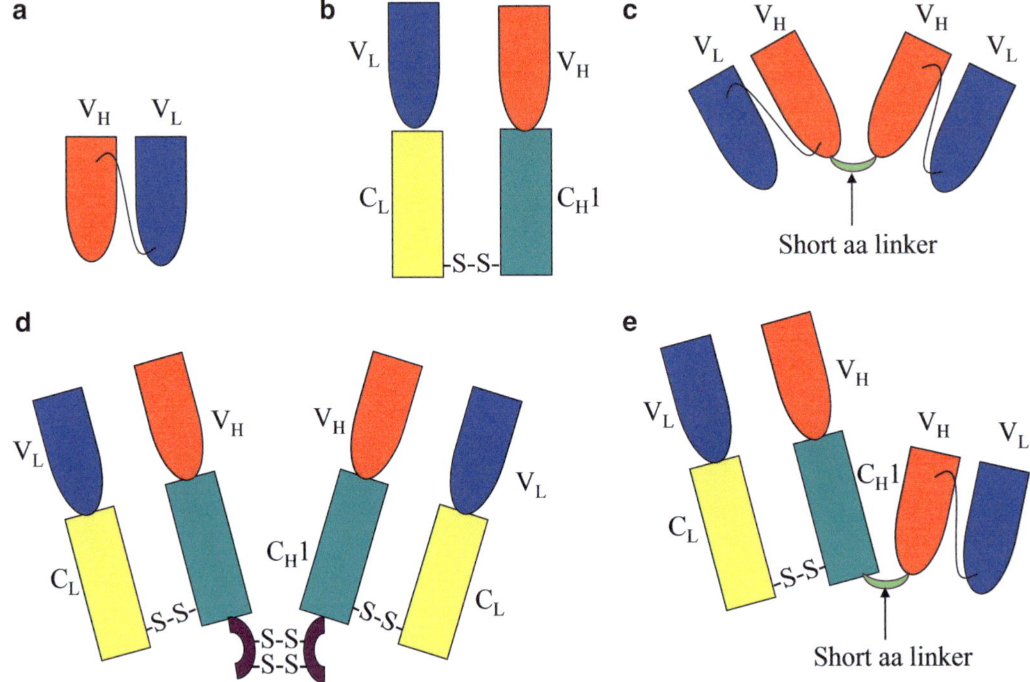

Fig. 2 Structures of a scFv (**a**), a Fab fragment (**b**), a $(scFv)_2$ (**c**), a $F(ab')_2$ (**d**), and a Fab-scFv (**e**). V_H = variable region of antibody heavy chain; V_L = variable region of antibody light chain; C_L = constant region of antibody light chain; C_H1 = constant domain one of antibody heavy chain; S–S = disulfide bond; and aa = amino acid

linker [9] (*see* Fig. 2c). A $F(ab')_2$ fragment contains two Fab fragments linked by disulfide bonds. It can be obtained by cleaving whole immunoglobulins using the enzyme pepsin below the hinge region (*see* Fig. 2d). The Fab-scFv fusion antibody is formed by a Fab and an scFv via a short amino acid linker (*see* Fig. 2e).

Acknowledgement

This work is supported by Science Foundation Ireland under CSET Grant No. 05/CE3/B754 and 10/CE/B1821.

References

1. Meffre E, Casellas R, Nussenzweig MC (2000) Antibody regulation of B cell development. Nat Immunol 1:379–85
2. Chailyan A, Tramontano A, Marcatili P (2012) A database of immunoglobulins with integrated tools: DIGIT. Nucleic Acids Res 40:1230–4
3. Schroeder HW Jr, Cavacini L (2010) Structure and function of immunoglobulins. J Allergy Clin Immunol 125:S41–52
4. Shaker GH (2010) Evaluation of anti-diphtheria toxin nanobodies. Nanotechnol Sci Appl 3: 29–35

5. Stanfield RL, Dooley H, Verdino P, Flajnik MF, Wilson IA (2007) Maturation of shark single-domain (IgNAR) antibodies: evidence for induced-fit binding. J Mol Biol 367:358–72, Epub 2006 Dec 22
6. Warr GW, Magor KE, Higgins DA (1995) IgY: clues to the origins of modern antibodies. Immunol Today 16:392–8
7. Townsend S, Finlay WJJ, Hearty S, O'Kennedy R (2006) Optimizing recombinant antibody function in SPR immunosensing. The influence of antibody structural format and chip surface chemistry on assay sensitivity. Biosens Bioelectron 22:268–74
8. Burton DR (2001) Antibody library. In: Phage Display a Laboratory Manual, Barbas 3rd CF, Burton DR, Scott JK and Silverman GJ (eds). Cold Spring Harbor Laboratory: New York, p. 1–3
9. Byrne H, Conroy PJ, Whisstock JC, O'Kennedy RJ (2013) A tale of two specificities: bispecific antibodies for therapeutic and diagnostic applications. Trends Biotechnol 31:621–32

Chapter 3

Prediction of Antigenic B and T Cell Epitopes via Energy Decomposition Analysis: Description of the Web-Based Prediction Tool BEPPE

Claudio Peri, Oscar C. Solé, Dario Corrada, Alessandro Gori, Xavier Daura, and Giorgio Colombo

Abstract

Unraveling the molecular basis of immune recognition still represents a challenging task for current biological sciences, both in terms of theoretical knowledge and practical implications. Here, we describe the physical-chemistry methods and computational protocols for the prediction of antibody-binding epitopes and MHC-II loaded epitopes, starting from the atomic coordinates of antigenic proteins (PDB file). These concepts are the base of the Web tool BEPPE (Binding Epitope Prediction from Protein Energetics), a free service that returns a list of putative epitope sequences and related blast searches against the Uniprot human complete proteome. BEPPE can be employed for the study of the biophysical processes at the basis of the immune recognition, as well as for immunological purposes such as the rational design of biomarkers and targets for diagnostics, therapeutics, and vaccine discovery.

Key words PPV, Antigen–antibody recognition, MHC-II, Epitope prediction, Energy decomposition, BEPPE, Web server

1 Introduction

The increasing availability of experimentally solved protein structures and the advances in theoretical knowledge and computational resources permitted to obtain a deeper understanding of proteins and their physicochemical properties. This in turn paved the way for the development of dedicated computational tools for the analysis and the functional prediction of these properties. In this context, BEPPE (Binding Epitope Prediction from Protein Energetics) is a Web-based tool for the prediction of antibody binding sites and MHC-II loaded epitopes, based on the analysis of protein internal energetics. BEPPE is rooted in a computational biophysics approach, defined MLCE, for the prediction of antigenic substructures in isolated proteins [1].

Gunnar Houen (ed.), *Peptide Antibodies: Methods and Protocols*, Methods in Molecular Biology, vol. 1348, DOI 10.1007/978-1-4939-2999-3_3, © Springer Science+Business Media New York 2015

The method was initially applied to the prediction of B-cell epitopes, i.e., epitopes that can be recognized and bound by antibodies [2–4]. The predictive performance [5] of the method, originally benchmarked on 19 antibody–antigen interactions, yielded 0.46 percentage in sensitivity and 0.84 specificity, with a PPV (positive predicted value, also referred as precision) of 0.32 and AUC values (Area Under the Curve) [6] of 0.71, at the variation of the prediction cutoff. Now we have expanded the possibilities of BEPPE to include the prediction of MHC-II epitopes. Our latest benchmark composed of 13 antigens and 57 non-redundant MHC-II epitopes, indicate a prediction performance scoring sensitivity, specificity and PPV of 0.41, 0.83, and 0.31, respectively, which is very close to the predictivity observed for antibody-binding sites (Table 1). One obvious use of the results is the design of synthetic epitope sequences. Moreover, our methods can be exploited for the design of optimized antigens. This can be

Table 1
Overall performance analysis of BEPPE in the task of MHC II epitopes prediction

MHC-II Epitopes Prediction performance analysis							
PDB ID	**Cutoff**	**Sensitivity**	**Specificity**	**Precision**	**Accuracy**	**MCC**	***p*-value**
1CB0.A	0.25	0.55	0.63	0.06	0.63	0.07	1.42E-059
1D3B.A	0.15	0.20	0.94	0.33	0.84	0.17	5.05E-003
1EA3.A	0.10	0.14	1.00	1.00	0.36	0.20	4.00E-009
1HA0.A	0.25	0.00	0.88	0.00	0.85	−0.06	7.16E-065
1I7Z.A	0.10	0.44	0.90	0.16	0.88	0.22	7.32E-021
1OVA.D	0.25	0.27	0.59	0.19	0.51	−0.12	3.41E-022
2GIB.B	0.25	0.38	1.00	1.00	0.59	0.41	9.80E-008
2JK2.A	0.15	0.87	0.86	0.28	0.86	0.44	6.36E-028
2VB1.A	0.25	0.34	0.83	0.62	0.60	0.19	1.60E-001
2WA0.A	0.20	0.70	0.82	0.16	0.82	0.27	4.27E-031
3BZH.A	0.25	0.89	0.74	0.18	0.75	0.33	1.35E-026
3FEY.C	0.25	0.50	0.74	0.08	0.73	0.11	1.28E-078
3HLA.A	0.15	0.00	0.78	0.00	0.59	−0.25	1.09E-020
Average	0.2	0.41	0.82	0.31	0.69	0.15	

For every protein (PDB code) is reported the optimal cutoff for prediction and the classification performance parameters sensitivity, specificity, precision, and accuracy [5]. The Matthews Correlation Coefficient (MCC) [22] test and statistical significance of the analysis (*p*-value) are also reported

achieved, e.g., through the identification of antigen conformations that optimally present immunogenic regions to the recognition by the immune system molecules, and the identification of mutations that can block the protein in such conformation. BEPPE can be accessed as a free Web service at the following URL: http://bio-inf.uab.es/BEPPE

2 Materials

BEPPE is largely composed of code written in-house, yet relies on some public free and commercial packages to perform the Energy Decomposition Analysis (as described by Tiana et al. [7]) and BLAST search [8]. Here we describe in detail the conditions and parameters integrated in the automated analysis procedure.

2.1 MM-GBSA Calculation Parameters

The input structure is treated in implicit solvent by means of the MM-GBSA calculation approach (*Molecular Mechanics Generalized Born Surface Area*) included in the AMBER (Assisted Model Building with Energy Refinements) software package [9].

Molecular Mechanics (MM) parameters for interactions are described by the forcefield *ff03* [10, 11]. The MM calculations are carried out with the *sander* module.

The polar solvation term is approximated with the Generalized Born (GB) model [12, 13] and OBC re-scaling [14] for solvation energy. The dielectric constant is set to 80 for the bulk (water as solvent) and 1 for the protein and internal cavities. The GB approach used by AMBER takes into account the shielding effect; we have adopted a physiological salt concentration (0.1 M).

The nonpolar solvation term is calculated through the evaluation of the solvent accessible surface area (SA) using the icosahedra approximation. The non polar contribution to the free energy of solvation is calculated as Enp = surften*SA (no offset correction, SURFOFF = 0) and the surface tension (surften) used is 0.0072 kcal/mol/A^2.

2.2 Energy Minimization and Residue Pairwise Decomposition

Input structures undergo a preliminary energy minimization procedure before any subsequent analysis, consisting of 200 minimization steps with the steepest descent optimization algorithm. Short range and long-range non-bonded interactions are considered within a 12 Å cutoff.

The interaction terms are used to build a matrix, describing the interaction energy between every residue pair in the sequence. This matrix is simplified through Principal Component Analysis decomposition. The resulting main eigenvectors and eigenvalues are used to generate a simplified matrix recapitulating the most relevant stabilizing interactions within the protein structure.

2.3 Matrix of Local Coupling Energy (MLCE) and Epitope Prediction

The simplified energy matrix reconstructed after PCA is used as the input for epitope prediction. BEPPE intersects the energy matrix with information on the protein topology, expressed as a contact matrix (6 Å cutoff from beta carbons): the result is the Matrix of Local Coupling Energies (MLCE) [1]. From this, the algorithm selects the list of contacting amino acids with a minimal coupling to the rest of the protein residues. This list forms the final epitope prediction, expressed in the form of patches. The selection of the residues carrying the weakest coupling energy depends upon a cutoff, defined as the percentage of most-uncoupled pairs over all possible ones in the structure. The cutoff is set by default to 15 % of all contributions.

2.4 BLAST Search of Epitopes

As reported in the bibliography, immunoreactive bacterial epitopes should not match sequences present in the human proteome [15]. To check for this, predicted epitope sequences are blasted against the human proteome, downloaded from Uniprot ftp [16] (this file is updated every 4 weeks). NCBI blast+ version 2.2.23 is executed with parameters adjusted for short input sequences using a strategy file downloaded from a 16-residue query NCBI blast search result. This file contains parameters optimized for short sequences as described in the blast help [17]. Blast results are provided with the prediction output file.

3 Methods

This section explains in detail the procedures and calculation steps to run a prediction job. In Fig. 1 we illustrate the individual modules of the server and the job pipeline.

1. Access the Web page for BEPPE at URL http://bioinf.uab.es/BEPPE
2. Upload your antigen pdb file or pdb file bundle in zip format. The program is compatible with single and multiple chain files (*see* **Note 1**). Upon submitting the job, each structure is passed to the core module for analysis and prediction (*see* **Note 2**).
3. The "check and fix" module is composed of a Python script, analyzing the PDB for common format anomalies, such as missing residues, presence of selenomethionines (MSE), duplicate residues (alternative sidechain position and mutations), and alternative atom positions. The script will automatically convert any MSE to MET, and remove any atom/residue duplicate, keeping only the first one indicated in the PDB file. In case of chain holes the calculation of the pairwise energetic contributions may be dramatically affected. In this case the program will be terminated and an error returned via email (*see* **Notes 3** and **4**).

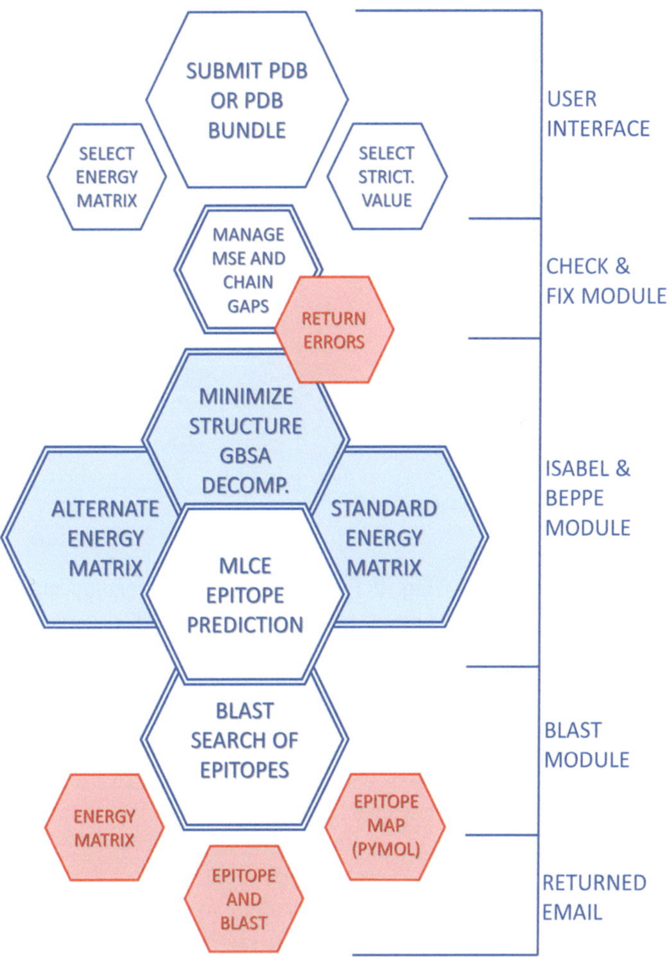

Fig. 1 Layout of the Web tool BEPPE: each individual PDB file is passed by the User Interface to the prediction pipeline (*double line hexagons*). The main modules processing the structure, existent even in stand-alone format, consist of a preliminary input check script, the package ISABEL for the energy decomposition and analysis, the program BEPPE for the prediction of epitopes, and the BLAST script. All results (and eventually all errors) are gathered in a unique text file accompanied by supplementary data, and sent to the user via email (*red, shaded background, single line hexagons*)

To avoid any unexpected inconvenience, we strongly encourage the user to manually check the coordinates file before submitting.

4. Once the PDB file is accepted, it is passed to a comprehensive automated pipeline, converting the PDB file format to AMBER standard, launching the energy minimization and the energy decomposition, as described in the Materials section. This independent code layer is named "ISABEL" (ISabel is Another BEppe Layer). After decomposition, the interaction energy

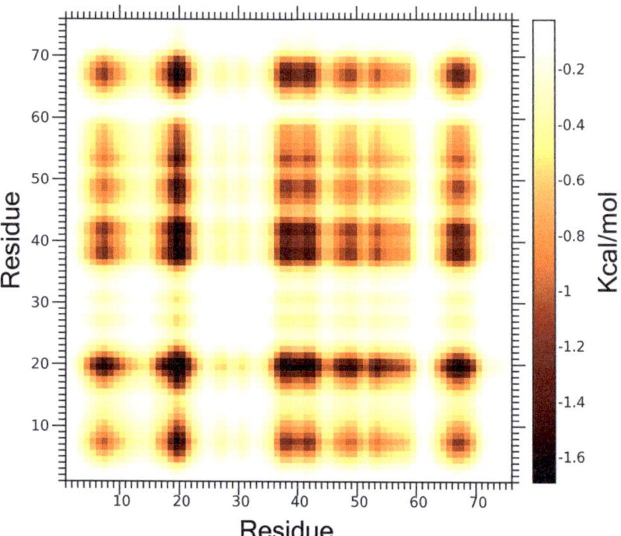

Fig. 2 Example of simplified energy matrix of the human cyclin-dependent kinases regulatory subunit 1 (CKS1B), reconstructed from the first eigenvector of the non-bonded energy matrix, after diagonalization. The information is expressed on a residue pair basis, so each value represent the coupling energy of two amino acids on the protein sequence (*x* and *y* axis). The weaker the signal, the more intense is the color, highlighting the most energetically uncoupled residues. Conversely, those areas that carry the strongest energy of stabilization are depicted here as light bands

matrix is diagonalized into its eigenvalues and related eigenvectors. By default only the first eigenvector will be taken into account [7]. Alternatively, it is possible to select a subset of the most representative eigenvectors according to the method proposed by Genoni et al. [18]. On the basis of the eigenvector(s) selected, a simplified interaction energy matrix will be generated (Fig. 2); such a procedure is intended to emphasize the more relevant (from an energetic point of view) non-bonded interactions from the initial "raw" interaction energy matrix, as described in detail in refs. [1, 19, 20]. The user can select the alternative energy matrix by checking the box "alternative energy matrix" before submission. The use of the alternative matrix is limited to large, multidomain proteins, or in presence of multi-chain complexes. The default energy matrix should be preferable in the vast majority of cases (*see* **Note 5**).

5. The newly built energy matrix is passed to the module BEPPE, which builds the MLCE matrix as the intersection of energy and contact matrices. The energetic cutoff for the selection of the epitope patches is set by default to 15 % of all contributions, which generally scored best for antibody–antigen interfaces

[1], but other cutoffs may be chosen depending on the user requirements. A stricter cutoff will limit the results to the most uncoupled residues, usually located in very small patches. Conversely, loosening the cutoff will produce larger patches. The optimal range for prediction goes from 5 % to 25 % of all low-energy contacts (*see* **Note 6**). The cutoff can be manually adjusted from the drop-down menu "Strictness of prediction". Note that the actual percentages have been substituted in the Web page with a much simpler notation, being 1 for a strict prediction (5 %), and 5 for a soft one (25 %). Clearly, in this simple values notation, the standard cutoff is set to 3.

6. Finally, the patches are blasted against the human proteome for similar motifs. Given the fact that most of the applications of BEPPE, including the discovery of biomarkers for diagnostics and targets for vaccines and therapeutics, would require immunogenic epitopes with no resemblance of preexistent, endogenous proteins, we provide the end-user with a quick outlook on the sequence alignment of the results.
7. The output provided includes the prediction at the residue level in text format subdivided in patches, along with the blast search results for each one. In attachments, the system provides a pictorial representation of the energy matrix as regenerated after decomposition and diagonalization (such as shown in Fig. 2), and a Pymol script [21] for an automatic, easy visualization of the results mapped on the protein structure.

4 Notes

1. The program is compatible with PDB files including multiple chains. If the user needs to run a prediction on individual subunits of a protein complex, or different conformations of the same protein, it is required that each structure is submitted as an individual PDB file. In a multi-chain file, the total energy of the system will be estimated, guiding the predictor towards the most uncoupled patches of the whole assembly. Such an analysis may be indicated in the case of complexes undergoing obligate interactions.
2. BEPPE is a Web tool returning one prediction for each input PDB file. One can use the reference structure obtained by diffraction techniques or NMR solution, but we would like to stress the possibility of using multiple conformations of the same protein to draw a consensus prediction out of different snapshots (e.g., 3 files), to help reduce the noise and focus on the significant patches. These conformations may be different fits of an NMR bundle, or representative frames from a Molecular Dynamics simulation.

3. The script ignores any missing residue as commonly annotated in the PDB remarks, since occasionally chain gaps are present within the protein structure even if not indicated. For similar reasons, it neither fully relies on the annotated numbering of the amino acids. The program progressively evaluates the distance of alpha carbons of consecutive residues. If large distances are found (>2.8 Å) the chain gap is evident, even if not indicated, and the job for that file is aborted. If smaller distances are found, the annotated chain gap may be just nominal and the chain integrity is likely to be maintained. In this case, the PDB file will be renumbered, since an uneven numbering may produce artifacts. The renumbered PDB file is accepted and a warning is returned (take into account that any prediction will follow the numbering of this new file, not the original one!).
4. The program automatically rejects any input structure containing chain gaps, since the extent of the energy perturbations and their effect on the predictivity have never been addressed in our benchmarks. Nevertheless, if you wish to run a prediction on a protein structure containing chain holes, try submitting it as a multiple chain file.
5. Usually, the first eigenvector is sufficient to reconstruct the energetic signature of a protein [7]. However, when attempting to reconstruct the energy matrix of proteins composed of independent domains (or subunits, considering complexes), the first eigenvector may be insufficient for a complete protein coverage. The user can be aware of this issue by looking at the graphic representation of the energy matrix (provided as an email attachment. An example is shown in Fig. 2). When the energetic information (the colored clusters) is not spread along the whole energy matrix (i.e., in the presence of large, white bands in correspondence of one domain), this may be due to the inability of the first eigenvector to represent the full matrix. The alternative method reconstructs the matrix with the specific task of maximizing coverage while considering only the most cohesive and representative energy modules from the top-ranked principal eigenvectors. This method is to be intended as a recovery system for those areas that would be left undescribed, since its efficacy has never been validated for the purpose of epitope prediction.
6. BEPPE is particularly sensitive to extremely low-coupling energies. Meaning that unstructured regions extended towards the solvent, thus bearing minimal contacts with the rest of the protein (such as long, flexible loops), will inevitably mark the local energy matrix (MLCE) with a strong uncoupled signal. Albeit these regions may actually include an epitope, the presence of such strong signals may mask the presence of other

regions worth of attention. Such an event can be easily acknowledged by looking at your reference structural data, and comparing it with the energy matrix provided as an email attachment. If the energy matrix shows an overwhelming signal in correspondence of such motifs, we strongly recommend to use a very low cutoff (values 1–2 from the drop-down menu) for your primary prediction (returning the core of the uncoupled motif/region), and run a subsequent prediction with a loose cutoff (value 5 from the drop-down menu). The second run can extend the prediction to include other, distal significant patches. In this case, one can ignore the "bleaching" noise effect produced at high cutoffs around the primary patch (due to the extremely strong signal) and focus on the core patch identified with the first prediction and the "secondary" patches highlighted at higher cutoff.

Acknowledgments

All authors contributed equally to this new version of the Web tool BEPPE.

This project was supported by EU's FP6 ("BacAbs", ref. LSHB-CT-2006-037325) and the Cariplo Foundation ("GtA", ref. 2009-3577).

References

1. Scarabelli G, Morra G, Colombo G (2010) Predicting interaction sites from the energetics of isolated proteins: a new approach to epitope mapping. Biophys J 98:1966–1975
2. Lassaux P, Peri C et al (2012) A Structure-Based Strategy for Epitope Discovery in Burkholderia pseudomallei OppA Antigen. Structure 21:167–175
3. Peri C, Gagni P et al (2013) Rational Epitope Design for Protein Targeting. ACS Chem Biol 8:397–404
4. Gourlay LJ, Peri C et al (2013) Exploiting the Burkholderia pseudomallei Acute Phase Antigen BPSL2765 for Structure-Based Epitope Discovery/Design in Structural Vaccinology. Chem Biol 20:1147–1156
5. Ponomarenko JV, Bourne PE (2007) Antibody-protein interactions: benchmark datasets and prediction tools evaluation. BMC Struct Biol 7:64
6. Fawcett T (2006) An introduction to ROC analysis. Pattern Recognitt Lett 27:861–974
7. Tiana G, Simona F, De Mori GM et al (2004) Understanding the determinants of stability and folding of small globular proteins from their energetics. Protein Sci 13:113–124
8. Altschul SF, Gish W, Miller W et al (1990) Basic local alignment search tool. J Mol Biol 5:403–410
9. http://ambermd.org/
10. Duan Y, Wu C, Chowdhury S et al (2003) A point-charge force field for molecular mechanics simulations of proteins based on condensed-phase quantum mechanical calculations. J Comput Chem 24:1999–2012
11. Lee MC, Duan Y (2004) Distinguish protein decoys by using a scoring function based on a new Amber force field, short molecular dynamics simulations, and the generalized Born solvent model. Proteins 55:620–634
12. Hawkins GD, Cramer CJ, Truhlar DG (1995) Pairwise solute descreening of solute charges from a dielectric medium. Chem Phys Lett 246:122–129

13. Hawkins GD, Cramer CJ, Truhlar DG (1996) Parametrized models of aqueous free energies of solvation based on pairwise descreening of solute atomic charges from a dielectric medium. J Phys Chem 100:19824–19839
14. Onufriev A, Bashford D, Case DA (2004) Exploring protein native states and large-scale conformational changes with a modified generalized Born model. Proteins 55: 383–394
15. Amela I, Cedano J, Querol E (2007) Pathogen proteins eliciting antibodies do not share epitopes with host proteins: a bioinformatics approach. PLoS One 2(6):e512
16. The UniProt Consortium (2014) Activities at the Universal Protein Resource (UniProt). Nucleic Acids Res 42:191–198
17. http://www.ncbi.nlm.nih.gov/BLAST/Why.shtml
18. Genoni A, Morra G, Colombo G (2012) Identification of Domains in Protein Structures from the Analysis of Intramolecular Interactions. J Phys Chem B 116:3331–3343
19. Corrada D, Morra G, Colombo G (2013) Investigating allostery in molecular recognition: insights from a computational study of multiple antibody-antigen complexes. J Phys Chem B 117(2):535–552
20. Corrada D, Colombo G (2013) Energetic and dynamic aspects of the affinity maturation process: characterizing improved variants from the bevacizumab antibody with molecular simulations. J Chem Inf Model 53(11):2937–2950
21. http://www.pymol.org
22. Mattews BW (1975) Comparison of the predicted and observed secondary structure of T4 phage Lysozyme. Biochim Biophys Acta 405:442–451

Chapter 4

Prediction of Antibody Epitopes

Morten Nielsen and Paolo Marcatili

Abstract

Antibodies recognize their cognate antigens in a precise and effective way. In order to do so, they target regions of the antigenic molecules that have specific features such as large exposed areas, presence of charged or polar atoms, specific secondary structure elements, and lack of similarity to self-proteins. Given the sequence or the structure of a protein of interest, several methods exploit such features to predict the residues that are more likely to be recognized by an immunoglobulin.

Here, we present two methods (BepiPred and DiscoTope) to predict linear and discontinuous antibody epitopes from the sequence and/or the three-dimensional structure of a target protein.

Key words Epitope, Linear epitope, Discontinuous epitope, Prediction

1 Introduction

The antibody-antigen recognition problem is one of the major yet unsolved topics in immunology and structural biology. Antibodies bind their targeted molecules with extreme precision, robustness, and specificity [1, 2].

Naïve, antigen-inexperienced antibodies are usually poly specific and cross-reactive and bind their target molecule with low affinity, typically in the micromolar range. These characteristics are achieved by enrichment in tyrosines, tryptophans, and serines, and to some degree by structural flexibility in the antigen-binding site (ABS) [3–5]. As soon as they recognize a suitable pathogenic molecule, they undergo an affinity maturation process in order to become more potent and specific [6–8]. This process consists in the diversification of the antibody repertoire through the introduction of random somatic mutations and the subsequent selection of B cells for the antigen affinity and against binding of self-molecules, resulting in a focused amplification of only the antibodies that have the highest affinity and specificity.

Gunnar Houen (ed.), *Peptide Antibodies: Methods and Protocols*, Methods in Molecular Biology, vol. 1348, DOI 10.1007/978-1-4939-2999-3_4,

Understanding the mode of binding of an immunoglobulin to its antigen has immense medical, industrial, and biological repercussions [9–11]. In the last few years, immunoglobulins have been subject to extensive studies, resulting in more than 200,000 sequences and close to 2000 solved structures of antibodies deposited in public databases as of November 2014 [12–16]. The analysis of such data has cast some light on the way an antibody recognizes its antigen. Nonetheless, it has to be underlined that, because of differences in the definition of the datasets used in different studies, there is a large variability in the statistical results presented in different works [17–22].

The average interaction area between an antibody and an antigen is typically between 500 and 1000 Å^2 and involves approximately 15 residues [21]. Traditionally B cell epitopes have been divided into two distinct classes: linear epitopes composed only by residues that are contiguous in the protein sequence, and conformational or discontinuous epitopes formed by protein regions that are brought in spatial proximity only when the protein is folded. However, the vast majority of characterized B cell epitopes are somewhat a mixture of the two with a continuous linear core constituting more than half of the epitope size [21]. Clearly, the majority of protein epitopes are composed by regions lying on the surface of the molecule. It has to be noted that, in a number of experimental assays, the analysis of the interaction between antigens and antibodies is performed by analyzing the binding of the latter molecules to the peptides that form the antigenic molecules rather than on the whole proteins in their native structural conformation [23, 24]. In this case, it appears that even residues that are scarcely exposed might be part of an epitope; it is not clear how much this fact might be relevant in vivo or rather be an experimental artifact [20].

Finally, the nature of the specific residues involved in the interaction has been analyzed in several works [25–27]. Again, there are some differences between the results reported in different works, but all the studies converge on two points: the interacting residues in antibody/antigen complexes are significantly different from the ones in other protein complexes, and aromatic residues and aromatic interactions in the paratope (i.e., the antibody region involved in the binding) have a primary role in antigen recognition.

The aforementioned characteristics have been exploited to develop methods that, from the sequence or the structure of a given protein, can predict the regions that are more likely to be potential epitopes [28, 29, 19, 30–41].

Here, we present and guide through the usage of two of such methods.

2 Materials

The methods presented take advantage of the antigen sequence and, if available, of its three-dimensional structure.

2.1 Sequence

The sequence data can be generated in-house or retrieved by querying publicly available databases such as NCBI protein database using the id or the name of the target protein (e.g., "H1N1 influenza virus hemagglutinin"). The results have to be saved in the FASTA format, which can be read and edited with text processing software tools such as Microsoft Word or Notepad. The sequence is used to predict linear antibody epitopes (Subheading 3.1).

2.2 Antigen-Solved Structure

It is possible to obtain either the solved structure or a reliable structural model of many antigenic proteins that can be used for the prediction of conformational epitopes (Subheading 3.2). To check whether an experimentally solved structure of the target antigenic protein exists, go to the PDB website (www.rcsb.org) [42]. Now, in the left-side menu click on "Search" and then on "Sequences." Copy the sequence in FASTA format you retrieved in the previous step and paste it in the "Paste Sequence" text box. Press "Run Sequence Search."

You will be redirected to the results page, in which all the significant hits (if any) are listed. Now you can inspect the similarity between the target protein and the matches by clicking, for each protein listed in the result page, on the "Display Full Alignment" link and eventually by inspecting the alignment to decide whether the protein is suitable for the aim of epitope prediction. Ideally, it should cover all your target sequence and the sequence identity between the experimentally solved protein and the query protein should be 100 %. Very often this is not the case, but it can still be possible to obtain a model of the antigen.

2.3 Antigen Model

The topic of producing reliable protein three-dimensional models exceeds the focus of the current tutorial, so for a more comprehensive study on this topic we invite the readers to refer to other papers and resources [43, 44].

CPHmodels [45] is an accurate yet intuitive tool to produce protein models.

In order to build a model of the antigen using CPHmodels, go to the webpage (http://www.cbs.dtu.dk/services/CPHmodels/) and paste in the text area the FASTA sequence of the target protein. Press submit and wait until your model will be produced.

When the final output page is produced, you will find, together with a number of details and information on the modeling process, a summary line and a line that contains the link "query.pdb" that allows you to download the model of the target protein.

As already stated, a comprehensive description of the modeling process and its result is out of the scope of this chapter; as a rule of thumb, we suggest the readers to use the modeled protein only if the identity percentage reported in the output page just before the alignment is greater than 50 %, the e-value (if reported in the summary line) is smaller than 1.0e–5, and the coverage is larger than 80 %. These thresholds are quite conservative and might exclude models that could potentially be used for epitope prediction purposes, but that require a more detailed analysis.

3 Methods

3.1 Linear Epitope Prediction

Go to the BepiPred [22] webpage (http://www.cbs.dtu.dk/services/BepiPred/) and paste the sequence of your protein in the text area. Leave all the other parameters in the page to their default value. Press "Submit" and wait (few seconds in general) until the result page appears. The final output page contains the predicted linear epitopes reported in the GFF format. For each sequence submitted to the server, the prediction will consist of a header (i.e., the lines reporting details such as the date of submission and the linear sequence among others) and of the prediction itself. Each line contains the sequence name (column 1), the prediction score (column 5), and the final result of prediction consisting of the residue id and an "E" for residues predicted to be part of a linear epitope and a "." for non-epitope residues.

3.2 Conformational Epitope Prediction

Go to the DiscoTope 2.0 [19, 20] webpage (http://www.cbs.dtu.dk/services/DiscoTope/). On this page, use the input option number 3 ("A file from your local disk containing your own structure in PDB format (not necessarily present in PDB)") and upload the protein model you have generated (Subheading 2.3). Leave all other parameters to the default values. Notice that, in case your antigen has an experimentally solved structure, you can skip the modeling step and use input options 1 or 2 in the DiscoTope server.

Wait until the result page appears—it might take seconds to few minutes.

The output consists of seven columns: the chain id, the residue number, the amino acid, the number of internal contacts, the propensity score, and the DiscoTope score. Finally, the residues predicted to be part of an epitope are marked with a "<=B" in the last column.

The user can download a modified pdb file in which the b-factor field has been substituted by the epitope prediction score. Most molecular viewers—such as PyMOL [46], Swiss PDB Viewer [47], and VMD [48]—offer the option to color a molecule according to its b-factor. Doing so, the user can have a visual representation of the epitope prediction results (Fig. 1, right panel, subheading 4.1-4).

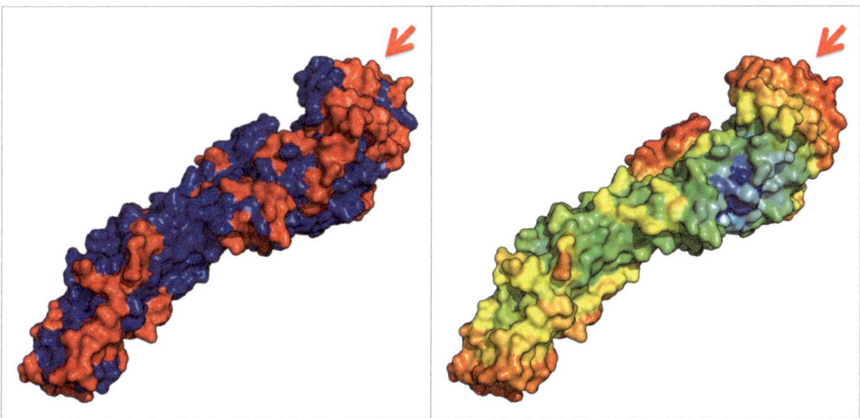

Fig. 1 The modeled structure of H1N1 influenza hemagglutinin colored according to the epitope scores predicted with BepiPred (*left*) and DiscoTope (*right*). Color scale goes from *blue* (low epitope probability) to *red* (high epitope probability). The *arrows* indicate the 98–110 epitope region

4 Final remarks

Even though the accuracy of antibody epitope predictions has considerably increased in the last few years, these computational results should be carefully inspected and interpreted to avoid meaningless or misleading conclusions.

Our overall understanding of the immune system and of the strategies it adopts to recognize foreigner molecules is still very limited.

New technologies are now available to study protein epitopes. Peptide arrays can be used to identify which peptides of a given protein react with the serum of an immunized host [23, 24]. However, we should underline that the ability of an antibody to recognize a given linear sequence in its original protein environment or when presented as a single peptide might be different. Thus, whenever possible, complementing the computational tools presented before with these experimental data can provide more robust results and give further details of some characteristics of the epitopes.

5 Notes

1. Some of the predicted epitopes might not be significant in vivo if the identified region is never exposed to antibodies, as in the case of patches that are involved in protein interaction or transmembrane regions. The user can rule out these cases either using annotations from experimental data or prediction tools [49–51].

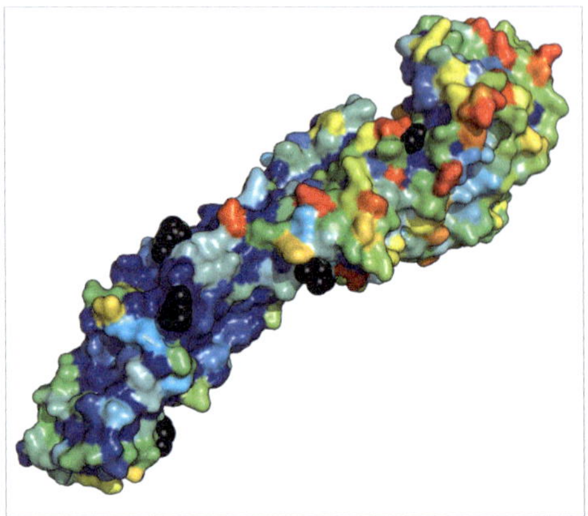

Fig. 2 Prediction of sequence variability and glycosylation sites on the H1N1 influenza hemagglutinin. Variability scores are colored from *blue* (more conserved) to *red* (more variable). N-linked glycosylation sites are depicted in *black*

2. Some posttranslational modifications (PTMs) can heavily modify the ability of antibodies to bind a given region of a protein [52–54] and should therefore be taken into account when predicting epitopes. Therefore, we encourage the users to perform a prediction of common PTMs, such as glycosylation and phosphorylation [55–58], on the antigenic protein of interest and to check whether these might interfere with the binding.
3. In some cases it can be useful to analyze the sequence conservation of the regions predicted to be epitopes. A common strategy used by pathogens to escape the host immune system is to have some regions of their proteins that are subject to very high mutation rates (hypermutation) [59, 60]. In this way, as soon as the immune system develops molecules targeting these regions, the mutations induced in the sequence allow the pathogen to avoid the immune response. Several servers can infer the conservation of each residue in a protein given its sequence and/or its structure [61, 62]. This information can be used for example to decide whether a given epitope can be used as a possible target for a vaccine or an antibody drug.
4. If the sequences of one or more antibodies targeting the antigenic protein are available, they can be used to improve the predictions and to gain further insights into the molecular mechanisms behind the binding. To this aim, the user should first build a three-dimensional model of the antigen—as explained in Subheading 2.3—and of the antibody, using one of the automatic modeling tools available [63–67]. Then, one of the automatic tools for antibody/antigen docking [25–27, 68–70] can be used to increase the reliability of the epitope predictions and to have a more detailed understanding of the binding at a molecular level.

5. An example of application:

   We predict the hemagglutinin B cell epitopes for a novel strain of H1N1 influenza virus (A/Helsinki/1454/2014 (H1N1)). We download the protein sequence from the NCBI protein database (accession code: AIP92890) and build its three-dimensional model using the CPHmodels server as previously explained.

   We submit the primary sequence to BepiPred [22] and the structural model to DiscoTope [20]. The predicted epitopes are displayed in Fig. 1. The two predictions only partially overlap. Even though the structure-based predictions obtained from the DiscoTope server are in general terms more reliable, we can focus first on the epitopes that are predicted by both methods. Both servers assign high scores to the protein region from residue 198 to residue 210 (HPSTTADQQSLYQ) that is proximal to the sialic acid-binding site. We can now check whether this region is potentially subject to hypermutation or glycosylation. We use the NetNGlyc server [56] to predict N-linked glycosylation. As apparent from Fig. 2, none of the predicted glycosylation sites is in close spatial proximity to the 198–210 region. The sequence conservation results obtained using the ConSurf server [61], however, show that this region contains residues that are more variable than the rest of the protein; therefore, as anticipated, the efficacy of a vaccine targeting this epitope might be lowered by hypermutations and sequence variability in general.

References

1. Carter PJ (2006) Potent antibody therapeutics by design. Nat Rev Immunol 6:343–357. doi:10.1038/nri1837
2. Davies DR, Cohen GH (1996) Interactions of protein antigens with antibodies. Proc Natl Acad Sci U S A 93:7–12
3. Birtalan S, Zhang Y, Fellouse FA, Shao L, Schaefer G, Sidhu SS (2008) The intrinsic contributions of tyrosine, serine, glycine and arginine to the affinity and specificity of antibodies. J Mol Biol 377:1518–1528. doi:10.1016/j.jmb.2008.01.093
4. Bond CJ, Marsters JC, Sidhu SS (2003) Contributions of CDR3 to V H H domain stability and the design of monobody scaffolds for naive antibody libraries. J Mol Biol 332: 643–655
5. Sethi DK, Agarwal A, Manivel V, Rao KV, Salunke DM (2006) Differential epitope positioning within the germline antibody paratope enhances promiscuity in the primary immune response. Immunity 24(4):429–438. doi:10.1016/j.immuni.2006.02.010
6. Daugherty PS, Chen G, Iverson BL, Georgiou G (2000) Quantitative analysis of the effect of the mutation frequency on the affinity maturation of single chain Fv antibodies. Proc Natl Acad Sci U S A 97:2029–2034. doi:10.1073/pnas.030527597
7. Neuberger MS, Ehrenstein MR, Rada C, Sale J, Batista FD, Williams G, Milstein C (2000) Memory in the B-cell compartment: antibody affinity maturation. Phil Transact Royal Soc London Ser B Biol Sci 355:357–360. doi:10.1098/rstb.2000.0573
8. Wabl M, Cascalho M, Steinberg C (1999) Hypermutation in antibody affinity maturation. Curr Opin Immunol 11:186–189
9. Olimpieri PP, Marcatili P, Tramontano A (2014) Tabhu: tools for antibody humanization. Bioinformatics. doi:10.1093/bioinformatics/btu667

10. Schrama D, Reisfeld RA, Becker JC (2006) Antibody targeted drugs as cancer therapeutics. Nat Rev Drug Discov 5:147–159. doi:10.1038/nrd1957
11. Shirai H, Prades C, Vita R, Marcatili P, Popovic B, Xu J, Overington JP, Hirayama K, Soga S, Tsunoyama K, Clark D, Lefranc MP, Ikeda K (2014) Antibody informatics for drug discovery. Biochim Biophys Acta 1844:2002–2015. doi:10.1016/j.bbapap.2014.07.006
12. Chailyan A, Tramontano A, Marcatili P (2012) A database of immunoglobulins with integrated tools: DIGIT. Nucleic Acids Res 40(Database issue):D1230–1234. doi:10.1093/nar/gkr806
13. Lefranc M (2014) Immunoglobulins: 25 Years of Immunoinformatics and IMGT-ONTOLOGY. Biomolecules 4:1102–1139. doi:10.3390/biom4041102
14. Vita R, Overton JA, Greenbaum JA, Ponomarenko J, Clark JD, Cantrell JR, Wheeler DK, Gabbard JL, Hix D, Sette A, Peters B (2014) The immune epitope database (IEDB) 3.0. Nucleic Acids Res doi:10.1093/nar/gku938
15. Dunbar J, Krawczyk K, Leem J, Baker T, Fuchs A, Georges G, Shi J, Deane CM (2014) SAbDab: the structural antibody database. Nucleic Acids Res 42(Database issue):D1140–1146. doi:10.1093/nar/gkt1043
16. Allcorn LC, Martin AC (2002) SACS–self-maintaining database of antibody crystal structure information. Bioinformatics 18:175–181
17. Blythe MJ, Flower DR (2005) Benchmarking B cell epitope prediction: underperformance of existing methods. Protein Sci 14:246–248. doi:10.1110/ps.041059505
18. Greenbaum JA, Andersen PH, Blythe M, Bui HH, Cachau RE, Crowe J, Davies M, Kolaskar AS, Lund O, Morrison S, Mumey B, Ofran Y, Pellequer JL, Pinilla C, Ponomarenko JV, Raghava GP, van Regenmortel MH, Roggen EL, Sette A, Schlessinger A, Sollner J, Zand M, Peters B (2007) Towards a consensus on datasets and evaluation metrics for developing B-cell epitope prediction tools. J Mol Recognit 20:75–82. doi:10.1002/jmr.815
19. Haste Andersen P, Nielsen M, Lund O (2006) Prediction of residues in discontinuous B-cell epitopes using protein 3D structures. Protein Sci 15:2558–2567. doi:10.1110/ps.062405906
20. Kringelum JV, Lundegaard C, Lund O, Nielsen M (2012) Reliable B cell epitope predictions: impacts of method development and improved benchmarking. PLoS Comput Biol 8, e1002829. doi:10.1371/journal.pcbi.1002829
21. Kringelum JV, Nielsen M, Padkjaer SB, Lund O (2013) Structural analysis of B-cell epitopes in antibody:protein complexes. Mol Immunol 53:24–34. doi:10.1016/j.molimm.2012.06.001
22. Larsen JE, Lund O, Nielsen M (2006) Improved method for predicting linear B-cell epitopes. Immunome Res 2:2. doi:10.1186/1745-7580-2-2
23. Buus S, Rockberg J, Forsstrom B, Nilsson P, Uhlen M, Schafer-Nielsen C (2012) High-resolution mapping of linear antibody epitopes using ultrahigh-density peptide microarrays. Mol Cell Proteom 11:1790–1800. doi:10.1074/mcp.M112.020800
24. Hansen LB, Buus S, Schafer-Nielsen C (2013) Identification and mapping of linear antibody epitopes in human serum albumin using high-density Peptide arrays. PLoS One 8, e68902. doi:10.1371/journal.pone.0068902
25. Brenke R, Hall DR, Chuang GY, Comeau SR, Bohnuud T, Beglov D, Schueler-Furman O, Vajda S, Kozakov D (2012) Application of asymmetric statistical potentials to antibody-protein docking. Bioinformatics 28:2608–2614. doi:10.1093/bioinformatics/bts493
26. Krawczyk K, Baker T, Shi J, Deane CM (2013) Antibody i-Patch prediction of the antibody binding site improves rigid local antibody-antigen docking. Protein Eng Des Sel 26:621–629. doi:10.1093/protein/gzt043
27. Krawczyk K, Liu X, Baker T, Shi J, Deane CM (2014) Improving B-cell epitope prediction and its application to global antibody-antigen docking. Bioinformatics 30:2288–2294. doi:10.1093/bioinformatics/btu190
28. Chen J, Liu H, Yang J, Chou KC (2007) Prediction of linear B-cell epitopes using amino acid pair antigenicity scale. Amino Acids 33:423–428. doi:10.1007/s00726-006-0485-9
29. Costa JG, Faccendini PL, Sferco SJ, Lagier CM, Marcipar IS (2013) Evaluation and comparison of the ability of online available prediction programs to predict true linear B-cell epitopes. Prot Pept Lett 20:724–730
30. Kulkarni-Kale U, Bhosle S, Kolaskar AS (2005) CEP: a conformational epitope prediction server. Nucleic Acids Res 33 (Web Server issue):W168–171. doi:10.1093/nar/gki460
31. Odorico M, Pellequer JL (2003) BEPITOPE: predicting the location of continuous epitopes and patterns in proteins. J Mol Recognit 16:20–22. doi:10.1002/jmr.602
32. Ponomarenko J, Bui HH, Li W, Fusseder N, Bourne PE, Sette A, Peters B (2008) ElliPro: a new structure-based tool for the prediction of antibody epitopes. BMC Bioinformatics 9:514. doi:10.1186/1471-2105-9-514
33. Ponomarenko JV, Bourne PE (2007) Antibody-protein interactions: benchmark datasets and

prediction tools evaluation. BMC Struct Biol 7:64. doi:10.1186/1472-6807-7-64

34. Reimer U (2009) Prediction of linear B-cell epitopes. Methods Mol Biol 524:335–344. doi:10.1007/978-1-59745-450-6_24
35. Sollner J (2006) Selection and combination of machine learning classifiers for prediction of linear B-cell epitopes on proteins. J Mol Recognit 19:209–214. doi:10.1002/jmr.770
36. Sollner J, Mayer B (2006) Machine learning approaches for prediction of linear B-cell epitopes on proteins. J Mol Recognit 19:200–208. doi:10.1002/jmr.771
37. Sun J, Wu D, Xu T, Wang X, Xu X, Tao L, Li YX, Cao ZW (2009) SEPPA: a computational server for spatial epitope prediction of protein antigens. Nucleic Acids Res 37 (Web Server issue):W612–616. doi:10.1093/nar/gkp417
38. Wang HW, Lin YC, Pai TW, Chang HT (2011) Prediction of B-cell linear epitopes with a combination of support vector machine classification and amino acid propensity identification. J Biomed Biotechnol 2011:432830. doi:10.1155/2011/432830
39. Wang HW, Pai TW (2014) Machine learning-based methods for prediction of linear B-cell epitopes. Methods Mol Biol 1184:217–236. doi:10.1007/978-1-4939-1115-8_12
40. Wee LJ, Simarmata D, Kam YW, Ng LF, Tong JC (2010) SVM-based prediction of linear B-cell epitopes using Bayes Feature Extraction. BMC Genomics 11 Suppl 4:S21. doi:10.1186/1471-2164-11-S4-S21
41. Yang J, Liu N, Zhang T, Zhao S, Qiang L, Su B, Kang A, Li Z (2013) (Prediction and identification of B-cell linear epitopes of hepatitis B e antigen). Nan fang yi ke da xue xue bao (J Southern Med Univ 33:253–257
42. Berman HM, Westbrook J, Feng Z, Gilliland G, Bhat TN, Weissig H, Shindyalov IN, Bourne PE (2000) The Protein Data Bank. Nucleic Acids Res 28:235 242
43. Bordoli L, Kiefer F, Arnold K, Benkert P, Battey J, Schwede T (2009) Protein structure homology modeling using SWISS-MODEL workspace. Nat Protoc 4:1–13. doi:10.1038/nprot.2008.197
44. Webb B, Sali A (2014) Comparative Protein Structure Modeling Using MODELLER. Curr Protocols Bioinformatics (Baxevanis AD et al Eds) 47:5 6 1–5 6 32. doi:10.1002/0471250953.bi0506s47
45. Nielsen M, Lundegaard C, Lund O, Petersen TN (2010) CPHmodels-3.0--remote homology modeling using structure-guided sequence profiles. Nucleic Acids Res 38 (Web Server issue):W576–581. doi:10.1093/nar/gkq535
46. Schrodinger, LLC (2010) The PyMOL Molecular Graphics System, Version 1.3r1.
47. Guex N, Peitsch MC (1997) SWISS-MODEL and the Swiss-PdbViewer: an environment for comparative protein modeling. Electrophoresis 18:2714–2723. doi:10.1002/elps.1150181505
48. Humphrey W, Dalke A, Schulten K (1996) VMD: visual molecular dynamics. J Mol Graphics 14(33–38):27–38
49. Bernsel A, Viklund H, Hennerdal A, Elofsson A (2009) TOPCONS: consensus prediction of membrane protein topology. Nucleic Acids Res 37 (Web Server issue):W465–468. doi:10.1093/nar/gkp363
50. Krogh A, Larsson B, von Heijne G, Sonnhammer EL (2001) Predicting transmembrane protein topology with a hidden Markov model: application to complete genomes. J Mol Biol 305:567–580. doi:10.1006/jmbi.2000.4315
51. Zhang QC, Petrey D, Garzon JI, Deng L, Honig B (2013) PrePPI: a structure-informed database of protein-protein interactions. Nucleic Acids Res 41(Database issue):D828–833. doi:10.1093/nar/gks1231
52. Ekiert DC, Bhabha G, Elsliger MA, Friesen RH, Jongeneelen M, Throsby M, Goudsmit J, Wilson IA (2009) Antibody recognition of a highly conserved influenza virus epitope. Science 324:246–251. doi:10.1126/science.1171491
53. Sternberger LA, Sternberger NH (1983) Monoclonal antibodies distinguish phosphorylated and nonphosphorylated forms of neurofilaments in situ. Proc Natl Acad Sci U S A 80:6126–6130
54. Wei X, Decker JM, Wang S, Hui H, Kappes JC, Wu X, Salazar-Gonzalez JF, Salazar MG, Kilby JM, Saag MS, Komarova NL, Nowak MA, Hahn BH, Kwong PD, Shaw GM (2003) Antibody neutralization and escape by HIV-1. Nature 422(6929):307–312. doi:10.1038/nature01470
55. Blom N, Gammeltoft S, Brunak S (1999) Sequence and structure-based prediction of eukaryotic protein phosphorylation sites. J Mol Biol 294:1351–1362. doi:10.1006/jmbi.1999.3310
56. Blom N, Sicheritz-Ponten T, Gupta R, Gammeltoft S, Brunak S (2004) Prediction of post-translational glycosylation and phosphorylation of proteins from the amino acid sequence. Proteomics 4:1633–1649. doi:10.1002/pmic.200300771
57. Hansen JE, Lund O, Tolstrup N, Gooley AA, Williams KL, Brunak S (1998) NetOglyc: prediction of mucin type O-glycosylation sites based on sequence context and surface accessibility. Glycoconj J 15:115–130

58. Julenius K, Molgaard A, Gupta R, Brunak S (2005) Prediction, conservation analysis, and structural characterization of mammalian mucin-type O-glycosylation sites. Glycobiology 15:153–164. doi:10.1093/glycob/cwh151
59. Brunham RC, Plummer FA, Stephens RS (1993) Bacterial antigenic variation, host immune response, and pathogen-host coevolution. Infect Immun 61:2273–2276
60. McKenzie GJ, Rosenberg SM (2001) Adaptive mutations, mutator DNA polymerases and genetic change strategies of pathogens. Curr Opin Microbiol 4:586–594
61. Armon A, Graur D, Ben-Tal N (2001) ConSurf: an algorithmic tool for the identification of functional regions in proteins by surface mapping of phylogenetic information. J Mol Biol 307:447–463. doi:10.1006/jmbi.2000.4474
62. Ashkenazy H, Erez E, Martz E, Pupko T, Ben-Tal N (2010) ConSurf 2010: calculating evolutionary conservation in sequence and structure of proteins and nucleic acids. Nucleic Acids Res 38 (Web Server issue):W529–533. doi:10.1093/nar/gkq399
63. Marcatili P, Olimpieri PP, Chailyan A, Tramontano A (2014) Antibody structural modeling with prediction of immunoglobulin structure (PIGS). Nat Protoc 9:2771–2783. doi:10.1038/nprot.2014.189
64. Marcatili P, Rosi A, Tramontano A (2008) PIGS: automatic prediction of antibody structures. Bioinformatics 24:1953–1954. doi:10.1093/bioinformatics/btn341
65. Sircar A, Kim ET, Gray JJ (2009) RosettaAntibody: antibody variable region homology modeling server. Nucleic Acids Res 37 (Web Server issue):W474–479. doi:10.1093/nar/gkp387
66. Weitzner BD, Kuroda D, Marze N, Xu J, Gray JJ (2014) Blind prediction performance of RosettaAntibody 3.0: grafting, relaxation, kinematic loop modeling, and full CDR optimization. Proteins 82:1611–1623. doi:10.1002/prot.24534
67. Yamashita K, Ikeda K, Amada K, Liang S, Tsuchiya Y, Nakamura H, Shirai H, Standley DM (2014) Kotai Antibody Builder: automated high-resolution structural modeling of antibodies. Bioinformatics 30:3279–3280. doi:10.1093/bioinformatics/btu510
68. Lyskov S, Gray JJ (2008) The RosettaDock server for local protein-protein docking. Nucleic Acids Res 36 (Web Server issue):W233–238. doi:10.1093/nar/gkn216
69. Sircar A, Gray JJ (2010) SnugDock: paratope structural optimization during antibody-antigen docking compensates for errors in antibody homology models. PLoS Comput Biol 6, e1000644. doi:10.1371/journal.pcbi.1000644
70. Sivasubramanian A, Sircar A, Chaudhury S, Gray JJ (2009) Toward high-resolution homology modeling of antibody Fv regions and application to antibody-antigen docking. Proteins 74:497–514. doi:10.1002/prot.22309

Chapter 5

Fmoc Solid-Phase Peptide Synthesis

Paul R. Hansen and Alberto Oddo

Abstract

Synthetic peptides are important as drugs and in research. Currently, the method of choice for producing these compounds is solid-phase peptide synthesis. In this nonspecialist review, we describe the scope and limitations of Fmoc solid-phase peptide synthesis. Furthermore, we provide a detailed protocol for Fmoc peptide synthesis.

Key words Fmoc solid-phase synthesis, Resins, Linkers, Coupling reagents

1 Introduction

Synthetic peptides are important as therapeutics [1, 2] and for a number of research purposes including cancer diagnosis and treatment [3], antibiotic drug development [4], epitope mapping [5], production of antibodies [6], and vaccine design [7]. Chemical synthesis of biologically active peptides may provide enough material for additional study, structure-activity relationships, or result in new analogues with improved properties.

Large peptides and small proteins of more than 50 residues may be produced by recombinant methods [8]. However, most researchers prefer the solid-phase method (SPPS) for the chemical synthesis of peptides less than 50 residues. Using the SPPS approach, peptides may be synthesized in virtually any scale, synthesis may be carried out by a fully automated peptide synthesizer, and non-proteinogenic amino acids or posttranslational modifications may be introduced during synthesis.

In solid-phase peptide synthesis (Scheme 1) the *C*-terminal amino acid residue, masked with a temporary protecting group on the α-amino group and a semipermanent protecting group on the side chain, is coupled through its C^{α}-carboxylic acid group to a resin. The resin is a polymer which swells in organic solvents and expands as the peptide grows. Furthermore, the resin is equipped with a bifunctional linker, which allows for the cleavage of the final

Gunnar Houen (ed.), *Peptide Antibodies: Methods and Protocols*, Methods in Molecular Biology, vol. 1348, DOI 10.1007/978-1-4939-2999-3_5, © Springer Science+Business Media New York 2015

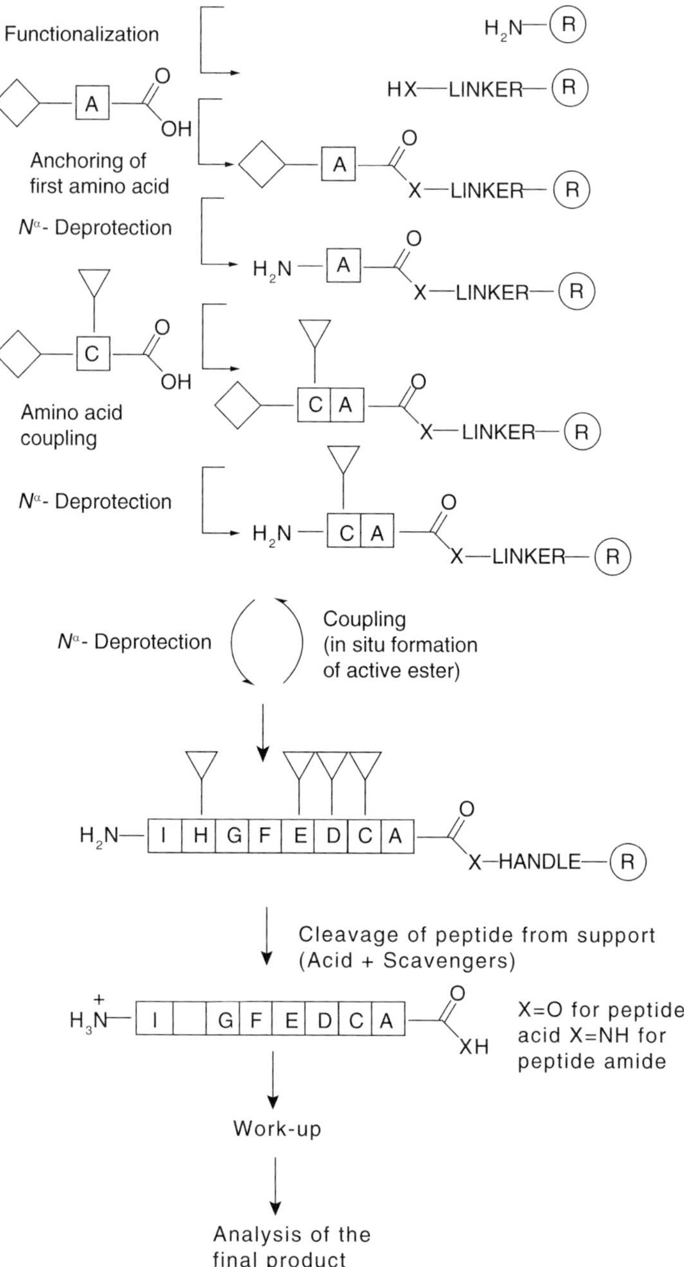

Scheme 1 Overview of solid-phase peptide synthesis

product following synthesis. The linker most often provides either a peptide acid or a peptide amide as the end product.

Following coupling of the first amino acid to the resin, the temporary protecting group masking the α-amino group is removed. The synthesis cycle is continued from the *C*-terminus towards the *N*-terminus until the desired protected peptide is obtained.

By convention, the resin is placed to the right in Scheme 1 which is very important for correct interpretation of sequence and stereochemistry. The product is then cleaved from the resin concurrently with the semipermanent side-chain protecting groups, isolated, and characterized.

The advantages of SPPS over solution synthesis are as follows: (1) All reactions are carried out in a single vessel; (2) excess amino acid and reagents are used to drive reactions to completion and can subsequently be filtered from the system, eliminating the need for purification of intermediates after each step; and (3) the solid-phase approach consists of many repetitive steps making automation possible. The only drawback of the method is that removal of by-products during synthesis is impossible.

Two strategies for the solid-phase synthesis of peptides exist, the Boc/benzyl and the Fmoc/tBu strategy.

In the Boc/benzyl strategy, which was introduced by R.B. Merrifield in 1963 [9], the α-amino group is protected by the *tert*-butyloxycarbonyl (Boc) group **1** and the side-chain functional groups by benzyl-based protecting groups. The Boc group is removed by trifluoroacetic acid (TFA) in dichloromethane (DCM) while the semipermanent side-chain protecting groups require hydrogen fluoride (HF) for removal. For a recent review on Boc SPPS *see* ref. 10.

In the Fmoc/tBu strategy, which was developed by Atherthon and Sheppard [11], the α-amino group is protected by the base-labile 9-fluorenylmethyloxycarbonyl (Fmoc) group **2** and side-chain functional groups by the acid-labile *tert*-butyl **3** or trityl-based protecting groups **4**. The Fmoc group is removed by 20 % piperidine in dimethylformamide (DMF) and the side-chain protecting groups by TFA. This is a truly orthogonal protecting scheme as opposed to the Boc/Benzyl strategy which relies on gradual acid lability.

Since the Boc/Benzyl strategy requires the use of HF for final deprotection the Fmoc/tBu strategy is the most widely used. In this chapter we focus on the scope and limitations of the Fmoc SPPS strategy [12–14].

1.1 Resins

Currently, resins may be divided into three different groups: (1) poly-styrene based, (2) PEG(poly (ethylene glycol))-grafted polystyrene, and (3) PEG resins without polystyrene.

The classical polystyrene resin is a resin cross-linked with 1 % of divinylbenzene (DVB) which swells well in DCM less well in DMF and not at all in H_2O. However, this type of resin is not well suited for Fmoc SPPS of longer peptides.

The most widely used resin in Fmoc SPPS is the TentaGel® resin developed by Rapp and coworkers [15]. TentaGel® resins are grafted copolymers consisting of a low cross-linked polystyrene matrix on which approx. 50–70 % PEG (w/w) is grafted. The PEG-graft copolymer swells in almost all relevant solvent systems, including H_2O. The end of the PEG chain is functionalized with either a hydroxyl group or an amino group. A number of these resins are commercially available.

ChemMatrix® is a resin developed by Albericio and coworkers which is made exclusively from PEG [16]. The authors demonstrated that the highly complex α-amyloid 1–42 peptide could be synthesized in a crude purity of 91 % using this resin. Finally, Meldal developed PEGA which is a beaded polyethylene glycol dimethylacrylamide copolymer resin [17] that allows enzymes to penetrate [18] for on-resin assays.

1.2 Linkers

A number of different linkers are available and most of them release the peptide as a peptide acid or peptide amide when treated with TFA (*see* Fig. 1). For a review of linkers *see* ref. 19 and linkers for protected peptides [20]. The most popular peptide acid and amide linkers are the Wang **5** [21] and Rink **6** [22], respectively.

The 2-chlorotrityl linker 7 which most often comes attached to a polystyrene resin is well suited for the synthesis of peptide acids [23]. Furthermore, protected peptide acids may be cleaved

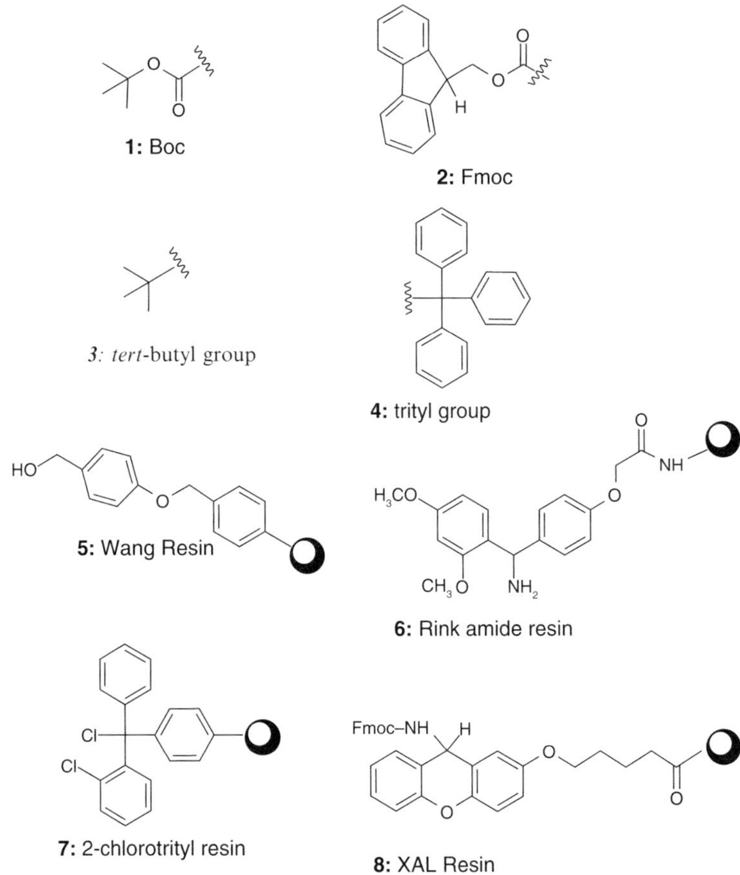

Fig. 1 α-Amino-protecting groups and commonly used linker-functionalized resins in Fmoc SPPS

from the resin using 1 % TFA or 10–20 % acetic acid, but Trt groups are not very stable in these conditions. This strategy is treated in other aspects of peptide synthesis. For protected peptide amides the XAL linker **8** may be used [24].

It should be noticed that in common practice resins are often identified by their linker. For example, the expression "Wang resin" refers to a polystyrene-based resin functionalized with the Wang linker. Other linkers are available which are cleaved by hydroxide ions [25], fluoride ions [26], 2 % hydrazine [27], or photolysis [28]. Finally, the backbone amide linker (BAL) should be mentioned. In this approach, the growing peptide is anchored through a backbone nitrogen, allowing the synthesis of peptide acids and of a number of modifications at the C-terminus [29]. The linker determines the way in which the first residue is anchored to the solid support: for the Rink amide linker (RAM), often used in combination with TentaGel® resins, this operation is straightforward as it is performed in the same way as a standard amino acid coupling; for the Wang and the 2-chlorotrityl linkers another protocol has to be followed, which ultimately makes it more convenient to buy a preloaded resin bearing the desired residue.

Manufacturers usually indicate a loading value (mmol/g) for their resins, which refers to the amount of functional groups (mmol) per gram of resin. This value is used to calculate the amount of reagents to use for each coupling. Low-loading resins offer a loading ≤0.25 mmol/g, while high-loading resins offer a functionalization ≥0.9 mmol/g. The latter are more cost effective and ensure a better ratio between the amount of resin and the coupling solution, but they also favor peptide aggregation and other undesired intermolecular reactions. A loading extent of 0.4–0.7 mmol/g is very versatile and suitable to most applications. In case no loading has been indicated, it is possible to determine it spectrophotometrically after the first Fmoc deprotection step [13].

Bead size is also often indicated either as particle diameter (μm) or mesh number. Most commercial resins are labeled either as 100–200 mesh (particle size = 74–149 μm) or 200–400 mesh (particle size = 37–74 μm). In our lab we prefer to use the 100–200 mesh size as bigger beads have a lower tendency to aggregate; this phenomenon can create pockets of resin that may be difficult to reach during the wash and deprotection procedures, leading to less pure products.

1.3 Peptide Bond Formation

Peptide synthesis is based on the formation of an amide (peptide) bond in a controlled, quantitative, and sequential fashion. If a carboxylic acid and an amine are mixed in solution, their acidic/basic properties predominate and only a proton transfer occurs (Fig. 2a). For this reason it is necessary to activate the carboxylic acid with a coupling reagent—and often an auxiliary nucleophile, too (Fig. 2b). Most of them result in a benzotriazyl ester **9** that reacts with the growing peptide chain on the resin, providing a nonacidic, efficient

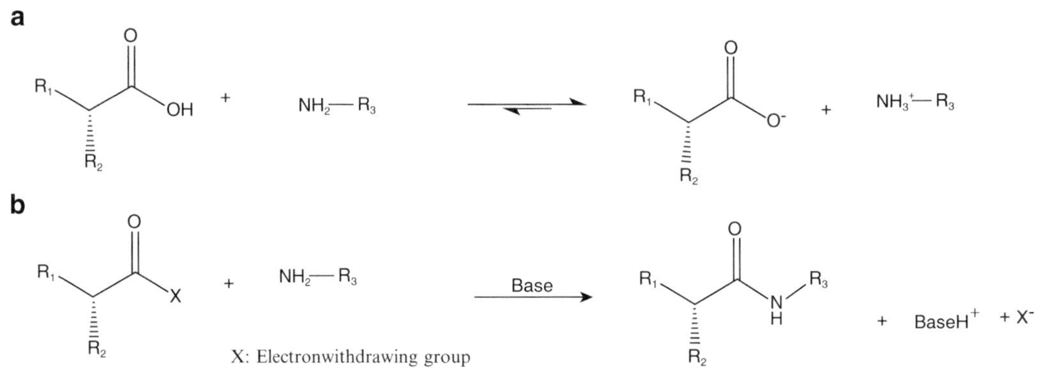

Fig. 2 Proton transfer (**a**) vs. activation of amino acids (**b**)

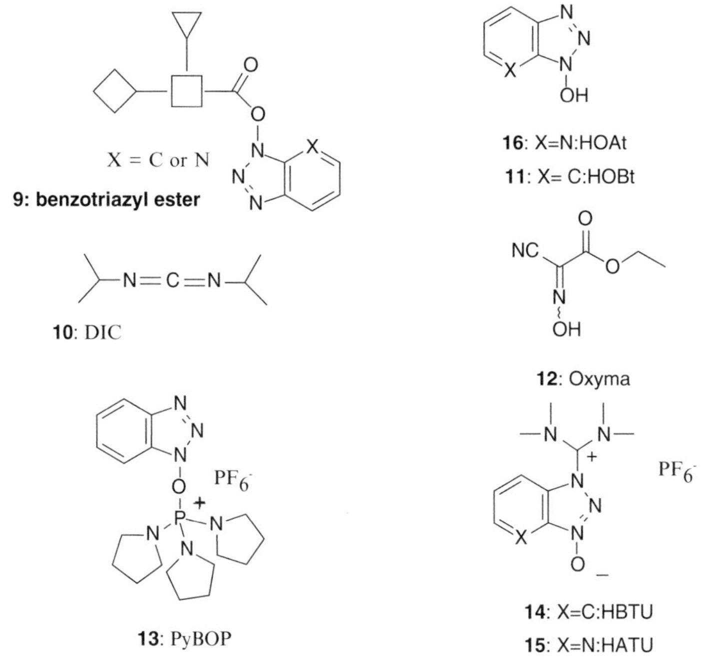

Fig. 3 Active ester generated in situ and coupling reagents

leaving group that ultimately allows the formation of the amide bond. This strategy typically results in a >99 % coupling efficiency.

Since the shelf stability of this type of compounds is very poor, the benzotriazyl ester is prepared in situ. A number of coupling reagents exist and they have been reviewed extensively [30] (*see* Fig. 3). Coupling protocols with DIC (*N,N*-diisopropylcarbodiimide) **10** and HOBt (1-hydroxybenzotriazole) **11** have been used for many years and work well for the synthesis of most short peptides (<15 amino acids) without clusters of branched or hydrophobic amino acids. The advantages of this approach are that the reagents are cheap and no base is used, eliminating the risk of epimerization at the α-carbon. In recent years, coupling protocols involving DIC and Oxyma (ethyl 2-cyano-2-(hydroxyimino)acetate) **12** [31] have gained increasing popularity. More difficult peptides require phosphonium or aminium type of coupling reagents such as PyBOP [32] (1-benzotriazolyloxy-tris-pyrrolidinophosphonium hexafluorophosphate) **13**, HBTU [33] (*N*-[(1H-benzotriazol-1-yl)(dimethylamino)methylene]-*N*-methylmethanaminium hexafluorophosphate *N*-oxide) **14**, or HATU [34] ((*N*-[(dimethylamino)-1H-1,2,3-triazole[4,5-b]pyridine-1-ylmethylene]-*N*-methylmethanaminium hexafluorophosphate *N*-oxide) **15**. These coupling reagents are typically used with HOBt or HOAt (7-aza-1-hydroxybenzotriazole) **16** and 2 equivalents (relative to the coupling reagent) of a tertiary amine such as diisopropylethylamine (DIPEA). Generally HOAt is a more effective additive than HOBt since it contains a nitrogen atom in the aromatic ring [34]. The coupling reaction is carried out in DMF or the less toxic NMP. The former is preferred in most labs since Fmoc-protected amino acids and triazole-based reagents are more soluble in this solvent.

In our lab we prefer to use 3 equiv. of HATU and HOAt and 6 equiv. of DIPEA relative to the resin loading (mmol/g resin) in DMF for 1.5 h. For couplings after sterically hindered amino acids, such as Arg, Ile, Val, and Leu, we do a 1-h recoupling.

Following synthesis, the peptide resin is then washed thoroughly with DMF, ethanol, and lyophilized. This step is important since residual DMF may add to the *N*-terminus of your peptide. Next, the peptide is cleaved from the resin concurrently with the protecting groups.

1.4 Protecting Groups

When the carboxylic acid is activated towards substitution, it can react with a variety of nucleophilic species. Besides the terminal α-amino group of the growing chain, nucleophiles can be present in the side chain of other residues in the sequence. From a synthetic perspective, this would lead to an unacceptable amount of impurities, hence the need for protecting groups.

The acid-labile protecting group *tert*-Bu is used to protect the side-chain functional groups of Ser, Thr, Tyr, Glu, and Asp [35]. The trityl (Trt) group [36] is employed for Cys, Asn, Gln, and His. For Arg 2,2,4,6,7-pentamethyl-dihydrobenzofuran-5-sulfonyl (Pbf) [37] is the most commonly used. Finally, the Boc group is preferred for Lys and Trp [38]. A number of protecting groups are also available for special purposes as described in the section on other aspects of SPPS, Table 1.

It is usually straightforward to remove the acid-labile protecting groups while cleaving the peptide from the solid support using a cocktail based on ~95 % TFA for 2 h. However, if a fully protected peptide is desired, particular attention has to be paid in order to ensure that the protecting groups can withstand the cleavage conditions. Particularly labile groups like Trt are never stable in acidic conditions, even when no TFA is used.

Table 1
Commonly used protecting groups for solid-phase peptide synthesis

Protecting group	Resistant to	Labile to	Removed with
Fmoc	Acids	Bases	10–20 % Piperidine
tBu	Bases	Acids	≥50 % TFA in DCM
Boc	Bases	Acids	≥50 % TFA in DCM
Trt	Bases	Acids	≥2 % TFA
Pbf	Bases	Acids	~95 % TFA

1.5 Cleavage Cocktails

An alphabet soup of cleavage cocktails is available which all contain TFA as the main component and scavengers that react with carbocations generated during cleavage. In our lab we use Reagent B′ TFA/H_2O/triisopropylsilane (95:2.5:2.5) (v/v) which works well for most peptides. We recommend that you do a test cleavage using approximately 5 mg of resin before choosing your cleavage cocktail.

Other popular cleavage cocktails include

Reagent K: TFA/thioanisole/H_2O/phenol/EDT (82.5:5:5:5:2.5) [39]

Reagent L: TFA/TIS/dithiothreitol/H_2O (88:2:5:5) [40]

Reagent R: TFA/thioanisole/ethanedithiol/anisole (90:5:3:2) [41]

Reagent B: TFA/phenol/H_2O/triisopropylsilane (88:5:5:2) [42]

For protected peptides: Acetic acid/TFE/dichloromethane (20:20:60) [43]

Following cleavage for 2 h, the solution is collected and TFA is evaporated by a stream of nitrogen down to a few hundred microliters. The peptide is precipitated in cold ether and centrifuged; the supernatant is then discarded and the washing procedure is repeated two times. Short and/or very lipophilic peptides can be considerably soluble in diethyl ether; in such cases they can be precipitated from fully apolar solvents such as hexane or heptane. For a successful wash it is important to obtain a fine peptide suspension; a short immersion in an ultrasound bath can help disintegrating undesired peptide aggregates. After drying, the peptide is dissolved in a mimimum volume of MeCN and diluted with water to approx. 20 %. (we actually use 10–20 % MeCN and 0.1 % TFA; solutions of acetic acid in water tend to defrost quickly in the freeze-dryer; for very insoluble peptides 100 % AcOH can be used—after considering compatibility with possible protecting groups) and lyophilized to give a fluffy white powder.

1.6 Analysis

Following synthesis, the product should be analyzed by liquid chromatography-MS (LC-MS) or analytical reverse-phase HPLC and MALDI-time-of-flight mass spectrometry. Analytical HPLC and preparative HPLC are most often done on a C-18 column using 0.1 % aqueous TFA as buffer A and CH_3CN/H_2O (9:1) as buffer B. For MALDI-TOF-MS, α-cyano-4-hydroxycinnamic acid is the most popular matrix. Alternatively, the HPLC UV and mass spectra may be generated by an LC-MS instrument. We are using an Agilent 6410 triple-quadrupole mass spectrometer instrument coupled to an Agilent 1200 HPLC system.

1.7 Other Aspects of SPPS

1.7.1 Modifications

A number of peptide modifications are possible. Some of the modifications include *C*->*N* termini cyclization [44], and Asp->Lys lactamization [45]; biotin [46], fatty acid [47], or fluorescent probe modification of the N-terminus [48]; secondary or tertiary amide, ester, aldehyde, or hydrazine modification of the C-terminus [29]; glycosylation [49], phosphorylation [50], or sulfation [51] of serine, threonine, or tyrosine; and introduction of non-proteinogenic amino acids such as citrulline for studies of autoimmune diseases [52], photoactivatable amino acids for peptide-protein conjugation [53], *N*-methylated [54] and peptoid residues [55, 56] for proteolytic stability, labeled amino acids for NMR studies [57], and branched lysine peptide constructs for antibody generation [58] and disulfide bond formation [59]. For these modifications special protecting groups are needed.

1.7.2 Special Protecting Groups

Lysine may be protected with a number of protecting groups (PGs) that can be cleaved in the presence of other PGs, while the peptide is still attached to the resin. This may be useful for selective introduction of modifications to the ε-amino group, e.g., biotin. The PGs include ivDde (2 % hydrazine) [60]; Mtt and Mmt (more acid-sensitive analogues of Trt, removed with ≤1.8 % TFA) [61]; Aloc (Pd + nucleophile) [62], or photolysis [63]. Similarly, C^{α}, Asp, and Glu may be protected with allyl ($PhSiH_3$ + $Pd(PPh_3)_4$) [64] or Dmab (2 % hydrazine) [65] for the synthesis of C-terminal-modified peptides. Strategies involving unusual protecting groups and anchoring of the side chain of Asn and Glu [66], Asp [64], Lys [65], or Cys [67] to the resin have been described. The latter is especially useful since Cys attached to the resin via an ester bond, but not amide bond, is prone to epimerization [68].

The acetamido group (Acm) [69] is often used for Cys when synthesizing peptides with two or more disulfide bonds. The Acm group is not cleaved by TFA but requires I_2 [70] or $Tl(Tfa)_3$ [71] for cleavage. Furthermore, the disulfide bonds can be formed on resin or in solution following cleavage [72].

1.7.3 Microwave Heating

Some peptides may form β-sheet-type structures during synthesis or involve many difficult couplings with β-branched amino acids. For such sequences microwave heating may lead to significant reductions in synthesis times and an increase in the crude peptide purity. However, conditions often need to be optimized for peptides containing Asp, Cys, and His. This is also true for phosphopeptides, glycopeptides, and *N*-methylated peptides since epimerization, aspartimide formation, and β-elimination may occur. For an excellent review on microwave heating in SPPS *see* ref. 73.

In the Fmoc strategy, the Fmoc group is cleaved by 20 % piperidine in DMF. However, the Fmoc group may also be cleaved by the non-nucleophilic reagent 1,8 diazobicyclo(5.4.0)-undec-7-ene (DBU) [74] for phosphorylated or glycosylated residues which are prone to β-elimination. Typically, a 2 % solution of DBU in DMF is used.

1.7.4 Synthesis of Long Peptides and Proteins

Synthesis of long peptides and proteins is most commonly carried out using segment condensation strategies [75] or native chemical ligation [76].

In the segment strategy, protected peptide fragments are synthesized on a 2-chlorotrityl resin, cleaved by 1 % TFA, purified, and coupled to another resin-bound peptide [23]. The HIV fusion inhibitor Fuzeon is synthesized using this strategy [77].

Native chemical ligation was introduced by Dawson and co workers [76]. This strategy involves chemoselective reaction of two unprotected peptide segments, one having a thioester at the C-terminal end, and the other a Cys residue at the N-terminus. Reaction gives an initial thioester-linked species. Spontaneous rearrangement gives a full-length product with a native peptide bond at the ligation site. Native chemical ligation has been used to make hundreds of proteins ranging in size up to more than 200 amino acids. For a review *see* ref. 78. Since thioesters are base labile they are more difficult to synthesize by Fmoc chemistry than Boc chemistry.

1.8 Concluding Remarks

Solid-phase peptide synthesis is a mature methodology which allows most peptides up to approximately 50 amino acids to be synthesized in good yield and purity. Furthermore, non-proteinogenic amino acids or other modifications may be incorporated, including cyclization, glycosylation, phosphorylation, fluorescent labelling, biotinylation, or disulfide bond formation. However, it is important to realize that certain peptide sequences are difficult to prepare even for experienced peptide chemists. In these cases careful attention should be paid to the peptide sequence, protecting groups, coupling reagents, heating protocol, and cleavage conditions as outlined in this chapter and the literature cited.

2 Materials and Preparations

General procedure for peptide synthesis in disposable syringes.

2.1 Choosing the Resin

This mostly depends on two variables:

(a) Is the goal to obtain a peptide acid or a peptide amide?

(b) Should the side-chain protecting groups be removed when the peptide is released from the solid support?

If a free peptide is desired (a), it is usual to employ a resin that releases the peptide after treatment with a 95 % TFA cocktail. A preloaded Wang resin would be the choice for peptide acids; a non-preloaded TentaGel®/Hypogel® resin functionalized with a RAM linker is very useful for the synthesis of peptide amides. If the peptide should be obtained without removing the side-chain protecting groups (b), a 2-chlorotrityl and Sieber resin are good choices for peptide acids and amides, respectively.

2.2 Synthesis Scale

The amount of resin to be used depends on how much peptide is desired. This can be calculated using the following formula:

$$g\ of\ Resin = \frac{mmol\ of\ Peptide\,(desired)}{Resin\ loading}$$

As this is a purely theoretical value (i.e., not factoring impurities and other losses), it is convenient to multiply it by 1.5 or 2.

2.3 Reactors

The reactor is the vessel where reactions take place. For peptide synthesis it is important that the reactor has one upper and one lower opening, and one filter (usually made of polytetrafluoroethylene, PTFE) to allow solvents to flow while retaining the resin (*see* **Note 1**). This protocol is based on the use of 5 ml syringes equipped with a PTFE filter as reactors for 30–120 mg. of resin and assumes the availability of a suction device (*see* **Note 2**). The suction device is however not strictly necessary. The use of pipette tips of 200 μl capacity is advisable, as these provide a disposable interface between the syringe and the solvents, reagents, and suction plate. It is often necessary to trim the wide end of such tips to achieve a good fit on the syringes.

2.4 HOAt and HATU Solutions

Three equivalents of HOAt and HATU are needed for each coupling, based on the synthesis scale calculated before (*see* **Note 3**).

It is convenient to calculate the total number of couplings and prepare a 0.4 M stock solution of both in dimethylformamide (DMF):

$$mg\ HOAt = 135.12 * 3 * \left[synthesis\ scale(mmol)\right] * \left[number\ of\ couplings\right]$$

$$mg\ HATU = 380.23 * 3 * \left[synthesis\ Scale(mmol)\right] * \left[number\ of\ couplings\right]$$

The respective total amounts are then to be dissolved in DMF up to a volume calculated as follows:

$$V\ Solution(ml) = \frac{3 * \left[synthesis\ scale(mmol)\right] * \left[number\ of\ couplings\right]}{0.4\,M}$$

The HATU solution should be divided in as many individual tubes as the total number of couplings. The HOAt solution shall be used to dissolve the amino acid samples (*see* next point). It is advisable to prepare a standard template using Microsoft® Excel™ to perform these calculations quickly and automatically.

2.5 Amino Acid Solutions and N, N-Diisopropylethylamine

In our lab we prepare 0.4 M solutions of our amino acids by dissolving them in an appropriate volume of the HOAt stock solution prepared in Subheading 2.4. One individual sample tube containing three equivalents of amino acid and HOAt should be prepared for each coupling.

The required amount of amino acid for each coupling should be weighed out in an individual tube:

$$mg\ Amino\ acid = Molecular\ weight\ of\ Amino\ acid * 3 * \left[synthesis\ scale(mmol)\right]$$

Equally divide the HOAt stock solution among the amino acid samples.

By doing so, a tube containing the amino acid and HOAt solution and a tube containing the HATU solution have been prepared for each coupling. For overnight or medium-term storage, place the tubes in a −20 °C freezer. Let them warm up again to room temperature for at least 20 min before use.

During coupling of each amino acid *N*,*N*-diisopropylethylamine (DIEA) is added to the solution. The amount for each coupling is calculated as follows:

$$\mu\text{L DIEA} = \frac{\left[129.24\frac{g}{mol} * 6 * \left[synthesis\ scale(mmol)\right]\right]}{0}.742\frac{g}{ml}$$

2.6 Piperidine Solution

This solution is needed to remove the Fmoc group between one coupling and the following. To prepare a 20 % v/v solution of piperidine in DMF, simply measure both volumes (e.g., 40 ml and 160 ml, respectively) separately in a graduated cylinder and mix them together in the final container. The final container should be a brown bottle with a screw cap.

3 Methods

3.1 Initiating the Synthesis

1. Weigh out the resin directly inside the reactor (*see* **Note 4**), reinsert the piston, and push it all the way down.
2. Transfer 3 ml of DMF into a beaker and draw it into the syringe.
3. Close the bottom end with a pressure cap and let it swell in DMF for at least 30 min (preferably overnight).

3.2 Quick Wash

1. Remove the piston and place the syringe on the suction plate.
2. Wash down with DMF any resin residue on the piston head.
3. Wash three times with DMF.

3.3 Amino Acid Coupling (See Note 5)

1. Add the HATU solution to the HOAt/amino acid solution.
2. Add DIEA (Sigma-Aldrich) and shake briefly: the solution should turn intense yellow.

3. Draw the coupling solution into the syringe avoiding to introduce too much air.
4. Cover the syringe with tin foil, leave it on a shaker for 1:30 h, and then discharge the coupling solution.
5. If a double coupling is desired, do a *Quick wash* (Subheading 3.2) and then repeat **steps 1–4**.

3.4 Full Wash

1. Remove the piston and place the syringe on a suction plate.
2. Wash down with DMF any resin residue on the piston head.
3. Wash with DMF (3×), DCM (Sigma-Aldrich) (3×), and DMF again (5×); fill the reactor all the way up at least a couple of times in order to wash the walls.

3.5 Fmoc Group Removal

1. Transfer 2–3 ml of piperidine solution into the syringe and leave standing.
2. After 4 min drain the solution and do a *Quick wash* (Subheading 3.2).
3. Repeat **steps 1** and **2** twice (thus a total of three deprotection cycles; for peptides longer than ten residues, extend the time of the last two deprotection cycles to 7 min each).
4. Put the piston back in place and change the syringe tip.

3.6 Full Wash (as in Subheading 3.4)

3.7 Reiteration (as in Subheadings 3.3–3.6)

Final Fmoc Group Removal (as in Subheadings 3.5–3.6)

1. Do as in Subheading 3.5, and then wash as in subheading 3.4, and finally wash the resin five times with ethanol (rinse the whole reactor and the piston, too).
2. Insert the piston and push it halfway down, then discard the tip, and leave the reactor in a lyophilizer overnight or at least for 2 h.

3.8 Cleavage

1. Freshly prepare at least 6 ml of TFA/H_2O/triisopropylsilane (95:2.5:2.5) (v/v) cleavage cocktail (*see* **Note 6**) for each peptide.
2. Push the piston all the way down.
3. Transfer 3.5 ml of cleavage cocktail in a beaker and draw it into the syringe. Pay attention to drops leaking out.
4. Carefully cap the lower end of the syringe with a pressure cap.
5. Place on a shaker for at least 2 h. Place the reactor vertically on a shaker and shake gently for 2 h.
6. Using the piston, push the cleavage solution into a 5 ml cryotube without touching the resin with the piston head.
7. Remove the piston.
8. Wash the resin twice pouring ~1 ml of cleavage cocktail from above collecting the eluate in the cryotube.

9. Push the piston all the way down to collect the last drops of solution into the cryotube.
10. Evaporate the solution with a gentle stream on N_2.
11. When 300 μl or less are left, add 4 ml of cold (−20 °C) diethyl ether (*see* **Note 7**). Put the cap on and shake gently.
12. Centrifuge at 2000 rpm for 6 min.
13. Using a pipette, carefully remove and discard the supernatant.
14. Resuspend the solid into another 3–4 ml of cold ether.
15. Repeat **steps 12–14** for a total of three washes. The last centrifuge run should be at 4000 rpm.
16. Leave the cryotube standing open overnight to let the residual ether evaporate.
17. Dissolve the crude product in 90 % water and 10 % acetonitrile +0.1 % TFA and freeze-dry to obtain fluffy white crystals (*see* **Note 8**).

4 Notes

1. Standard polypropylene 5 ml syringes can be fitted with a PTFE filter and used for this purpose. In this case, pressure caps from Sigma-Aldrich (Cat# Z120979) to close the narrow end are useful. The syringe piston head must not have a rubber O-ring, as this could be unstable in the reaction and cleavage conditions. It is also possible to buy 5 ml polypropylene reactors equipped with a PTFE filter and a bottom and a top cap from Thermo Scientific (Cat# 29922). The main difference between the two approaches is that, in the case of syringes, the bottom opening serves both as inlet and outlet and the piston can be used to drive the flow; with other reactors the inlet is the top opening and the outlet is the bottom one, and a suction device is needed to drive the flow. If no suction device is available, syringes must be used: washing solvents and the piperidine solution must be transferred in a beaker, and then drawn into (and ejected from) the syringe using the piston.
2. This is basically an in-house-made Teflon® plate featuring five holes with a diameter of ~4 mm. This size allows to accommodate syringes, 200 μl pipette tips, and the Thermo Scientific reactors. The plate and the vacuum pump are connected to a 2.5 L glass reservoir using Omnifit® fittings (Fig. 4). We drill a hole in a cork stopper to let the drain pipe pass through, and then we insert it in a rubber ring adapter (the sort used for filtration) to ensure a tight fit with the reservoir neck. A three-way flow switch is needed to prevent the piperidine solution from being drained from the reactors during deprotection (Fig. 4).

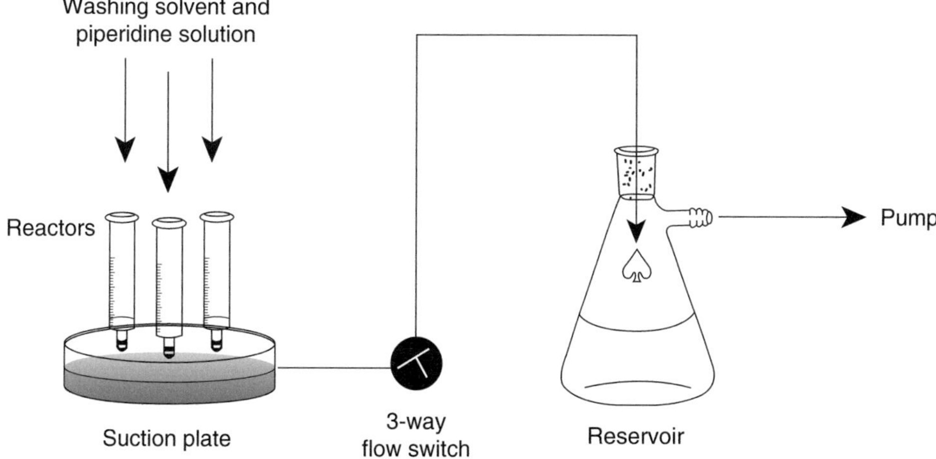

Fig. 4 Peptide synthesis setup

3. It is usual to acylate one amino functionality at a time, but when acylations are taking place at multiple positions, this should be accounted for. For example, when lysine-based branched peptides are synthesized by coupling on both amino groups simultaneously, this has the same effect as doubling the resin loading as far as stoichiometry is concerned.
4. When synthesizing multiple peptides at once, it is very convenient to label the reactors with tape of different colors.
5. If the amino functionality on the resin (preloaded or not) is Fmoc protected, **steps 5** and **6** need to be completed before the first coupling can take place. Please note that except for 2-chlorotrityl resins, most other resins (preloaded or not) are Fmoc protected.
6. Cleavage cocktails based on TFA and other caustic substances are highly corrosive and need to be handled with care. Remember that acids are always added to water and never the other way around. Always use glass containers and glass Pasteur pipettes to prepare and transfer the cocktail. These cocktails cannot be stored and have to be freshly prepared before use. Triisopropylsilane dissolves slowly in TFA and persists as oily spots on the solvent surface; make sure that dissolution is complete before exposing the resin to the cocktail.
7. Short (≤6 AA) and lipophilic peptides can be considerably soluble in diethylether; in this case use n-hexane instead. If in doubt, the supernatant can be saved and kept at −20 °C overnight. In the morning, check for the presence of precipitate.
8. Note that the peptide will be obtained as a TFA salt.

References

1. Fosgerau K, Hoffmann T (2015) Peptide therapeutics: current status and future directions. Drug Discovery Today 20:122–128
2. Craik DJ, Fairlie DP, Liras S et al (2013) The future of peptide-based drugs. Chem Biol Drug Des 81:136–147
3. Okarvi SM (2008) Peptide-based radiopharmaceuticals and cytotoxic conjugates: Potential tools against cancer. Cancer Treat Rev 34:13–26
4. Mercer DK, O'Neil DA (2013) Peptides as the next generation of anti-infectives. Future Med Chem 5:315–337
5. Gori A, Longhi R, Peri C et al (2013) Peptides for immunological purposes: design, strategies and applications. Amino Acids 45:257–268
6. Trier NH, Hansen PR, Houen G (2012) Production and characterization of peptide antibodies. Methods 56:136–144
7. Robinson JA (2013) Max Bergmann lecture Protein epitope mimetics in the age of structural vaccinology. J Pept Sci 19:127–140
8. Johnson IS (1983) Human insulin from recombinant DNA technology. Science 213: 632–637
9. Merrifield RB (1963) Solid phase peptide synthesis.I. The synthesis of a tetrapeptide. J Am Chem Soc 85:2149–2154
10. Pedersen SW, Armishaw CJ, Strømgaard K (2013) Synthesis of peptides using tert -butyloxycarbonyl(Boc) as the α-amino protection group. In: Jensen KJ, Shelton PT, Pedersen SL (eds) Peptide synthesis and applications. Humana Press, New York, pp 65–80
11. Atherthon E, Fox H, Harkiss D, et al. (1978) A Mild procedure for Solid Phase Peptide Synthesis: Use of Fluorenylmethyloxycarbonylamino-Acids. J Chem Soc Chem Commun 537–539
12. Jensen KJ, Shelton PT, Pedersen SL (2013) Peptide synthesis and applications. Humana Press, New York
13. Fields GB, Lauer-Fields JL, Liu R-Q et al (2002) Principle and practice of solid-phase peptide synthesis. In: Grant G (ed) Synthetic peptides: a user's guide. Oxford University Press, Oxford, pp 93–219
14. Fields GB (1997) Solid-phase peptide synthesis. Methods Enzymol 289
15. Rapp W, Zhang L, Habich R et al (1989) Polystyrene-polyoxyethylene graft copolymers for high speed peptide synthesis. In: Bayer E, Jung G (eds) Peptides 1988: proceedings of the 20th european peptide symposium. De Gruyter, Walter, Inc, Berlin, pp 199–201
16. García-Martín F, Quintanar-Audelo M, García-Ramos Y et al (2006) ChemMatrix, a poly(ethylene glycol)-based support for the solid-phase synthesis of complex peptides. J Comb Chem 8:213–220
17. Meldal M (1992) PEGA: a flow stable polyethylene glycol dimethyl acrylamide copolymer for solid phase synthesis. Tetrahedron Lett 33:3077–3080
18. Meldal M, Auzanneau F-I, Hindsgaul O, et al. (1994) A PEGA Resin for use in the Solid-phase Chemical-Enzymatic Synthesis of Glycopeptides. J Chem Soc Chem Commun 1849–1850
19. Songster MF, Barany G (1997) Handles for solid-phase peptide synthesis. Methods Enzymol 289:126–174
20. Góngora-Benítez M, Tulla-Puche J, Albericio F (2013) Handles for fmoc solid-phase synthesis of protected peptides. ACS Comb Sci 15: 217–228
21. Wang SS (1973) p-alkoxybenzyl alcohol resin and p-alkoxybenzyloxycarbonylhydrazide resin for solid phase synthesis of protected peptide fragments. J Am Chem Soc 95:1329–1333
22. Rink H (1987) Solid-Phase synthesis of protected peptide fragments using a tri-alkoxy-diphenyl-methylester resin. Tetrahedron Lett 28:3787–3790
23. Barlos K, Chatzi O, Gatos D et al (1991) 2-chlorotrityl chloride resin. Int J Pept Protein Res 37:513–520
24. Han Y, Bontems S, Hegyes P et al (1996) Preparation and applications of xanthenylamide (XAL) handles for solid-phase synthesis of C-terminal peptide amides under particularly mild conditions. J Org Chem 61: 6326–6339
25. Atherthon E, Sheppard RC (1989) Solid phase peptide synthesis. A practical approach. Oxford University Press, Oxford
26. Mullen DG, Barany G (1988) A New fluoridolyzable anchoring linkage for orthogonal solid-phase peptide synthesis: design, preparation, and application of the (N-3 or 4)-((4-(hydroxymethyl) phenoxy)-tert-butylphenylsilyl)phenyl pentanedioic acid monamide Pbs handle. J Org Chem 53:5240–5248
27. Chhabra SR, Parekh H, Khan AN et al (2001) A Dde-based carboxy linker for solid-phase synthesis. Tetrahedron Lett 42:2189–2192
28. Chumachenko N, Novikov Y, Shoemaker RK et al (2011) A dimethyl ketal-protected benzoin-based linker suitable for photolytic release of unprotected peptides. J Org Chem 76:9409–9416
29. Jensen KJ, Alsina J, Songster MF et al (1998) Backbone amide linker (BAL) strategy for

solid-phase synthesis of C-terminal-modified and cyclic peptides. J Am Chem Soc 123: 5441–5452

30. El-Faham A, Albericio F (2011) Peptide coupling reagents, more than a letter soup. Chem Rev 111:6557–6602
31. Subiros-Funosas R, Prohens R, Barbas R et al (2009) Oxyma: an efficient additive for peptide synthesis to replace the benzotriazole-based HOBt and HOAt with a lower risk of explosion. Chemistry 15:9394–9403
32. Coste J, LeNguyen D, Castro B (1990) PyBoP®: a New peptide coupling reagent devoid of toxic by-product. Tetrahedron Lett 31:205–208
33. Knorr R, Trzcieak A, Bannwarth W et al (1989) New coupling reagents in peptide chemistry. Tetrahedron Lett 30:1927–1930
34. Carpino LA (1993) 1-hydroxy-7-azabenzotriazole. An efficient peptide additive. J Am Chem Soc 115:4397–4398
35. Callahan FM, Anderson GW, Paul R et al (1963) The tertiary butyl group as a blocking agent for hydroxyl, sulfhydryl and amido functions in peptide synthesis. J Am Chem Soc 85:201–207
36. Sieber P, Riniker B (1991) Protection of carboxamide functions by the trityl residue. Application to peptide synthesis. Tetrahedron Lett 32:739–742
37. Carpino LA, Shroff H, Triolo SA et al (1993) The2,2,4,6,7-pentamethyldihydrobenzofuran-5-sulfonyl group (Pbf) as arginine side chain protectant. Tetrahedron Lett 34:7829–7832
38. McKay FC, Albertson NF (1957) New amine-masking groups for peptide synthesis. J Am Chem Soc 79:4686–4690
39. King DS, Fields CG, Fields GB (1990) A cleavage method which minimizes side reactions following fmoc solid phase peptide synthesis. Int J Pept Protein Res 36:254–266
40. Nielsen SL, Frimodt-Moller N, Kragelund BB et al (2007) Structure activity study of the antibacterial peptide fallaxin. Protein Sci 16:1969–1976
41. Albericio F, Kneib-Cordonier N, Biancalana S et al (1990) Preparation and application of the PAL handle for the solid-phase peptide synthesis of C-terminal peptide amides under mild conditions. J Org Chem 55:3730–3743
42. Solé NA, Barany G (1992) Optimization of solid-phase peptide synthesis of [Ala8]-dynorphin. J Org Chem 57:5399–5403
43. Napolitano A, Rodriquez M, Bruno I et al (2003) Synthesis, structural aspects and cytotoxicity of the natural cyclopeptides yunnanins A, C and phakellistatins 1, 10. Tetrahedron 59:10203–10211
44. Davies J (2003) The cyclization of peptides and depsipeptides. J Pept Sci 9:471–501
45. Houston ME Jr, Gannon CL, Kay CM et al (1995) Lactam bridge stabilisation of α-helical peptides: ring size. Orientation and positional effects. J Pept Sci 1:274–282
46. Natarajan S, Festin SM, Hedberg A et al (1992) Site-specific biotinylation. A novel approach and its application to Endothelin-1 analogs and PTH-analogs. Int. J. Pept Protein Res 40:567–574
47. Chicharro C, Granata C, Lozano R et al (2001) N-terminal fatty acid substitution increases the leishmanicidal activity of CA(1-7)M(2-9), a cecropin-melittin hybrid peptide. Antimicrob Agents Chemother 45:2441–2449
48. Fernandez-Carneado J, Giralt E (2004) An efficient method for the solid-phase synthesis of fluorescently labelled peptides. Tetrahedron Lett 45:6079–6081
49. Jensen KJ, Hansen PR, Venugopal D et al (1996) Synthesis of 2-acetamido-2-deoxy-D-glucopyranose O-glycopeptides from N-dithiasuccinoyl-protected derivatives. J Am Chem Soc 118:3148–3155
50. Valerio RM, Bray AM, Maeji NJ et al (1995) Preparation of O-phosphotyrosine-containing peptide by fmoc solid-phase synthesis: evaluation of several fmoc-Tyr(PO3R2)-OH derivatives. Lett Pept Sci 2:33–40
51. Musiol H-J, Escherich A, Moroder L (2002) Synthesis of sulfated peptides. Synthesis of Peptides and Peptidomimetics E 22b: 425–453
52. Trier NH, Leth ML, Hansen PR et al (2012) Cross-reactivity of a human IgG1 anticitrullinated fibrinogen monoclonal antibody to a citrullinated profilaggrin peptide. Protein Sci 21:1929–1941
53. Strynadka NC, Redmond MJ, Parker R et al (1998) Use of synthetic peptides to Map the antigenic determinants of glycoprotein D of herpes simplex virus. J Virol 62:3474–3483
54. Chatterjee J, Gilon C, Hoffman A et al (2008) N-methylation of peptides: a new perspective in medicinal chemistry. Acc Chem Res 41: 1331–1342
55. Miller SM, Simon RJ, Ng S et al (1994) Proteolytic studies of homologous peptide and N-substituted glycine. Bioorg Med Chem Lett 4:2657–2662
56. Jahnsen RD, Sandberg-Schaal A, Vissing KJ et al (2014) Tailoring cytotoxicity of antimicrobial peptidomimetics with high activity against multidrug-resistant escherichia coli. J Med Chem 57:2864–2873
57. Schneider T, Kruse T, Wimmer R et al (2010) Plectasin, a fungal defensin, targets the bacterial

cell wall precursor lipid II. Science 328: 1168–1172

58. Tam JP, Zavala F (1989) Multiple antigenic peptide. A novel approach to increase detection sensitivity of synthetic peptides in solid-phase immunoassays. J Immunol Methods 124:53–61
59. Andreu D, Albericio F, Solé NA et al (1994) Formation of disulfide bonds in synthetic peptides and proteins. Methods Mol Biol 35:91–169
60. Chhabra SRBH, Evans DJ, White PD et al (1998) An appraisal of new variants of Dde amine protecting group for solid phase peptide synthesis. Tetrahedron Lett 39:1603–1606
61. Li D, Elbert DL (2002) The kinetics of the removal of the N-methyltrityl (Mtt) group during the synthesis of branched peptides. J Pept Res 60:300–303
62. Loffet A, Zhang HXM (1993) Allyl-based groups for side-chain protection of amino-acids. Int J Pept Protein Res 42:346–351
63. Rusiecki VK, Warne SA (1993) Synthesis of Nα-Fmoc-Nε-Nvoc-Lysine and Use in the preparation of Selectively Functionalized Peptides. Bioorg Med Chem Lett 3:707–710
64. Salvati M, Cordero FM, Pisaneschi F, et al. (2006) New cyclic Arg-Gly-Asp pseudopenta-peptide containing the β-turn mimetic GPTM. Synlett, 13:2067–2070
65. Berthelot T, Goncalves M, Laın G et al (2006) New strategy towards the efficient solid phase synthesis of cyclopeptides. Tetrahedron Lett 62:1124–1130
66. Albericio F, Van Abel R, Barany G (1990) Solid-phase synthesis of peptides with C-terminal asparagine or glutamine. Int J Pept Protein Res 35:284–286
67. Barany G, Han Y, Hargittai B et al (2003) Side-chain anchoring strategy for solid-phase synthesis of peptide acids with C-terminal cysteine. Biopolymers 71:652–666
68. Han Y, Albericio F, Barany G (1997) Occurrence and minimization of Cysteine racemization during step-wise solid-phase synthesis. J Org Chem 62:4307–4312
69. Veber DF, Milkowski JD, Varga SL et al (1972) Acetamidomethyl. A Novel Protection Group for Cystein. J Am Chem Soc 94:5456–5461
70. Goulas S, Gatos D, Barlos K (2006) Convergent solid-phase synthesis of hirudin. J Pept Sci 12:116–123
71. Fujii N, Otaka A, Funakoshi S et al (1987) Studies on peptides. CLI. Syntheses of cysteine-peptides by oxidation of S-protected cysteine-peptides with thallium(III) trifluoroacetate. Chem Pharm Bull 35:2339–2347
72. Munson M, Barany G (1993) Synthesis of a-Conotoxin-SI, a bicyclic tridecapeptide amide with Two disulfide bridges: illustration of novel protection schemes and oxidation strategies. J Am Chem Soc 115:10203–10210
73. Pedersen SL, Tofteng AP, Malik L et al (2012) Microwave heating in solid-phase peptide synthesis (2012). Chem Soc Rev 41:1826–1844
74. Wade JD, Bedford J, Sheppard RC et al (1991) DBU as an N-a deprotection reagent for the fluorenyl methoxycarbonyl group in continuous flow SPPS. Pept Res 4:194–199
75. Lloyd-Williams P, Albericio F, Giralt E (1993) Convergent solid-phase peptide synthesis. Tetrahedron 48:11065–11133
76. Dawson PE, Muir TW, Clark-Lewis I et al (1994) Synthesis of proteins by native chemical ligation. Science 266:776–779
77. Schneider S, Bray B, Mader C et al (2005) Development of HIV fusion inhibitors. J Pept Sci 11:744–753
78. Kent SBH (2009) Total chemical synthesis of proteins. Chem Soc Rev 38:338–351

Chapter 6

Peptide-Carrier Conjugation

Paul R. Hansen

Abstract

To produce antibodies against synthetic peptides it is necessary to couple them to a protein carrier. This chapter provides a nonspecialist overview of peptide-carrier conjugation. Furthermore, a protocol for coupling cysteine-containing peptides to bovine serum albumin is outlined.

Key words Peptide-carrier conjugation, Bovine serum albumin, m-Maleimidobenzoyl-*N*-hydroxy-succinimidyl ester (MBS)

1 Introduction

In order to produce antibodies against synthetic peptides with a molecular weight less than 1500 Da it is necessary to couple them to a protein carrier [1]. This is because both a T-cell epitope and B-cell epitope are required for a strong immune response and small synthetic peptides cannot provide both. For excellent reviews on peptide-carrier conjugation see Muller [2] and Hermanson [3].

The choice of carrier molecule is determined by the number of functional groups available for conjugation, immunogenicity of the carrier protein, cost, and most importantly whether the produced peptide-protein conjugate is water soluble. Commonly used carriers are BSA (MW 67,000 Da), KLH (MW 4.5×10^5 to 1.3×10^7 Da), thyroglobulin (MW 660,000 Da), and ovalbumin (MW 43,000 Da) [3].

It is not well established how many peptide molecules per carrier protein are necessary to generate a good antibody response. However, Hodges et al. have shown that two peptides per BSA molecule are enough to generate antibodies against peptides related to herpes simplex virus [4]. The peptide-carrier conjugation ratio may be determined by amino acid analysis [5], spectrophotometry [6], or mass spectrometry [7].

The most widely used strategy for preparation of peptide-carrier conjugates is solution conjugation.

Gunnar Houen (ed.), *Peptide Antibodies: Methods and Protocols*, Methods in Molecular Biology, vol. 1348,
DOI 10.1007/978-1-4939-2999-3_6, © Springer Science+Business Media New York 2015

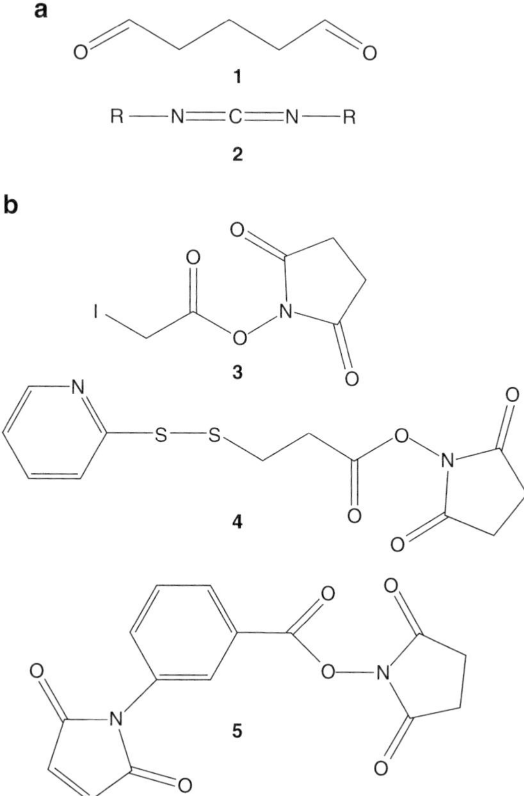

Fig. 1 Peptide-protein coupling reagents. (**a**) One-pot reagents. (**b**) Two-step reagents (require Cys in the peptide sequence)

The solution strategy may be divided into a (1) one-pot or (2) two-step conjugation method (Fig. 1). The majority of them react with the ε-amino group of lysine and/or the thiol group of cysteine.

In the one-pot method, the cross-linking reagent and peptide are added to the carrier protein followed by dialysis to remove unwanted compounds.

In the two-step method, two separate reactions are used to form the covalent bond between carrier protein and peptide. In the first reaction the cross-linking reagent is attached to the carrier protein via an amide bond. Gel filtration or dialysis is often used to remove excess of the coupling reagent, followed by conjugation of the peptide.

1.1 One-Pot Method

One of the oldest one-step cross-linking reagents is glutaraldehyde **1** which reacts with amino groups on the carrier protein and peptide to form Schiff bases (Fig. 2a). However, the reagent is very nonspecific and known also to form peptide and protein aggregates [8].

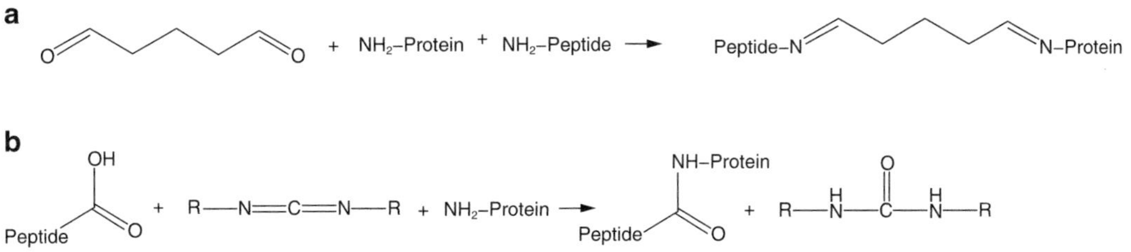

Fig. 2 One-pot peptide-protein coupling reagents. (**a**) Glutaraldehyde. (**b**) Carbodiimides

The most popular one-step reagents for peptide-protein conjugation are the carbodiimides **2**, which were first used to generate antibodies against bradykinin and angiotensin [9]. As shown in Fig. 2b, this class of reagents activates a C-terminal carboxylic acid group in the peptide which in turn reacts with an amino group in the protein to form an amide bond.

1.2 Two-Step Method

The most widely used two-step reagents are iodoacetic acid *N*-hydroxysuccinimide ester (IAAOSu) **3** [10], *N*-succinimidyl 3-(2-pyridyldithio) propionate (SPDP) **4** [6], and m-maleimidobenzoyl-*N*-hydroxysuccinimidyl ester (MBS) **5** [11]. These three coupling reagents all require that a Cys residue is present in the peptide sequence. If this is not the case it may be added to the *N*-terminus during solid-phase synthesis. The described two-step reagents all react in the same way in the initial step (Fig. 3a) First they form an amide with amino groups on the carrier protein via its NHS ester. The cysteine residue of the peptide then reacts with either the alkyliodide (Fig. 3b), the 2-pyridyldisulfide group (Fig. 3c), or m-maleimidobenzoyl group (Fig. 3d), resulting in a covalent bond between carrier and peptide.

IAAOSu: Rector et al. introduced iodoacetic acid N-hydroxysuccinimide ester to produce well-defined protein-protein conjugates using ovalbumin and IgG as test proteins [10]. Houen and coworkers demonstrated that IAAOSu also works well for peptide carrier conjugation. The authors conjugated several peptides, including CGHEKEGFMEAEQC and glutathione to ovalbumin which gave high antibody titers [12].

SPDP: *N*-succinimidyl 3-(2-pyridyldithio)-propionate **4** was first described by Carlsson and coworkers in 1978. Since pyridine-2-thione is released during the second step of the conjugation reaction, the peptide-protein carrier ratio may be estimated spectrometrically [6].

MBS: MBS **5** was first reported by Kitagawa and Aikawa who conjugated insulin to β-d-galactosidase [11]. Since then MBS has been used to conjugate a number of peptides to proteins including peptides of two regions of the surface protein VP1 of foot-and-mouth disease virus to KLH [13]. When using MBS as

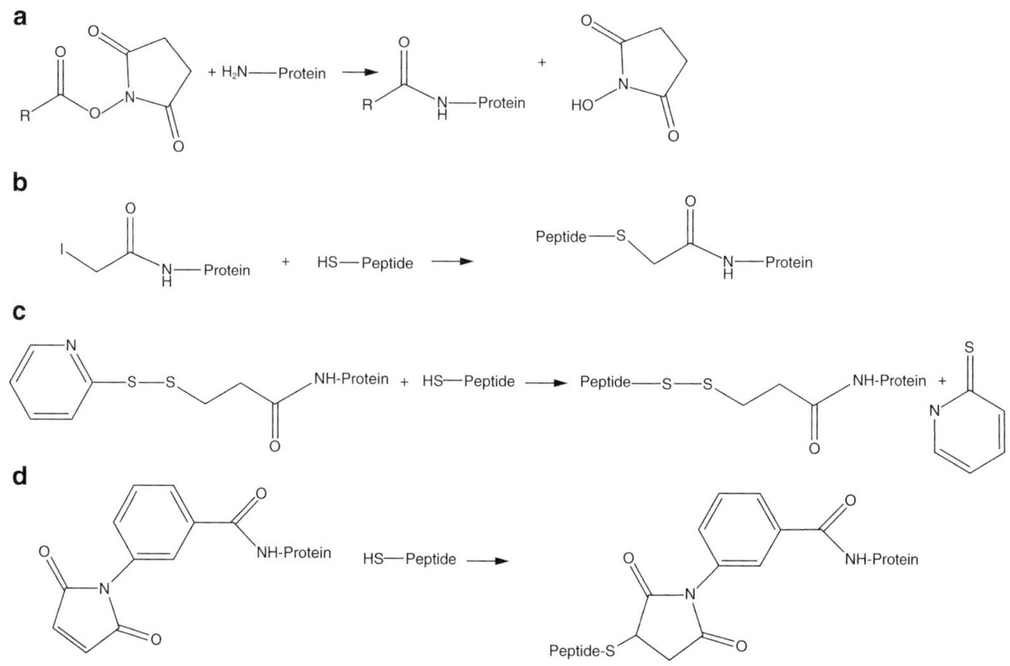

Fig. 3 Peptide-protein coupling reagents. (**a**) General acylation step for the NHS esters LAAOSu, SPDP, and MBS. Second coupling step for IAAOSu, SPDP, and MBS (require Cys in the peptide sequence). (**b**) IAAOSu. (**c**) SPDP. (**d**) MBS

coupling reagent, the carrier protein is typically dissolved in phosphate-buffered saline (PBS) followed by the addition of MBS in dimethylformamide (DMF). After 1 h, the MBS-protein conjugate is passed through a PD-10 desalting column. The peptide is then dissolved in DMF and mixed with MBS-conjugated carrier protein. Following overnight reaction, 0.1 M bicarbonate is added and the peptide-carrier conjugate lyophilized.

Lateef and coworkers successfully raised antipeptide antibodies using the above-described procedure [7]. Two test peptides, EMVAQLRNSSEPAKKC and RNTKGKRKGQGRPSPLAPC, were conjugated to BSA and the products analyzed by MALDI-TOF MS. The authors found that between 1 and 13 peptides were conjugated to BSA. A detailed peptide-protein conjugation protocol using MBS is provided in Subheading 3.

Most published peptide-protein conjugation procedures work quite well. However, there are pitfalls. In an excellent study by Briand et al., nine peptides, three carrier proteins, and four coupling reagents were used to prepare a number of peptide-protein conjugates [5]. The authors examined some of the problems that may be encountered when peptide-carrier conjugates are prepared for immunochemical assay such as (1) generation of specific antibodies that react with unrelated carrier proteins treated with the same coupling agent, (2) instability of the conjugate, and (3) alteration of the antigenic properties of the peptide moiety. Finally, some guidelines as to which control experiments are suitable are given.

1.3 Other Approaches

Multiple antigenic peptides: An alternative to traditional peptide-carrier conjugates is multiple antigenic peptides (MAPs) which were introduced by Tam [14] and have been used extensively in diagnostics and vaccine research [15, 16]. MAPs are a branched lysine core matrix carrying four copies of the antigenic peptide which is synthesized entirely by solid-phase synthesis. The advantage of this approach is that single chemically defined peptide antigens are produced, which yields good immune responses. The drawback is that MAP constructs containing hydrophobic peptide antigens may be difficult to synthesize and analyze.

2 Materials

1. Beaker 50 ml.
2. Bovine serum albumin (BSA).
3. Cysteine-containing peptide (*see* **Note 1**).
4. Dialysis tubing average flat width 43 mm, molecular weight cutoff 12,000 Da with dialysis tubing closures 50 mm.
5. Dimethylformamide (DMF).
6. Eppendorf tubes (1 and 5 mL).
7. MBS (*see* **Note 2**).
8. PBS (0.01 M), pH 7.4 (*see* **Note 3**).
9. Sodium azide.

3 Methods

3.1 Procedure for Coupling a Peptide to BSA Using MBS [2, 7]

3.1. Dissolve BSA (5 mg) in 0.5 mL of 0.01 M phosphate buffer (pH 7).

3.2. Prepare a solution of 10 mg/ml MBS in DMF.

3.3. Add 100 μL of the MBS solution dropwise and stir for 30 min.

3.4. To remove excess of MBS, transfer the conjugation solution to a dialysis tubing and close with dialysis tubing closures.

3.5. Put the dialysis tubing in a 50 ml beaker with 30 ml of 0.01 M phosphate buffer (pH 7.4) for 2 h in a fridge. Repeat twice and leave the last dialysis overnight.

3.6. Dissolve the cysteine-containing peptide (5 mg) in 100 μL DMF.

3.7. Transfer the solution in the dialysis tubing to a 5 ml Eppendorf tube and add the peptide solution.

3.8. Following overnight reaction at RT, dialyze the conjugation mixture as described in **steps 3.4–3.5**.

3.9. Store aliquots in a freezer or in a fridge with 0.02 % sodium azide *(see* **Note 4**).

4 Notes

1. The peptide may be produced as described by Hansen and Oddo (Chapter 5) or obtained commercially. A number of companies offer custom peptide synthesis. It is recommended that you order a purity of >90 %.
2. MBS conjugation kits are commercially available from several suppliers.
3. To prepare a 0.01 M PBS solution mix 8.064 g NaCl and 0.201 g KCl in a 1 L volumetric flask and H_2O to the calibration mark.
4. Ready-to-use PBS with azide buffer is commercially available from several suppliers.

References

1. Lee B-S, Huang J-S, Jayathilaka GDL et al (2010) Production of antipeptide antibodies. Methods Mol Biol 657:93–108
2. Muller S (1999) Peptide Carrier Conjugation In: van Regenmortel MVH and Muller S (eds) Synthetic Peptides as Antigens, Elsevier, Amsterdam pp. 79–130
3. Hermanson GT (2013) Bioconjugate techniques, 3rd edn. Academic, New York
4. Strynadka NC, Redmond MJ, Parker R et al (1988) Use of synthetic peptides to map the antigenic determinants of glycoprotein D of herpes simplex virus. J Virol 62:3474–3483
5. Briand JP, Muller S, van Regenmortel MHV (1985) Synthetic peptides as antigens: pitfalls of conjugation methods. J Immunol Methods 78:59–69
6. Carlsson J, Drevin H, Axén R (1978) Protein thiolation and reversible protein-protein conjugation. SPDP, a New heterobifunctional reagent. Biochem J 173:723–737
7. Lateef SS, Gupta S, Jayathilaka LP et al (2007) An improved protocol for coupling synthetic peptides to carrier proteins for antibody production using DMF to solubilize peptides. J Biomol Tech 18:173–176
8. Plaué S, Muller S, Briand JP et al (1990) Recent advances in solid-phase peptide synthesis and preparation of antibodies to synthetic peptides. Biologicals 18:147–157
9. Goodfriend, T. L., Levine, L. Fasman, G. D. (1964). Antibodies to bradykinin and angiotensin: a use of carbodiimides in immunology. Science 144:1344–1346
10. Rector ES, Schwenk RJ, Tse KS et al (1978) A method for the preparation of protein-protein conjugates of predetermined composition. J Immunol Methods 24:321–336
11. Kitagawa T, Aikawa T (1976) Enzyme coupled immunoassay of insulin using a novel coupling reagent. J Biochem 76:233–236
12. Houen G, Olsen DT, Hansen PR et al (2003) Preparation of bioconjugates by solid-phase conjugation to ion exchange matrix-adsorbed carrier proteins. Bioconjug Chem 1376:75–79
13. Schaaper WMM, Lankof H, Puijk WC et al (1989) Manipulation of antipeptide immune response by varying the coupling of the peptide with the carrier protein. Mol Immunol 26: 81–85
14. Tam JP (1988) Synthetic peptide vaccine design: synthesis and properties of a high-density

multiple antigenic peptide system. Proc Natl Acad Sci U S A 85:5409–5413

15. Kowalczyk W, Monsó M, de la Torre BG et al (2011) Synthesis of multiple antigenic peptides (MAPs)—strategies and limitations. J Peptide Sci 17:247–251

16. Tam JP, Lu Y-A (1989) Vaccine engineering: enhancement of immunogenicity of synthetic peptide vaccines related to Hepatitis in chemical defined models consisting of T- and B-cell epitopes. Proc Natl Acad Sci U S A 86: 9084–9088

Chapter 7

Solid-Phase Peptide-Carrier Conjugation

Gunnar Houen and Dorthe T. Olsen

Abstract

Conjugation to carrier proteins is necessary for peptides to be able to induce antibody formation when injected into animals together with a suitable adjuvant. This is usually performed by conjugation in solution followed by mixing with the adjuvant. Alternatively, the carrier may be adsorbed onto a solid support followed by activation and conjugation with the peptide by solid-phase chemistry. Different reagents can be used for conjugation through peptide functional groups (-SH, $-NH_2$, -COOH) and various carrier proteins may be used depending on the peptides and the intended use of the antibodies. The solid phase may be an ion-exchange matrix, from which the conjugate can subsequently be eluted and mixed with adjuvant. Alternatively, the adjuvant aluminum hydroxide may be used as the solid-phase matrix, whereupon the carrier is immobilized and conjugated with peptide. The resulting adjuvant-carrier-peptide complexes may then be used directly for immunization.

Key words Peptide, Carrier, Conjugation, Solid phase, Antibodies

1 Introduction

Peptides of smaller size usually do not elicit production of high-affinity antibodies (Abs) due to a lack of T cell epitopes and an inherently low immunogenicity and in order to induce specific Abs they need to be conjugated covalently to a larger carrier protein and to be injected together with a suitable adjuvant [1–3]. Conjugation can be done in solution by one- or two-step procedures with a number of reagents [3–7]. Alternatively, conjugation may be performed by solid-phase chemistry using principles developed for peptide synthesis. Here, convenient procedures for conjugation to different solid phase-immobilized carrier proteins are described (Fig. 1). The solid phase can be an ion-exchange matrix, from which the resulting conjugates can be eluted and mixed with adjuvant [8]. Alternatively, the adjuvant aluminum hydroxide can be used as the solid phase, whereupon the carrier is immobilized [7].

Gunnar Houen (ed.), *Peptide Antibodies: Methods and Protocols*, Methods in Molecular Biology, vol. 1348, DOI 10.1007/978-1-4939-2999-3_7, © Springer Science+Business Media New York 2015

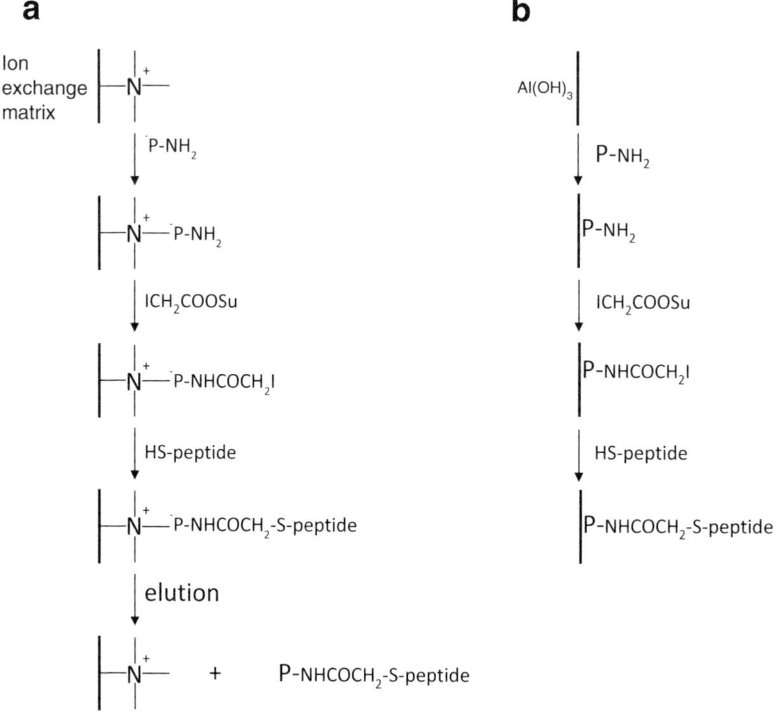

Fig. 1 (**a**) Conjugation to ion-exchange matrix-adsorbed carrier protein. (**b**) Conjugation to aluminum hydroxide adjuvant-adsorbed carrier protein

In this case, the resulting adjuvant-carrier-peptide complexes may be used directly for immunization. The advantages of these methods include improved control of conjugation chemistry and peptide orientation and ease of the different steps involved.

2 Materials

1. Ion-exchange matrix (strong quaternary cation exchanger) (*see* **Notes 1** and **2**).
2. Aluminum hydroxide (aqueous suspension, e.g., 2 % by dry matter (determined as Al_2O_3)).
3. Carrier proteins (e.g., ovalbumin, bovine serum albumin, keyhole limpet hemocyanin).
4. Sodium phosphate buffer (PB), 10 mM, pH 9.0 (*see* **Note 3**).
5. NaOH, 1 M stock, dilute to 0.1, 0.01, 0.001 M with water.
6. Dimethylformamide (DMF).
7. Mercaptoethanol.

8. Iodoacetic acid *N*-hydroxysuccinimide ester (IAAOSu), e.g., synthesized as described by Hampton et al. [9].
9. Synthetic peptide(s) with cysteine added at the N- or C-terminus as appropriate for conjugation (*see* **Note 4**).
10. Amino acid analysis (AAA) apparatus or access to this service, e.g., as described by Barkholt and Jensen [10].
11. Chromatographic equipment.
12. End-over-end rotator.
13. Centrifuge.

3 Methods

3.1 Conjugation to Ion-Exchange Matrix-Adsorbed Protein (Fig. 1a)

3.1.1 Adsorption of Carrier Protein

The required amount of ion-exchange matrix (here 10 mL settled matrix) is washed with 10 mM PB, pH 9.0.

The carrier (10 mL, 1–2 mg/mL in PB) (e.g., ovalbumin) is mixed with the prewashed ion-exchange matrix and incubated with gentle end-over-end agitation for 1 h or overnight at 5 °C. The matrix is settled by centrifugation (2000 × *g*) and the amount of carrier protein bound is estimated by determining the UV absorption of the first supernatant (*see* **Note 5**).

The matrix is washed three times with 10 mM PB (centrifugation in between at 2000 × *g*).

3.1.2 Activation

The settled matrix with bound carrier is resuspended with 10 mL 10 mM PB.

1 mL iodoacetic acid *N*-hydroxysuccinimide ester in DMF (10 mg/mL) is added and incubated at 5 °C for 1 h.

The activated carrier matrix is washed three times with 10 mM PB as above.

3.1.3 Conjugation

Resuspend the activated matrix in 10 mL 10 mM PB and add the cysteine-containing peptide (*see* **Note 4**) dissolved in 1 mL PB or DMF as required (*see* **Note 6**).

Incubate overnight at 5 °C with gentle end-over-end agitation.

Wash three times with PB (centrifugation in between at 2000 × *g*). The amount of conjugated peptide may be determined by AAA (*see* **Note 7**).

At this stage, the peptide-carrier matrix may be used directly for immunization with or without added adjuvant. Alternatively, the peptide-carrier complexes may be eluted, characterized, and then used for immunization.

3.1.4 Elution

The matrix with peptide-conjugated carrier is packed in a small column and eluted with a gradient of increasing NaCl concentration (0–1 M) in 10 mM PB. The absorbance at 280 nm is recorded. Conjugates elute according to their isoelectric points (pIs), which are determined by the composition and number of conjugated peptides together with the composition of the carrier.

The eluted conjugates may be dialyzed or used directly for immunization. The conjugation ratios may be determined by AAA as above.

3.1.5 Mixing with Adjuvant

Mix 1:1 with 2 % aluminum hydroxide (recommended) or another adjuvant.

3.1.6 Immunization

Mice are injected subcutaneously with 0.1 mL and rabbits with 1 mL, e.g., every 2 weeks until a desired titer is obtained (as determined by ELISA or other method) (*see* **Notes 8** and **9**).

3.2 Peptide Conjugation to Aluminum Hydroxide-Adsorbed Carrier Protein (Fig. 1b)

3.2.1 Adsorption of Carrier Protein to Aluminum Hydroxide

Ten mL 2 % aluminum hydroxide in water is mixed with 10 mg carrier protein (e.g., ovalbumin) in 1 mL water (*see* **Note 10**) and incubated with end-over-end rotation overnight at 5 °C.

The adjuvant-carrier matrix is washed three times with 10 mL dilute base (0.001–0.01 M NaOH) (it is important to control the pH, which must remain above 7 at all times.

3.2.2 Activation

Add 1 mL iodoacetic acid N-hydroxysuccinimide ester (10 mg/mL) in DMF and incubate overnight at 5 °C with end-over-end agitation.

The activated carrier-adjuvant matrix is collected by centrifugation (2000 × *g*).

Wash three times with dilute base (0.001–0.01 M NaOH) (centrifugation at 2000 × *g* in between).

Resuspend in 10 mL dilute base (0.001–0.01 M NaOH).

3.2.3 Conjugation

Add peptide containing a cysteine (*see* **Note 4**) dissolved in 1 mL water, dilute base (0.001–0.01 M NaOH), DMF, or mixtures of these as required (*see* **Note 6**).

Incubate overnight at 5 °C with end-over-end agitation. The pH may be checked in between to assure that it is high enough for successful conjugation (*see* **Note 11**).

Excess of reactive groups may be blocked by addition of 5 μL mercaptoethanol but this is not mandatory.

Wash three times with water or dilute base (0.001–0.01 M NaOH) (*see* **Note 12**).

The amount of conjugated peptide may be determined at this stage by AAA (*see* **Note 7**).

The adjuvant-carrier-peptide conjugates are now ready for immunization.

3.2.4 Immunization

Mice are injected subcutaneously with 0.1 mL and rabbits with 1 mL, e.g., every 2 weeks until a desired titer is obtained (as determined by ELISA or other method) (*see* **Notes 8** and **9**).

4 Notes

1. A strong anion-exchange matrix (quaternary amine) is chosen for carrier proteins with a low isoelectric point and a strong cation-exchange matrix for carrier proteins with a high pI.
2. The nature of the matrix restricts the types of conjugation chemistry that may be used.
3. A 0.1 M stock can be made and diluted afterwards. Make 0.1 M stocks of NaH_2PO_4 and Na_2HPO_4 and mix to obtain the desired pH.
4. The cysteine is added preferably at the N-terminus (for C-terminal peptide(s) or C-terminus (for N-terminal peptide(s) but any position may in principle be used, e.g., naturally occurring cysteine residues.
5. Use Lambert-Beer's equation: $A_{280} = \varepsilon \, l \, c$.
6. 0.1–1 mg/mL depending on peptide availability and solubility.
7. If the conjugation is successful, carboxymethyl-cysteine will be formed by the reaction of the cysteine side chain with the iodoacetic acid. Furthermore, the ratios of the individual amino acids can be used to calculate a conjugation number, provided that the amino acid compositions of the carrier and the peptide are known.
8. For mAb production mice are injected intraperitoneally the first, intermittent, and last times or intravenously the last time (only eluted conjugates without adjuvant for intravenous injection).
9. Animals are followed daily for adverse reactions. If required, animals may have to be treated or sacrificed according to ethical guidelines.
10. Depending on the availability and the degree of saturation wanted, smaller or larger amounts of carrier protein may be used.
11. The reactivity of the cysteine side-chain SH group is optimal at pH 8–9.
12. Washing with DMF or mixtures of DMF and water or dilute base may be required to remove excess of very hydrophobic peptides. However, the final washes must be without organic solvent.

References

1. Shinnick TM, Sutcliffe JG, Green N, Lerner RA (1983) Synthetic peptide immunogens as vaccines. Annu Rev Microbiol 37:425–446
2. Naz RK, Dabir P (2007) Peptide vaccines against cancer, infectious diseases, and conception. Front Biosci 12:1833–1844
3. Harlow E, Lane D (1988) Antibodies—a laboratory manual. Cold Spring Harbor Laboratory Press, New York
4. Hermanson GT (1996) Bioconjugate Techniques. Academic, San Diego
5. Carter JM (1994) Techniques for conjugation of synthetic peptides to carrier molecules. Methods Mol Biol 36:155–191
6. Houen G, Jensen OM (1995) Conjugation to preactivated proteins using divinylsulfone and iodoacetic acid. J Immunol Methods 181:187–200
7. Houen G, Jakobsen MH, Svaerke C, Koch C, Barkholt V (1997) Conjugation to preadsorbed preactivated proteins and efficient generation of anti peptide antibodies. J Immunol Methods 206:125–134
8. Houen G, Olsen DT, Hansen PR, Petersen KB, Barkholt V (2003) Preparation of bioconjugates by solid-phase conjugation to ion exchange matrix-adsorbed carrier proteins. Bioconjug Chem 14:75–79
9. Hampton A, Slotin LA, Chawla RR (1976) Evidence for species-specific substrate-site-directed inactivation of rabbit adenylate kinase by N-6-(6-iodoacetamido-n-hexyl)adenosine 5′-triphosphate. J Med Chem 19:1279–1283
10. Barkholt V, Jensen AL (1989) Amino acid analysis: determination of cysteine plus half-cystine in proteins after hydrochloric acid hydrolysis with a disulfide compound as additive. Anal Biochem 177:318–322

Chapter 8

Analysis of Peptides and Conjugates by Amino Acid Analysis

Peter Højrup

Abstract

Amino acid analysis is a highly accurate method for characterization of the composition of synthetic peptides. Together with mass spectrometry, it gives a reliable control of peptide quality and quantity before conjugation and immunization.

Peptides are hydrolyzed, preferably in gas phase, with 6 M HCl at 110 °C for 20–24 h and the resulting amino acids analyzed by ion-exchange chromatography with post-column ninhydrin derivatization. Depending on the hydrolysis conditions, tryptophan is destroyed, and cysteine also, unless derivatized, and the amides, glutamine and asparagine, are deamidated to glutamic acid and aspartic acid, respectively. Three different ways of calculating results are suggested, and taking the above limitations into account, a quantitation better than 5 % can usually be obtained.

Key words Peptides, Amino acid analysis

1 Introduction

Since the conception of the automatic amino acid analyzer in the 1950s by Moore and Stein [1], amino acid analysis (AAA) has transformed analytical biochemistry, enabling the quantitative analysis of proteins, peptides, and free amino acids.

The result of an amino acid analysis is the absolute amount of each amino acid in the sample. These data can be used for determining the composition of the sample and/or the total amount of sample present. If your purpose is only quantitation (typically when analyzing complex samples like proteomics) you have a multitude of choices (e.g., Bradford, BCA, Lowry); however, most of these methods are based on derivatization of amines with a chromophore followed by comparison to a standard. Although simple and fast, a main problem of this type of analysis is that you are measuring amino groups, not just amino acids. Another method, which is nondestructive, is the measurement of concentration by UV absorbance; but the result can easily be disturbed by UV-absorbing substances.

Gunnar Houen (ed.), *Peptide Antibodies: Methods and Protocols*, Methods in Molecular Biology, vol. 1348, DOI 10.1007/978-1-4939-2999-3_8, © Springer Science+Business Media New York 2015

These methods can be quite accurate if used in a controlled environment; that is, you always have the same buffer, concentration range, type of protein, etc., but if these are not under total control, quantitation is often wrong by a significant amount.

On the other hand, when using amino acid analysis you measure on the level of the individual amino acid, and most of the problems with colorimetric methods can readily be recognized and compensated for. However, the price to pay is an increase in cost and a longer analysis time. Amino acid analysis usually takes place by chromatographic separation of the amino acids, followed by determination of the exact quantity of each residue. As the amino acids do not have any absorbance in themselves, except for a rather low absorbance in the 250–280 nm UV range for the aromatic amino acids, they have to be derivatized prior to quantitation. Derivatization of amino acids can take place either prior to or after separation, also known as pre- and post-column derivatization. A multitude of methods are available (e.g., 2, 3), with the pre-column derivatization clearly showing higher sensitivity than post-column. However, two things speak in favor of post-column derivatization, one is stability, as the ion-exchange column used for separation is extremely stable, and the second is that the derivatization step in pre-column analysis is much more sensitive towards contaminations. This can be particularly important in a core lab where you do not have control of the samples you receive, and thus the following description will be based on a post-column derivatization system with separation taking place by cation-exchange chromatography.

Prior to AAA, polypeptides have to be hydrolyzed in order to obtain the free amino acids for the analysis. Since the early inception of AAA, acid hydrolysis using 6 N HCl at 110 °C for 20–24 h has been the method of choice, mainly due to the fact that hydrochloric acid is easily evaporated and produces few artifacts. The volatility of hydrochloric acid also enables gas-phase hydrolysis, thus further decreasing contaminations. A negative aspect of the acid hydrolysis method is that tryptophan is destroyed, and cysteine also, unless derivatized, and the amides, glutamine and asparagine, are deamidated to glutamic acid and aspartic acid, respectively (*see* **Note 1**). However, taking these limitations into account, the accuracy obtained by AAA is generally unmatched by other methods.

2 Materials

Hydrolysis solution: Amino acid analysis-grade 6 N HCl containing 0.1 % phenol, 0.1 % 2-thioglycolic acid.

Mini-inert valve (VICI PS-614163) (*see* **Note 2**).

40 ml screw-cap bottle (e.g., number ND24, EPA, Mikrolab Aarhus A/S, Århus, Denmark).

Large-gauge syringe needle.

Stock solution of amino acids (e.g., Pierce amino acid standard) appropriately diluted.

Norleucine and sarcosine to be used as internal standards.

Ninhydrin and chromatographic running buffers.

Unless otherwise noted, chemicals are best in analytical grade.

3 Methods

3.1 Sample Preparation

Sample preparation is essential for correct amino acid analysis, as you are analyzing the total content of your sample. In many other forms of analysis, you either do not see the contaminants, or, like in mass spectrometry, you can perform a micro purification prior to analysis.

If you are mainly interested in the composition, you should by all means transfer your sample to a salt-free buffer, using solid-phase extraction, dialysis, gel filtration, or similar. However, in most cases you are also interested in the quantity and often have a limited amount of sample. In these cases you have to consider contaminants, particularly in the last preparation steps.

3.1.1 Contaminations

Three kinds of contaminants are common in amino acid analysis:

Salts are to be avoided, as they may inhibit hydrolysis. If you have a large amount of salts, you may have to forsake gas-phase hydrolysis and add a drop of HCl to the sample. Salts may also disturb the chromophore derivatization, and the final chromatographic separation—post-column derivatization after ion-exchange separation is much less affected than pre-column derivatization. A special case is amine-containing buffers, in particular urea and guanidinium chloride. For pre-column derivatization they exhaust the chromophore leading to under-derivatization, and for post-column methods, they obscure the arginine peak. Again, ion exchange with post-column derivatization is less affected.

Amino acids can often be contaminants. Glycine is often used as a buffer, and a drop of spilled buffer may dry, become airborne, and contaminate other samples.

Finally, contaminating *proteins* will skew the results. Keratin is a typical contaminant arising from hair and skin, which gives rise to increased serine and glycine levels. If the purpose of the analysis is quantitation, a 5 % contamination may be in the same range as the accuracy of the method and thus be compensated for, but a similar contamination may lead to erroneous conclusions of the composition if analyzing a supposedly “pure” protein.

3.1.2 Amounts

When considering the amount to submit for amino acid analysis, two things have to be taken into consideration: the sensitivity of the analyzer and the sensitivity of the hydrolysis. Pre-column derivatization, particularly when carried out with fluorescent detection, can be made extremely sensitive (about 1 pmol/aa), while post-column detection using ninhydrin is difficult to drive below 100 pmol/aa—using fluorescent detection you can get 10× higher sensitivity. However, when preparing samples routinely, and particularly when running samples in a core-like facility, it is quite difficult to ensure total cleanliness in the preparation of samples. This typically results in contaminations, particularly of the small amino acids, Gly, Ala, Ser, and Thr, at a level of 10–30 pmol. This means that in practice it is difficult to obtain reliable results of less than 1 μg of the starting material. As people almost always overestimate the amount they are working with, you should always ask for 2 μg samples. If you have a high-sensitivity system, you can then run multiple technical replicates.

3.2 Gas-Phase Hydrolysis

The classical way of performing hydrolysis calls for pre-treated glass tubes, but for most practical purposes 500 μl polypropylene tubes (Eppendorf tubes) can be used (*see* **Note 3**).

The samples are placed in 500 μl polypropylene tubes, 2–3 holes are made in the lid with a wide-gauge needle, and the samples are subsequently dried in a vacuum centrifuge.

Place 200–300 μl hydrolysis solution in the bottom of a 40 ml screw-cap bottle, followed by 3–5 polypropylene tubes with samples. Be careful not to stack them directly on top of each other, as condensing HCl may drip into the tubes below.

The glass vial is flushed with argon (alternatively nitrogen) before being closed with a mini-inert valve (*see* **Note 4**). The glass vial is then evacuated to less than 10 Torr and placed at 110 °C for 20–24 h (*see* **Note 5**).

After hydrolysis, the mini-inert valve is opened in a fume hood, pointing away from you (acid fumes!). The valve is then removed and the sample tubes withdrawn, dried on the outside, and then dried in a vacuum centrifuge. Be careful to have a −80 °C trap between the centrifuge and the vacuum pump to catch residual HCl vapor.

3.3 Amino Acid Analysis

After removal of all traces of HCl, the samples are dissolved in loading buffer (sodium citrate pH 2.20) spiked with an internal standard (see below). The autosampler of the BioChrom 30 amino acid analyzer is unable to load and inject the entire sample, but setting the autosampler to "microliter-pickup," you can inject 60 of 70 μl if you use vials with tapered inserts.

The analysis is most easily done using BioChrom standard program and solvents, and by using the accelerated analysis buffers the cycle time per analysis is about 60 min.

In order to verify the running of the system, and to calibrate for loss of ninhydrin efficiency, every seventh sample run is a standard (in our case 1400 pmol, but should be adjusted to your standard running conditions).

3.4 Data Interpretation

After each analysis, the data are automatically integrated according to the current calibration and saved to a single data file containing the results in text format. This can be read by Excel or a dedicated program (e.g., MyAAA, *see* **Note 6**).

3.4.1 Calibration and Standards

Typical results from an analysis are shown in Fig. 1. Standards can be included in several positions to improve accuracy.

Initially you have to make certain that the system is calibrated for each residue, using at least a 3-point calibration curve. On our ninhydrin-based system, this is carried out whenever a new bottle of ninhydrin is put on the system.

To improve the results, you may include a standard during the hydrolysis. A typical standard is norleucine, which is stable to acid hydrolysis, and usually separates nicely from the other amino acids during the chromatography. The amount recovered can thus be used to compensate for handling losses during the hydrolysis and loading.

Secondly, a standard (norleucine, sarcosine) may be included in the loading buffer to compensate for injection errors. Standards also help to identify errors when you suddenly find yourself with a blank run: Is this caused by no sample or an error in the machine?

In our lab, every seventh sample is a standard, which is used for checking the chromatographic separation and, when integrated, to compensate for the loss of quality of ninhydrin over time (*see* **Note 7**).

By analyzing a number of pure standard proteins, a hydrolysis compensation factor may be calculated for each amino acid (*see* **Note 8**).

3.4.2 Quantitation

After compensating the measured amounts according to the outline above, the total amount of protein can be calculated as follows:

Quantitation—the fast and easy way. Add the amount of each residue together. Remember to omit the internal standard (if included) and to include Pro, which for ninhydrin systems is calculated at 440 nm.

The summed value is multiplied by 110 (the average mass of a residue), and the result is the total amount of protein in picogram (divide by 10^6 to get µg), if the results of the analysis are reported in picomoles.

This method usually agrees within 5 % with methods 2 and 3 below if the protein composition is "average." If the protein composition is skewed (deviating strongly from the composition in Table 1, e.g., collagen contains approximately 1/3 Gly), this method is unsuitable.

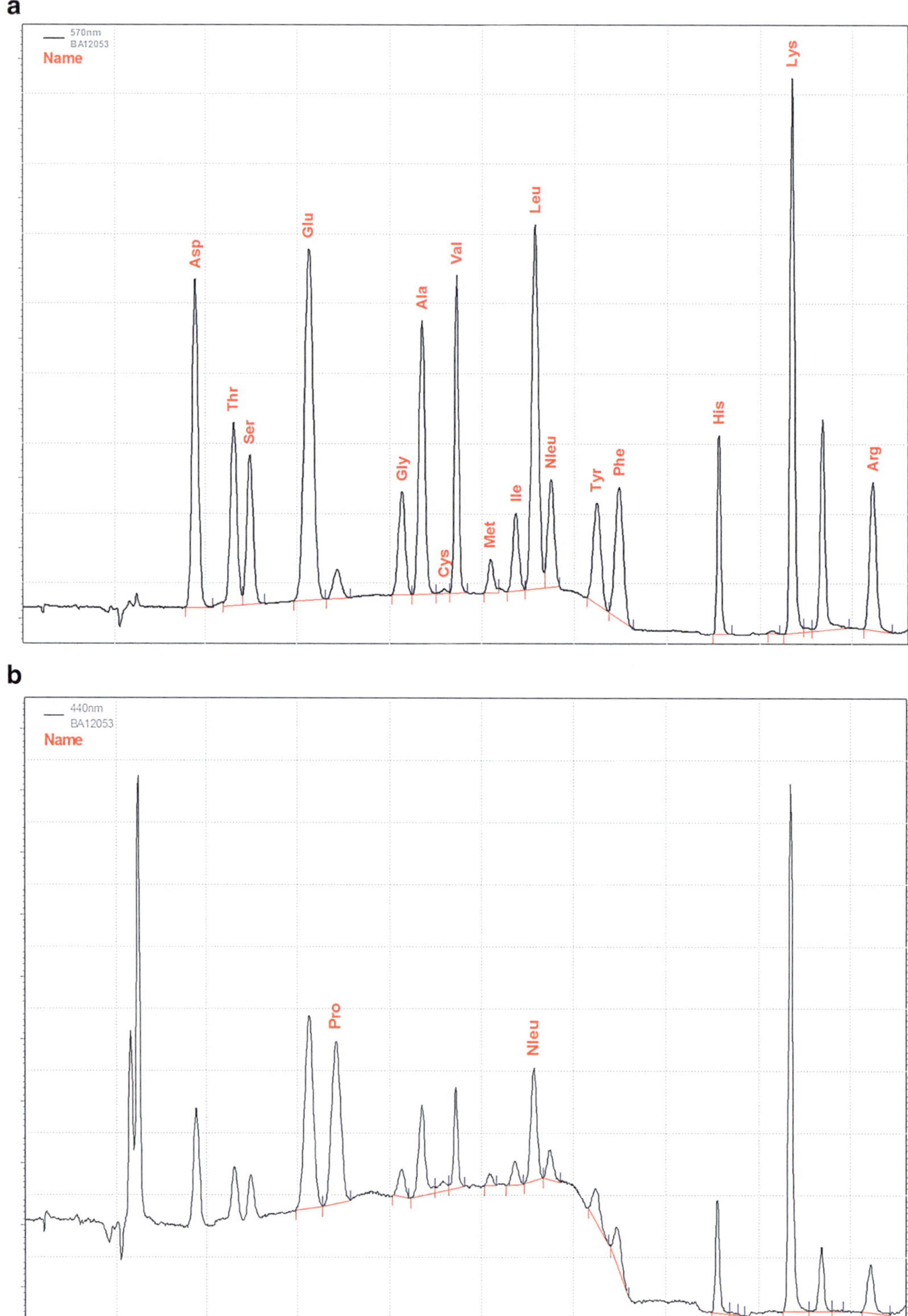

Fig. 1 Amino acid analysis chromatogram of hydrolyzed bovine serum albumin. The amino acids were separated by cation-exchange chromatography and detection was after post-column derivatization with ninhydrin. *Top* chromatogram shows 560 nm and *below* 440 nm trace. Only proline is calculated from the 440 nm chromatogram

Table 1
Composition in percent of the 20 common amino acids in the UniProt protein database. Data are from http://www.ebi.ac.uk/uniprot/TrEMBLstats

Ala	(A)	8.52	Gln	(Q)	4.04	Leu	(L)	9.92	Ser	(S)	6.45
Arg	(R)	5.17	Glu	(E)	6.22	Lys	(K)	5.51	Thr	(T)	5.54
Asn	(N)	4.27	Gly	(G)	6.99	Met	(M)	2.51	Trp	(W)	1.23
Asp	(D)	5.39	His	(H)	2.20	Phe	(F)	4.08	Tyr	(Y)	3.15
Cys	(C)	1.12	Ile	(I)	6.36	Pro	(P)	4.40	Val	(V)	6.80

Quantitation—the slightly more complicated way. The amount of each residue is multiplied by the mass of each residue (remember to multiply by the residue mass, not the amino acid mass—e.g., Gly is calculated as 57 Da, not 75 Da), as the amino acids are originally part of a protein, not free. You then sum the resulting values and get a result in picogram as above. For both methods 1 and 2 you will underestimate slightly, as Trp and Cys are not included in the calculations. If you are analyzing mixtures of proteins, you can usually get a more accurate estimate by adding 3.1 % mass due to the average content of Cys and Trp (Table 1).

Quantitation—the best, but most complicated, way. This method only applies to the analysis of pure proteins with a known sequence.

Start by calculating the composition of the given protein. Remember that Asn and Gln are deamidated during acid hydrolysis and are to be calculated as Asp and Glu, respectively. Then divide each measured amount with the given residue composition. This calculates amount in pmol per residue. In an ideal world this value would be identical for each residue; however, due to losses (mainly hydrolytic) there will be some variation. If a value is obviously wrong (too high—contamination; too low—losses) you can omit it before taking the average of the rest.

This average value represents the amount of one residue in the sequence, and thus the total amount of protein analyzed. Multiply it with the molecular mass of the protein (based on the sequence or the mass determined by, e.g., mass spectrometry), and you have the total mass of the protein. This has the advantage that you can compensate for contaminants and you can include "missing" residues (Trp, Cys) and posttranslational modifications like glycosylations in the mass.

Example 1. Table 2 shows the calculations related to the amino acid analysis chromatogram shown in Fig. 1. The first column shows the name of the amino acid in three-letter code. Note that Asx covers Asp and Asn like Glx covers Glu and Gln due to deamidation of the side-chain amide during hydrolysis. The next column shows the raw data as integrated by the amino acid analyzer. The third column has a compensation of 1.4 % added due to a decrease

Table 2
Amino acid composition of BSA as determined by amino acid analysis after acid gas-phase hydrolysis

	pmol	pmol	g/mol	pg	Residues	pmol/res
Asp	7087.6	7186.8	115.03	826,701	54	133.1
Thr	4188.0	4246.7	101.05	429,126	33	128.7
Ser	3257.0	3302.6	87.03	287,423	28	117.9
Glu	11,461.5	11,622.0	129.04	1,499,702	79	147.1
Gly	2142.9	2172.9	57.02	123,898	16	135.8
Ala	6045.0	6129.6	71.04	435,445	47	130.4
Val	4596.7	4661.0	99.07	461,770	36	129.5
Met	458.9	465.4	131.04	60,981	4	116.3
Ile	1654.3	1677.5	113.08	189,689	14	119.8
Leu	7769.6	7878.4	113.08	890,890	61	129.2
Tyr	2579.9	2616.0	163.06	426,560	20	130.8
Phe	3670.5	3721.8	147.07	547,370	27	137.8
His	2179.2	2209.7	137.06	302,864	17	130.0
Lys	7431.0	7535.1	128.09	965,168	59	127.7
Arg	3156.2	3200.4	156.1	499,577	23	139.1
Pro	3470.3	3518.9	97.05	341,510	28	125.7
Sum	71,148.62	72,144.70		8,288,675	546	2079.0
					Average	129.9
Amount	7.83 μg	7.94 μg		8.29 μg	Av.* 66,652	8.65 μg

*Average mass of intact BSA

in ninhydrin efficiency. The fourth column shows the residue mass of each amino acid, which is used to generate column five, which then shows the total mass of each residue. Finally column 6 lists the number of each residue in BSA, and the last column (number 7) shows the values in column 5 divided with column 6. The values in the last column should theoretically all be identical, but Ser and Met are lower than expected due to hydrolytic loss, while Glu is slightly higher due to a contamination. Taking the average residue amount (129.9 pmol/res) and multiplying with the mass of BSA (66652 Da) yield a total mass of 8.65 μg.

Comparing the calculated total mass shows that the initial 7.83 μg is increased to 7.94 μg by compensating for decreased ninhydrin sensitivity. This is again increased to 8.29 μg in column 5, mainly due to BSA having a lower than average content of small residues (e.g., Gly is only 2.7 % compared to an average of 7.0 %, Table 1).

Finally, when calculated based on the average pmol/residue, the total mass is calculated as 8.65 μg. The reason for this large increase in total mass is that BSA has two tryptophan residues and a large content of cysteine (6 %). For the average protein, the combined content of Trp and Cys is only 2.35 % (Table 1).

3.4.3 Calculating Protein/Peptide Composition

Calculating the composition of an unknown protein or peptide can be a little tricky. For this you need first to estimate the lowest common denominator. You start by summing the amount of all residues. As each residue should contribute equally to the composition, you will get the lowest common denominator by dividing the sum of all residue amounts with the number of residues. The number of residues can be estimated by dividing the mass of the protein by 110 (the mass of an average residue) taking into account the presence of Cys and Trp which are not part of the amino acid analysis.

If you do not know the mass of the protein, you will have to estimate the lowest common denominator. This is most easily done based on stable residues present in low amounts (e.g., His, Arg, Phe, Met).

Example 2. The example in Table 3 shows the composition of bovine ACBP. The column labeled pmol shows the amount of each residue with the sum below. The protein is 10 kDa and contains 86 residues of which two are known to be tryptophan and thus not seen in the analysis, leaving 84 residues. Dividing the sum of 65,070 pmol with 84 residues yields 774.6 pmol/residue. Dividing this value into each amount gives the number of residues in column 3 which when rounded to integer values are shown in column 4.

If you had no knowledge of the mass of the protein, you could start the search for the lowest common denominator by noting that the lowest value was Arg as 798 pmol, while His and Ser were double that at 1564 and 1643 pmol. Furthermore, at triple values around 2400 you find Val, Met, and Phe. Taking the average of these residues you would end up with 795 pmol/residue, which could then be used as starting point for the remainder of the amino acids. By making a least square fit to the data you can obtain a quite accurate value.

4 Notes

1. Although acid hydrolysis using HCl is the standard method, a number of methods have been developed for analysis of sensitive residues. Tryptophan can be recovered after hydrolysis in 2.5 M mercaptoethanesulfonic acid (170–185 °C for 12.5 min) or mixtures of HCl, mercaptoethanol, and phenol [4]. Cysteine can be determined after oxidation with performic acid [5] or sodium azide or by chemical derivatization before hydrolysis. Several not-so-stable residues can be recovered after basic

Table 3
Amino acid composition of ACBP as determined by amino acid analysis after acid gas-phase hydrolysis

	pmol	Residues	Integer
Asp	7588.8	9.8	10
Thr	3313.4	4.3	4
Ser	1643.4	2.1	2
Glu	10,613.0	13.7	14
Gly	3153.8	4.1	4
Ala	7137.9	9.2	9
Val	2487.3	3.2	3
Met	2386.8	3.1	3
Ile	3009.7	3.9	4
Leu	4095.7	5.3	5
Tyr	2988.3	3.9	4
Phe	2256.6	2.9	3
His	1564.0	2.0	2
Lys	9864.0	12.7	13
Arg	798.3	1.0	1
Pro	2169.1	2.8	3
Sum	65,070.0		
Residues	84		Total 84
Amount/res	774.64		

hydrolysis [6]. As these compounds are not volatile, you need a neutralization or purification step prior to analysis, thus increasing complexity and decreasing sensitivity.

2. The mini-inert valve is equipped with a rubber cylinder that enables the insertion of a needle into the vial. Before being taken into use, this rubber cylinder has to be pushed out, cut in the middle to allow airflow, and reinserted to close the hole. Be careful that the flow through the valve is not blocked.
3. For the classical hydrolysis you use 6×50 mm glass tubes. These are then pyrolyzed overnight at 400 °C and kept free of contaminants. For hydrolysis the sample is dried at the bottom

of the tube, and 5–6 vials can be placed in a 40 ml screw-cap vial for gas-phase hydrolysis. These are likely to contribute less to contamination when attempting high-sensitivity analysis; however, in our hands using >1 μg samples, the advantage is minimal compared to polypropylene tubes, as contaminations are more likely to originate from the sample.

4. Always check that the mini-inert valve has an unrestricted airflow when the green button is depressed and is closed when the red button is depressed, as the teflon bar may turn during use.
5. As the stability of amino acids towards acid hydrolysis varies (*see* **Note 8**), you can increase accuracy by making a time-course hydrolysis. If you hydrolyze for 24, 48, and 72 h you can extrapolate the quantity of the difficult-to-hydrolyze residues (Val, Ile) towards infinity, while the unstable ones (Ser, Thr, Tyr, Met) can be extrapolated to 0 h. Alternatively, if time is at a premium, you can increase the temperature and shorten the hydrolysis time. You can effectively halve the time for every 10 °C increase (i.e., 6 h at 130 °C). The results are not quite as accurate, but are usually acceptable. Note that some polypropylene tubes are not stable at high temperatures; please check beforehand.
6. MyAAA is an in-house-developed program for the analysis of data from a BioChrom 30 amino acid analyzer. Please contact the author for more information.
7. We typically observe a loss of 2–3 % ninhydrin intensity/week. This is compensated as a general factor.
8. Please observe that the yield of a given residue varies with the sequence, and absolute compensation factors for hydrolytic loss cannot be calculated reliably. The yield of beta-branched residues (Ile and Val) is thus reduced greatly when they are bound to each other (i.e., Ile-Val, Val-Val, Ile-Ile) due to steric hindrance. A yield lower than normal due to hydrolytic loss can also be expected for Ser (up to 10 %), Thr (up to 5 %), Tyr (dependent on phenol in the hydrolysis buffer), and Met (oxidation). Cysteine (unless derivatized) and tryptophan are completely lost during acid hydrolysis.

Acknowledgments

The Carlsberg Foundation is gratefully acknowledged for financial support.

References

1. Moor S, Spackman DH, Stein WH (1958) Automatic recording apparatus for use in the chromatography of amino acids. Fed Proc 17:1107–1115
2. Alterman MA, Hunziker P (eds) (2012) Amino acid analysis: methods and protocols, methods in molecular biology, vol 828. Springer Science+Business Media, New York, NY
3. Cooper C, Packer WK (eds) (2000) Amino acid analysis protocols, methods in molecular biology, vol 159. Springer Science+Business Media, New York, NY
4. Adebiyi AP, Jin DH, Ogawa T, Muramoto K (2005) Acid hydrolysis of protein in a microcapillary tube for the recovery of tryptophan. Biosci Biotechnol Biochem 69:255–257
5. Hirs CHW (1967) Performic acid oxidation. Methods Enzymol 11:177–199
6. Hirs CHW (1967) Detection of peptides by chemical methods. Methods Enzymol 11:325–329

Chapter 9

Characterization of Synthetic Peptides by Mass Spectrometry

Bala K. Prabhala, Osman Mirza, Peter Højrup, and Paul R. Hansen

Abstract

Mass spectrometry (MS) is well suited for analysis of the identity and purity of synthetic peptides. The sequence of a synthetic peptide is most often known, so the analysis is mainly used to confirm the identity and purity of the peptide. Here, simple procedures are described for MALDI-TOF-MS and LC-MS of synthetic peptides.

Key words Synthetic peptide, Mass spectrometry, MALDI-TOF-MS, LC-MS

1 Introduction

Mass spectrometry (MS) is well suited for analysis of the identity and purity of synthetic peptides [1]. The sequence of a synthetic peptide is almost always known, so the analysis mainly serves to confirm the identity and purity of the peptide before costly immunization and testing are initiated. In most cases, a matrix-assisted laser desorption ionization (MALDI) time-of-flight (TOF) mass spectrometry spectrum will yield sufficient information to proceed with the peptide for immunological studies. In case of problems (e.g., difficulties with the synthesis and/or cleavage of the peptide from the resin), analysis of peptide purity and sequence by MALDI-MS/MS or LC-MS or LC-MS/MS may be required.

MALDI is one of the most popular MS techniques, mainly because of its large *m/z* range of 500–100,000 Da. Furthermore, MALDI is one of the most tolerant MS techniques with respect to salts and solvents and has a very high sensitivity down to sub-millimolar concentrations [2]. Finally, peptides analyzed by MALDI-TOF-MS most often generate singly charged ions which make interpretation easier as compared to other mass spectrometry techniques such as LC-MS. In MALDI-TOF-MS, the peptide and matrix are

Gunnar Houen (ed.), *Peptide Antibodies: Methods and Protocols*, Methods in Molecular Biology, vol. 1348, DOI 10.1007/978-1-4939-2999-3_9,

co-crystallized on a metal plate. The matrix is most often α-cyano-4-hydroxycinnamic acid, sinapinic acid, or 2,5-dihydroxybenzoic acid. Following activation with UV laser energy, peptide, and matrix desorption, the mixture is transferred into the gas phase and ionized [3]. Although MALDI-TOF MS does not have the resolving power obtainable by LC-MS/MS, it has a significant edge in sample analysis time.

In recent years, liquid chromatography coupled with mass spectrometric detection (LC-MS or tandem MS, LC-MS/MS) has become a widely applied technique for the analysis of peptides up to 3000 Da. Furthermore, LC-MS instruments are very sensitive and can often achieve limits of detection in the pmol/L range [4]. Typically, an LC-MS/MS instrument consists of an (1) atmospheric pressure ionization source, (2) ion-inlet and focusing component, (3) first mass-filtering device, (4) collision chamber, (5) second mass-filtering device, and finally (6) an ion impact detector [5]. Many different mass spectrometer types are available including quadrupole, ion-trap, LTQ-Orbitrap, and time-of-flight mass analyzers. They may employ different ionization methods such as atmospheric pressure chemical ionization, electrospray ionization, or atmospheric pressure photoionization [6]. However, a detailed description of these techniques is beyond the scope of this chapter. Here, simple procedures for MALDI-TOF-MS and LC-MS are described. For more advanced procedures, see the cited literature and the following chapter for interpretation of MS/MS spectra.

2 Materials

2.1 MALDI-TOF-MS

Acetonitrile.

α-Cyano-4-hydroxycinnamic acid.

Trifluoroacetic acid (TFA).

MS target plate.

Bruker Microflex (TM).

2.2 LC-MS

Agilent 1100 series HPLC systems, Esquire 3000 plus ion trap mass spectrometry, Bruker Daltonics Esquire 5.3 software, Chromacol LC-MS glass vials.

Degassed mobile phases A and B.

Mobile phase A: 0.05 % formic acid in water.

Mobile phase B: 90 % acetonitrile in water.

3 Methods

3.1 MALDI-TOF-MS

1. Prepare a matrix solution of α-cyano-4-hydroxycinnamic acid dissolved in 1 mL acetonitrile:H_2O:trifluoroacetic acid (500:475:25) (10 mg/mL).

2. Dissolve the peptide in H_2O or acetonitrile:H_2O ($\geq$50 % H_2O) (*see* **Notes 1** and **2**).
3. The concentration should be within 0.01–0.1 mg/mL.
4. Mix a droplet of 1.0 μL of peptide with 1.0 μL of matrix in the lid of an eppendorf tube and transfer 1.0 μL of the mixed solution to a position on the target plate with a micropipette.
5. All samples are allowed to dry completely before a droplet (1.0 μL) of matrix solution is added on top of each sample spot.
6. Allow all spots to dry completely before proceeding.
7. The target plate is loaded into the MS instrument and the data generated using the computer program flexControl. The data is then processed using the computer program flexAnalysis (*see* **Note 3**).

3.2 LC-MS

1. Dissolve the peptide in phosphate buffer pH 7.0, to get a standard stock solution of 50 μM. If the peptide is not completely soluble, add acetonitrile (*see* **Notes 5** and 7).
2. The standard stock solution of peptide is diluted by methanol/water (50/50), such that the final concentration in a 100 μL aliquot is 25 μM. The sample must always be filtered using Millipore centrifugal filter units (0.25 μm) before injecting into the column (*see* **Notes 12–13**).
3. An amount of 20 μL of sample is injected into the LC and separated on an Agilent Poroshell C-18 column (5 μm, 2.1 × 75 mm, column temperature 40 °C). A linear gradient of (10–90 %) acetonitrile in water is used at a flow rate of 0.5 mL/min (*see* **Notes 8, 10,** and **11**).
4. The LC system is connected to the Esquire ion trap so that the eluted peptide is sprayed directly into it. Electrospray ionization (positive mode) is utilized and the ion trap is filled with 50,000 ions at a speed of 13,000 m/z per second. Scan range is set at 50–3000 m/z.
5. Once the data acquisition is complete, the data file can be imported into Esquire 5.3, wherein the masses can be analyzed and the chromatograms can be exported along with the mass spectra in a suitable format (e.g., TIFF) (Fig. 1) (*see* **Note 14**).
6. Note that with a UV detector located between the LC and the mass spectrometer there is a time lag between the UV and the MS signal corresponding exactly to the time for the flow to reach the MS detector after the UV detector (*see* **Note 9**).
7. For analysis of the resulting mass spectra, the suppliers of MS systems such as Agilent (Spectrum Mill) and Bruker Daltonics (BioTools) have their own software that can aid in sequencing and identification of your desired peptide. A general tool like

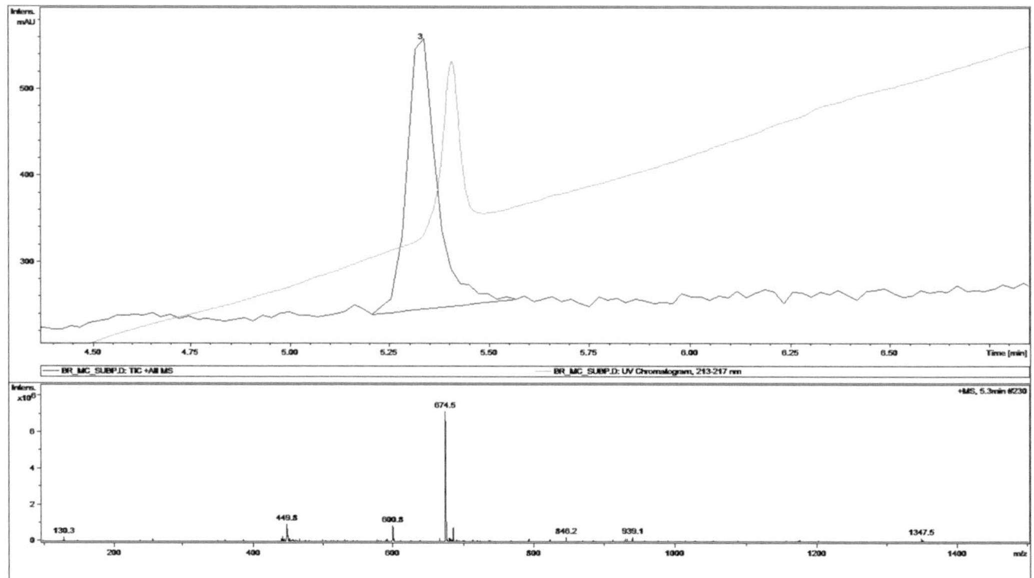

Fig. 1 LC-MS of Substance P (RPKPQQFFGLM amide), M.W. 1346.7 Da. The *green-colored* spectrum belongs to the UV and the *red* one to MASS. The lag time between the signals is discussed in LC-MS section point 13. The main peak is observed at $((M+2H)^{+}/2) = m/z$ 674.4

GPMAW (gpmaw.com) is also useful as well as free sources such as Protein Prospector (prospesctor.ucsf.edu) which is being widely used in our labs (*see* **Notes 4** and **6**).

4 Notes

1. Do not use organic solvents like dimethylsulfoxide to dissolve the peptide, since this will oxidize cysteines to disulfides, sulfoxides, or other oxidation products.
2. If the peptide is insoluble in water or acetonitrile water mixtures of TFA or HCOOH may be used (be aware of possible modification by trifluoroacetylation or formylation).
3. If the sample shows strong adduct ion formation (particularly alkali metal adducts, Table 1), it should be purified by micro solid-phase extraction (SPE) (e.g., [7]).
4. If quantification MS/MS spectra is desired you may need to perform multiple reaction monitoring (MRM) [8].
5. The standard solutions should be run in the same solvent/matrix as that of the sample to avoid matrix effects.
6. For quantifications, a calibration curve must be prepared with a selected target ion from MS/MS spectra.

Table 1
Adducts and chemical modifications typically observed when analyzing synthetic peptides

Mass	Composition	Description
−17	NH_3	Loss of ammonia, typically from N-terminal Gln
+16	O	Oxidation, typically of Met or Trp
+22	Na	Sodium adduct
28	CHO	Formyl
+38	K	Potassium adduct
42	C_2H_4O	Acetyl
56	C_4H_8	t-Butyl
71	C_3H_6NO	Acetamidomethyl from Cys
96	C_2F_3O	Trifluoroacetyl
242	$C_{19}H_{15}$	Trityl protecting group from Cys, Asn, Gln, or His
252	$C_{13}H_{16}O_3S$	Pbf-protecting group from Arg

7. All the solutions must be prepared with Milli-Q water (18 Ω, 25 °C) and stored at 4 °C.
8. For short peptides such as di- and tripeptides, the gradient of acetonitrile should be started with less organic content or in neat water.
9. The retention times may vary due to interactions of the peptides with the column; therefore it is always desirable to have an internal standard.
10. The column should not be overloaded, so the maximum concentration of the solution injected should not be more than 200 μM under any circumstances.
11. 0.1 % TFA can be added to mobile phase to improve peak shapes and retention times.
12. It is highly recommended to centrifuge the samples at 6500 rpm for 5 min before filtering them in order to get rid of any undesirable substances originating from the matrix.
13. Always make sure that a clear filtered sample is injected into the LC-MS.
14. In special cases SPE can be used to isolate the peptide of interest.

References

1. Kafka AP, Kleffmann T, Rades T et al (2011) The application of MALDI TOF MS in biopharmaceutical research. Int J Pharm 417:70–82
2. Schaiberger A, Moss J (2008) Optimized sample preparation for MALDI mass spectrometry analysis of protected synthetic peptides. J Am Soc Mass Spectrom 19:614–619
3. Montaudo G, Samperi F, Montaudo MS (2006) Characterization of synthetic polymers by MALDI-MS. Prog Polym Sci 31:277–357
4. Couchman L, Taylor DR, Krastins B et al (2014) LC-MS candidate reference methods for the harmonisation of parathyroid hormone (PTH) measurement: a review of recent developments and future considerations. Clin Chem Lab Med 52:1251–1263
5. Grebe SK, Singh RJ (2011) LC-MS/MS in the Clinical Laboratory Where to From Here? Clin Biochem Rev 32:5–31
6. Youdim KA, Saunders KC (2010) A review of LC-MS techniques and high-throughput approaches used to investigate drug metabolism by cytochrome P450s. J Chromatogr B Analyt Technol Biomed Life Sci 878: 1326–1336
7. Rappsilber J, Ishihama Y, Mann M (2003) Stop and go extraction tips for matrix-assisted laser desorption/ionization, nanoelectrospray, and LC/MS sample pretreatment in proteomics. Anal Chem 75:663–670
8. Sanda M, Pompach P, Brnakova Z et al (2013) Quantitative liquid chromatography-mass spectrometry-multiple reaction monitoring (LC-MS-MRM) analysis of site-specific glycoforms of haptoglobin in liver disease. Mol Cell Proteomics 12:1294–1305

Chapter 10

Interpretation of Tandem Mass Spectrometry (MSMS) Spectra for Peptide Analysis

Karin Hjernø and Peter Højrup

Abstract

The aim of this chapter is to give a short introduction to peptide analysis by mass spectrometry (MS) and interpretation of fragment mass spectra. Through examples and guidelines we demonstrate how to understand and validate search results and how to perform de novo sequencing based on the often very complex fragmentation pattern obtained by tandem mass spectrometry (also referred to as MSMS). The focus is on simple rules for interpretation of MSMS spectra of tryptic as well as non-tryptic peptides.

Key words Mass spectrometry, MS, De novo sequencing, Tandem mass spectra, Validation, Fragmentation

1 Introduction

Mass spectrometry (MS) is a technique that offers many diverse possibilities when the goal is to study and characterize polypeptides of any size, all the way from small peptides loaded onto MHC I to large proteins like immunoglobulins. Traditionally the determination of the primary structure has been carried out by a bottom-up approach where the protein in question has been cleaved into smaller peptides by a specific protease, whereupon these have been separated and individually analyzed by mass spectrometry (*see* **Note 1**). The purpose of the chapter is not only to demonstrate how sequences can be obtained and validated manually from tandem mass spectrometry data, but to introduce the reader to the complexity of MSMS spectra and demonstrate how simple guidelines can help to judge results determined from automated data analysis by studying specific features of the spectra. We do not give guidelines to automated data analysis but recommend a recent book chapter [1]. For a description of how to use mass spectrometry-based methods for other kinds of analysis of proteins like immunoglobulins we can recommend the following articles [2, 3].

Gunnar Houen (ed.), *Peptide Antibodies: Methods and Protocols*, Methods in Molecular Biology, vol. 1348, DOI 10.1007/978-1-4939-2999-3_10,

It is often possible to interpret an MSMS spectrum by hand, and we can start by asking the question: should you interpret all of your MSMS spectra manually? The answer would be: No, absolutely not. Today, a long list of software tools for interpreting tandem mass spectra can be generated from a simple Google search. The principles behind most of these tools are based on matching of the experimentally generated data against either theoretically generated data from peptide candidates from a database (e.g., 4, 5) or against already interpreted spectra (e.g., 6). Other tools try to automatically deduce the sequence directly from the spectra using for example the distance between the ion peaks. This strategy is called de novo sequencing. Then why should you spend time learning how to interpret an MSMS spectrum by hand?! For several reasons. First of all, dealing with peptides from more exotic species or from alternative splice variants may not be identified by traditional database-dependent searches if the sequence/variant is not present in the database. You may also find yourself in a situation where you work with nonspecifically cleaved peptides or modified peptides, where a simple search in a search engine may not be fruitful. Some of these situations may be solved by performing error-tolerant searches, but not always (*see* Subheading 3.1.3). Another—and maybe even more important—reason for learning the fragmentation behavior of peptides is to be able to evaluate the results given by the search engines or presented by other scientist (*see* Subheading 3.3). Unless you do a very stringent search in order to avoid any potential false-positive hits (and thereby also loosing weak but true peptide hits), you will never be able to avoid false-positive hits. Knowing how you should expect the peptide to fragment will help you evaluate and be critical towards such hits before spending months working on a hypothesis built on these false-positive hits. When performing automated de novo sequencing you will often be left with several possible partial peptide solutions that you have to judge and choose among and maybe even extend by hand. Here again it is essential to have knowledge on manual spectrum interpretation and validation.

In order to obtain detailed information on the primary sequence of a given protein, we often have to start by cutting the protein into smaller pieces of less than 30 residues, which is the size handled most efficiently by mass spectrometers. Trypsin is the most common choice due to a combination of high specificity, high activity, high stability, and low price. Trypsin cleaves C-terminal to lysine and arginine with high specificity, and therefore we expect the peptides to have a C-terminal lysine or arginine. These are both basic residues having an enhanced proton affinity and therefore a positive effect on the ionization event happening in the mass spectrometer. Often some of the peptides produced will be too long, too small, or simply too difficult to analyze by conventional MS methods. As a result, strategies involving only one enzyme will leave regions of the protein not analyzed. Full characterization of a protein therefore requires alternative strategies involving splitting the sample and using different enzymes for digestion. This will provide overlapping peptides covering the remaining part of the sequence, hopefully giving us full sequence coverage; Trypsin can be combined with more or less specific enzymes like chymotrypsin, Asp-N, Glu-C, and pepsin [6, 7] resulting in peptides which do not have the basic C-terminal characteristic of tryptic peptides. This is also the case if the goal is to analyze small naturally occurring peptides, which have arisen from unknown enzyme specificity, and therefore we have no prior knowledge of the terminal residues. This provides an extra challenge with respect to de novo sequencing, which we deal with in Subheading 3.2.

In order to analyze peptides by mass spectrometry, they need to be ionized. This takes place either by eletrospray ionization (ESI) [8] or matrix-assisted laser desorption/ionization (MALDI) [9]. Most peptides can be ionized using either of the two methods (*see* **Note 2**). The resulting fragmentation spectra may turn out quite different due to reasons that are discussed later in this chapter (*see* Subheading 3.3.3).

Inside the MS instrument, the ionized peptides are directed through a mass analyzer and sorted by their mass-to-charge ratio. When run in MSMS mode, peptide ions from a very narrow window of a pre-chosen mass-to-charge ratio are isolated and subjected to fragmentation, e.g., by colliding the peptide ions with gas molecules, like in collision induced dissociation (CID). The fragmentation pattern observed for a given peptide is dependent on the amino acid sequence of the peptide (not just the composition) as well as the number of charges carried by the peptide ion, i.e., the charge state of the peptide. Unfortunately, the exact structure of the resulting MSMS spectrum can only be partly predicted based on the sequence, as explained later.

The charge state of positive ions produced by ESI is dependent on the numbers of ionizable groups available, i.e., the number of basic amino acid residues (His, Lys, and Arg) and the N-terminal amino group. The ions will be observed in the spectrum as multiply charged molecular ions $(M+zH)^{z+}$ and often in more than one charge state (e.g., both as doubly and triply charged ions). In contrast, the ions produced in a MALDI-ion source in positive mode are almost always singly charged ions, i.e., only one proton is transferred pr. analyte molecule $(M+H)^{+}$. As we explain further, the difference in the number of charges is essential for the fragmentation behavior of the peptides in the two instrument types.

In the collision chamber of the MS instrument the ions will fragment through cleavage of one or more of the chemical bonds in the peptide, primarily in the peptide backbone. This gives rise to three distinct dissociation pathways as illustrated in Fig. 1 [10], resulting in several series of ions. The ion-series containing the

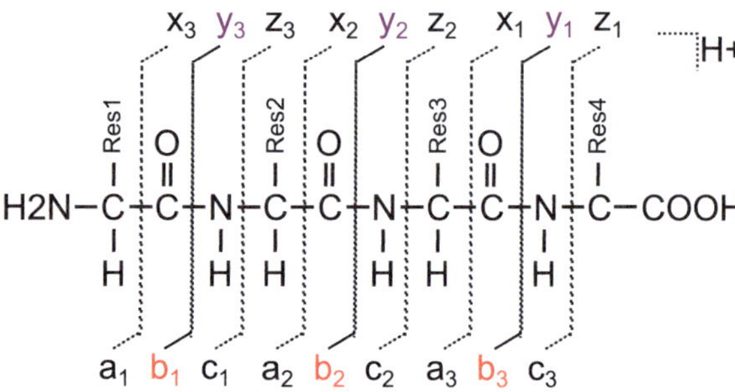

Fig. 1 Fragmentation pattern according to the Roepstorff notation. For CID, fragmentation primarily takes place in the peptide bond giving rise to b- and y-ions. Other fragmentation techniques can yield a, c, x, and z ions

N-terminal part of the peptide are denoted a*n*, b*n*, and c*n*, dependent on which backbone bond is involved in the cleavage. The subscript "*n*" denotes the number of residues in the resulting fragment ion (*see* Fig. 1). The corresponding ion-series containing the C-terminal part are denoted x*n*, y*n*, and z*n*. For CID fragmentation, the peptide ion will preferably undergo fragmentation in the amide bond, resulting in b- and y-ions depending on which terminus retains the positive charge. Each peptide ion will fragment only once or a few times. However, as we have isolated many copies of the same peptide ion, the resulting spectra will contain fragments representing a large part of the potential fragments (*see* Fig. 2). The exact distribution of the various ions (i.e., the peak height) is difficult to predict, and will depend both on the exact peptide sequence and on the distribution of positively charged residues.

As the mass difference between the fragment ions in an ion-series in the MSMS spectrum represents the mass of the individual amino acid residues (*see* Fig. 2), the complete peptide sequence can in principle be interpreted directly by calculating the distance/difference between the fragment ions. This is the fundamental principle behind de novo sequencing (*see* Subheading 3.2). However, the process is complicated considerably by several ion-series that may coexist in the same spectrum (e.g., y-, b-, and a-ions); some of the ions may be absent; some fragment are unstable and lose small neutral molecules like water or ammonia (18 and 17 Da respectively); some fragments result from two fragmentation events leaving us with fragments having neither the N- nor the C-terminal (internal fragments) and some peaks in the spectra may even be from co-isolated peptides. Some of these fragment types are illustrated in Fig. 2 and introduced in more detail in next section.

The reason why some of the ions are missing or of very low intensity is that cleavage is not equally likely to occur at each individual bond in the peptide; the bonds cleaved in a given

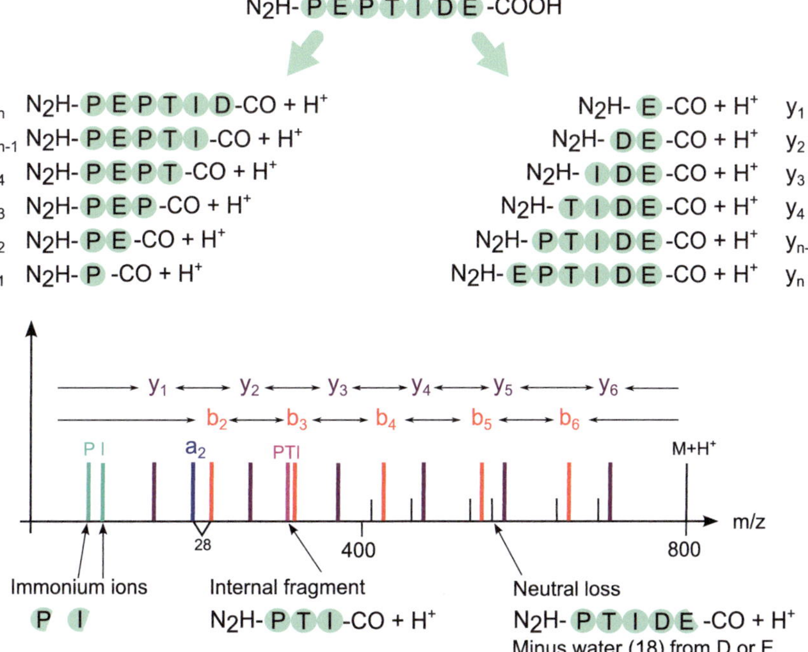

Fig. 2 Typical ions observed in a CID fragmentation mass spectrum. The peptide PEPTIDE gives rise to a number of primary fragmentation ions (the b- and y-ions) as well as satellite peaks arising from loss of water or ammonia. In addition, a few a ions like a2/b2 ion pairs, immonium ions (amino acid side chains), and internal fragments are usually observed

situation can be referred to as the "possible dissociation pathways/channels". Which pathways/channels are followed depends on various factors like; the activation method used; the amino acid side chain adjacent to the dissociation site; the number of protons available, and the rate at which the fragmentation takes place. The stability of the resulting ions will also have an influence on which fragments are observed, as unstable ions can dissociate further into smaller ions. As a result, we can think of some of the dissociation pathways being more populated than others and some will not be used at all, leading to differences in fragment intensities.

But can it then be concluded which of the pathways will be the preferred pathways and which will not be used at all? Yes and no. There are a few general rules for selectively enhanced fragmentation pathways. As we describe further, some of these rules depend on whether or not you have more protons on your peptide ion than you have arginine residues and are especially important for MALDI data. For ESI, the only clear rule involves the presence of the residue proline. These rules are described in the method section.

Most of the peptide ions will only be subjected to a single backbone cleavage. However, two successive cleavages may take place, resulting in either low-mass ions (e.g., a bn-ion fragments into a bn_{-1}–ion or an an-ion, especially the b_2 and a_2-ions are stable) or in internal fragment ions (containing neither the original C- nor N-terminus). For tryptic peptides, the enhanced proton affinity of the C-terminal residue, offer an explanation as to why the b-ions in the high mass range are typically of lower abundance (or missing) in a tryptic spectrum. Exceptions from this general finding are seen for peptides containing additional basic sites; b*n*-ions containing these residues will be more stable and thereby also more abundant (see examples later, Fig. 3c).

Fragment ions containing only a single residue are called immonium ions. These are of the structure HN=CH-R and range from *m/z* 30–159. They can be used as indicators of the presence of specific amino acid residues, *see* Fig. 2 and GPMAW Tables.

Satellite peaks connected to the individual fragment ions can be observed in an MS/MS spectrum (*see* Fig. 2) due to loss of small neutral molecules of ammonia (-17 Da, commonly from the side chains of Asn, Gln, Lys, and Arg), or water (-18 Da, either from the C-terminal COOH group or the side chains of Asp, Glu, Ser, or Thr) (for proposed mechanisms, *see* [11]). *See* **Note 3** for more neutral losses.

2 Materials

A large amount of programs, tools, and information is available through the Internet. For the manual interpretation of ms/ms spectra, we mention a few useful tools: Fragment Ion Calculator can be used directly at http://db.systemsbiology.net:8080/proteomicsToolkit/FragIonServlet.html and the software GPMAW (http://gpmaw.com) has a number of useful features, including the 'Fragment Analyzer' tool which is available as a free download. A number of tables can be downloaded from the GPMAW web page (http://gpmaw.com/html/ms-tables.html). In the following these tables are referred to as GPMAW Tables and comprise: elemental mass values, residue mass values, b1- and y1-ions, immonium ions, all a2–b2 ions and fragment ions of single, double and triple residues up to 300 Da.

3 Methods

3.1 Data Filtering, Database Searches and Automated De Novo Sequencing

Before investing a huge effort on identifying the analyzed peptides manually it is a good idea to filter away low-quality spectra and use relevant software to identify whatever can be easily identified by a database search. The following questions are often relevant for this process.

3.1.1 Is the Data of High Enough Quality?

It is always worthwhile to take a look at the quality of the data before starting the interpretation. Especially when dealing with large datasets it is advantageous to use software programs like

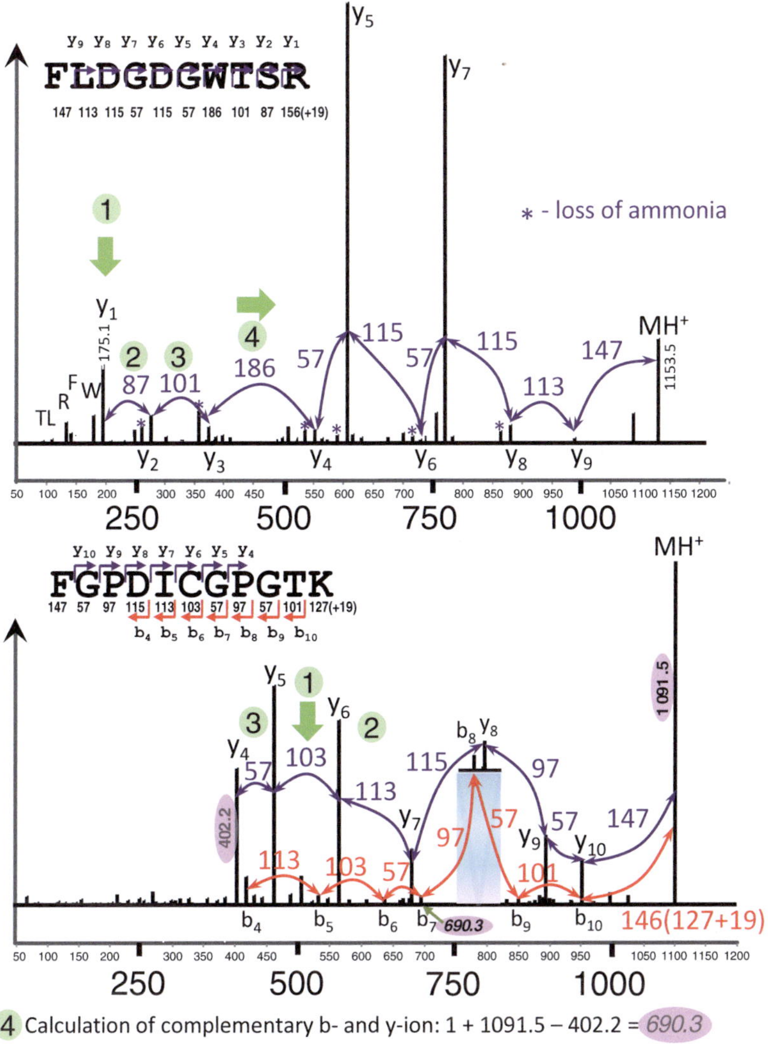

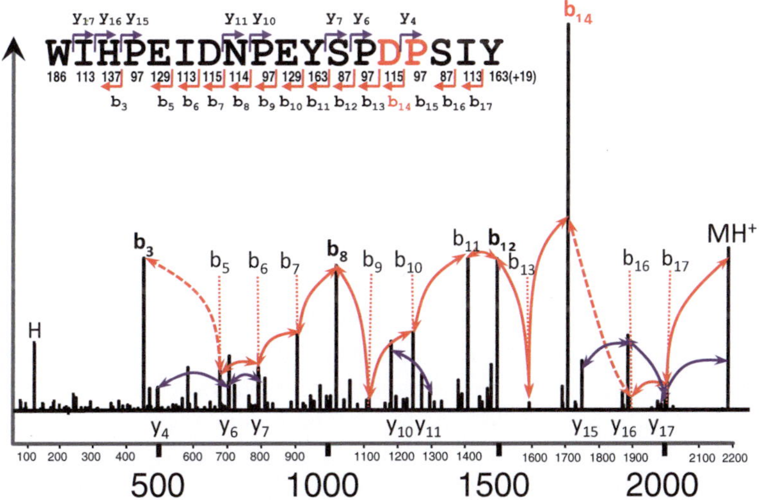

Fig. 3 Typical spectra of singly charged peptides obtained by CID ms/ms fragmentation. (a) Arginine-terminated peptide with typical Asp-induced base peaks. (b) Lysine-terminated peptide. (c) Non-tryptic peptide

MassAI to evaluate data quality (http://www.massai.dk). If the goal is to perform de novo sequencing (*see* Subheading 3.2) you can for example use the software to confidently exclude all MS/MS spectra with only 10 or less fragment peaks as these are less likely to contain enough information for de novo sequencing.

3.1.2 Can the Spectra Be Identified by a Simple Database Search?

Even if one expects that the peptides analyzed are not present in a public or personal database it is always a good idea to make a database-dependent search using one of the many available search engines. In this step you also have a fair chance to discover if your sample is contaminated with unexpected proteins, or if very close homologs are present in the database. See reference for a detailed description of how to perform such searches [1].

3.1.3 Does a Second Path Search Reveal Peptides with for Example Simple Unexpected Modifications?

For most search engines, it is also possible to make a second path search (sometimes referred to as error tolerant searches). The concept is that when you have identified the proteins present, you can repeat the search one more time, but this time the search will only be performed against the protein sequences identified in the first run. Your search is now performed in a much smaller search space and as a consequence you can allow many more modifications, missed cleavage sites and even common substitutions. However, be aware that this method often results in many false positives and manual validation (*see* Subheading 3.3) is often needed.

3.1.4 Are the Spectra Already Identified by Others?

An alternative to traditional database-dependent searches is the spectral library matching strategy [5]. Here an experimental spectrum is matched against a database of already obtained and assigned experimental spectra. It is then assumed that when a spectrum is nearly identical to an already identified spectrum in the database, then the peptides giving rise to these spectra are identical (*see* **Note 4**).

3.1.5 Can an Automated De Novo Sequencing Program Help?

Yes, but it should be stated that automated de novo sequencing is still in its infancy and manual evaluation of the obtained results are often needed (*see* **Note 5**). It is generally found that automated de novo software is having the hardest time predicting the terminals of the spectra, whereas most of them do an honorable job in the middle part of the peptide, where it often relies on complementary b- and y-ions (see later).

3.1.6 Can You Combine De Novo Sequencing and Database Searches?

Yes again. In programs like PEAKS the de novo sequencing strategy is combined with a database-dependent search strategy to increase the overall output and to cross-validate the results found by both methods [12]. As for database-dependent search strategies certain parameters need to be set for automatic de novo sequencing. These are typically enzyme specificity, number of missed cleavages, mass tolerance, modifications, instrument/fragmentation type and charge state.

3.2 Manual De Novo Sequencing

Equipped with spectra of sufficient quality, a calculator and relevant tables or with a software assisting tool like the Fragment Analyzer (*see* **Note 6**) from Lighthouse data you can train your ability to interpret spectra and sometimes identify spectra not identified by software search tools.

There are several ways to start the manual de novo sequencing. Three of these are described under Subheadings 3.2.1–3.2.3 and illustrated by the examples in Fig. 3a–c.

3.2.1 Look for the y1 Ion and Extend the y-Ion Series

This is particularly relevant for tryptic peptides where the C-terminal residue is expected to be either arginine (R), or lysine (K). Generally, the y1 ion can be found as the residue mass plus 19 (*see* GPMAW Tables).

1. Locate the potential y1 ion/ions. Each of these can be a starting point for the rest of the process. For tryptic peptides you should start from either *m/z* 175 (y1 ion for arginine-terminated peptides) or *m/z* 147 (y1 ion for lysine-terminated peptides). For the spectrum in Fig. 3a the y1 ion is easily located at *m/z* 175.
2. The y2 ion can now be found by calculating the mass distance between the y1 ion and the higher mass fragment ions (within a range of 57 and 186, if modified residues are ignored). For the spectrum in Fig. 3a, we find a fragment ion at *m/z* 262 (y1 + 87). This indicates that the second last residue in the peptide is a serine residue and we annotate this ion as the y2-ion.
3. When a potential y2-ion is found, the process can be repeated by calculating the distance between this ion and the sequence ions with higher mass in order to find the y3-ion. In the example given this could be the ion which is 101 Da heavier than the y2-ion, indicating a threonine as the next residue. As the y-ion series starts from the C-terminal of the peptide (*see* Fig. 1), the sequence we interpret from the y-ion series is the reverse peptide sequence. So for now, we have an arginine as the most C-terminal residue, a serine as the second-last residue and a threonine as the third-last residue. The most C-terminal part of the peptide is therefore found to be –TSR.
4. The process is repeated until the end or until no more y-ions can be found. In the latter case, you can try to look for larger distances that may represent a combination of two or three residues. See GPMAW Tables for such mass combinations.

Experience tells us that what is often found confusing is the fact that we are looking at both fragment masses and fragment-mass differences. In the example above, we looked for a fragment mass in **step 1**, wanting to find the y1-ion which we knew should be either 147 or 175 for tryptic peptides. For the rest of the steps we used mass differences between fragment ions to tell us the nature of the residues in the sequence.

3.2.2 Look for the a2–b2 Ion Pair and Extend the b-Ion Series

An alternative starting point could be to look for the a2–b2 ion pair (illustrated in Fig. 2). The difference between the a2 ion and the b2 ion is CO, i.e., a mass difference of 28 (*see* Fig. 1). As the a2 and b2 ions are quite stable, they are often of high intensity, whereas the b1 ion is often missing. The a2–b2 ion pair can therefore be used to kick start the b-ion series in the same way as we used the y1 ion to kick starting the y-ion series in the section above. As the b2-ion is composed of two residues, the mass region in which we can find the a2–b2 ion pair is limited by *m/z* 115 to *m/z* 373 (i.e., from Gly-Gly + H^+ to Trp-Trp + H^+).

1. Locate any potential a2–b2 ion pairs, i.e., two ions in the *m/z* 115–373 region spaced by 28. In case of several potential a2–b2 pairs, you are now advised to start with the most intense pair (see GPMAW Tables for relevant mass values of b2-ions). In the a2–b2 ion pair, the ion with the highest mass will be the b2 ion.
2. The b3 ion can be found by calculating the mass difference between the b2 ion and the higher-mass ions—this is parallel to the process described for y-ions above (*see* Subheading 3.2.1).
3. When a potential b3 ion is found the process can be repeated by calculating the mass distance to the ions with a higher mass in order to find the b4 ion etc.

Please notice that even though the y- and b-ions are read from each end of the peptide, the smaller (low number) ions (e.g., b2 and y2, each containing two residues) will be in the low-mass region for both ion types, whereas larger ions, like b10 and y10, will be in the higher-mass region, each containing 10 residues.

3.2.3 Start in the Middle/High Mass Area of the Spectrum and Extend in Both Directions

Especially if the low mass region is densely filled or contain only very few peaks (as in Fig. 3b), it may not be possible to locate the y1 ion or the a2/b2 ion pair. Instead, interpretation can be started in the middle or high-mass region of the spectrum. Here two relatively intense ions separated by the mass of a single residue can serve as a starting point. For tryptic peptides, it is likely that both peaks represent y-ions, as these often dominate in the high-mass area due to the basic residue in the C-terminal (*see* also Subheading 3.3.3). However, there is no guarantee that the peaks are from the same ion series (*see* **Note 7**).

1. Locate two peaks spaced by a distance that corresponds to a single amino acid residue (see GPMAW Tables). In the spectrum in Fig. 3b, two peaks (later determined to be y5 and y6) acted as such a starting point. The difference was found to be 103 corresponding to the mass of a cysteine residue (be aware that cysteine is often modified or part of a disulfide bridge).
2. From the fragment ion with the highest mass (y6 in Fig. 3b) you can measure the distance to higher mass peaks and find new members of the ion series. In the example in Fig. 3b,

a 113 Da heavier ion was found corresponding to the addition of leucine or isoleucine. From this ion there was a gap of 212 Da with only very low-intensity ions. According to the GPMAW Tables, this difference can correspond to either Asp-Pro or Pro-Asp. A low intensity ion was found indicating that the correct sequence was Asp-Pro. The ion series could then be extended all the way to the precursor ion (MH^+). The last difference is 147 indicating that the sequence has a terminal phenylalanine (and thus showed that we were analyzing the b-ion series).

3. From the initial lower-mass peak you can measure the distance to lower-mass peaks to locate members of the ion series containing fewer residues. In the present example, this was only repeated once as the low-mass ions of this spectrum were of very low intensity.

3.2.4 Finding the Corresponding b/y Ion Series

If you have succeeded in finding a partial or complete ion series, you can calculate the mass values of the ions in the complementary ion series. For example, having located the first three y-ions of a peptide of length n and mass MH^+, you can calculate the masses of the corresponding b-ions (i.e., b_{n-1}, b_{n-2}, and b_{n-3}) using the following formula

$$y_m + b_{n-m} = H^+ + MH^+$$

where H^+ is the mass of a proton in units, i.e., 1.007 u.

In the example given in Fig. 3b we can use this to calculate the mass of an ion from the complementary ion series. For this we use the mass of the peptide ion (MH^+, 1091.5) and the mass of the last ion used in the section above (402.2).

$$M(\text{ion}) = 1 + 1091.5 - 402.2 = 690.3$$

Using this value as a new starting point we extend the new ion series (the y-ions) in both directions. In the example in Fig. 3b this was not possible all the way due to lack of fragment ions in the low mass region. However, as the two ion series are overlapping, the combined information provides the entire peptide sequence.

Notice that the difference between the last b-ion and the precursor ion corresponds to the mass of the relevant residue plus 18. This is due to the C-terminal carboxyl acid group.

The sequence for the spectrum in Fig. 3c is also found using the strategy described under Subheadings 3.2.3–3.2.4. This sequence is a non-tryptic sequence and in a few positions we are missing both the relevant b- and y- ion (e.g., the complementary b4 and y14). We return to this example at the end of Subheading 3.3.3 and demonstrate how we use this knowledge on peptide fragmentation behavior to deduce the rest of the sequence.

3.3 Manual Validation

Most database-dependent search engines and automated de novo sequencing tools provide you with a confidence score based on how well the peptide matches the experimental data. As discussed in the introduction, it can be beneficial to be able to validate/study the results manually. The same is the case after manual de novo sequencing. The questions below can guide you through the process of manual evaluation, supported by the examples in the figures. The relevance of some of the questions depends on the ionization methods applied and the sequence of the peptide, i.e., whether it is a tryptic peptide having a basic residue in the C-terminal or a non-tryptic peptide, which may not even have any basic residues. These differences are explained as we go along.

3.3.1 Can the Majority of the Intense Peaks Be Explained from a Theoretical Fragmentation of the Suggested Peptide?

Assuming that the spectrum only represents a single peptide, we expect the majority of peaks to be explainable, even though the fragmentation behavior of peptides is not fully understood. Having an intense unexplained peak should make you suspicious. However, such peaks could be other sequence-relevant fragments, e.g., internal fragments, a lower charge state of the parent ion or an ion resulting from a neutral loss of for example water or ammonia (*see* Fig. 2). In Fig. 3a, loss of ammonia (indicated by *) explains some of the otherwise non-assigned fragments.

Calculation of the most common ions (a, b, c, x, y, and z) can be done using for example Fragment Ion Calculator. Here you type in the sequence and the mass values of the selected ion series will be calculated. For a more sophisticated calculation including ions like internal ions, modified residues, and satellite ions resulting from loss of water or ammonia, a software like GPMAW can be recommended (*see* Fig. 4). Be aware that we do not expect to see all potential calculated fragment peaks in the experimental spectrum.

Ions not assigned in the spectrum could also inspire you to check whether rearranging a few residues in the sequence could lead to a better result explaining these peaks. Alternatively, a large residue may mask two smaller residues like whether a tryptophan (W, residue mass 186) should instead be replaced by Ser+Val, Glu+Gly, or Asp+Ala and thus explain an otherwise unassigned peak. Looking for the corresponding immonium ions may guide you in the right direction in these relatively rare cases (*see* Subheading 3.3.4).

3.3.2 Is the Charge State of the Peptide as Expected from the Number of Basic Groups in the Peptide?

For MALDI data we only expect a single charge on each peptide, independent of the number of basic groups. For ESI we expect, as a rule of thumb, to observe one proton per basic residue (R, H and K) plus one for the N-terminal amino group. However, each peptide may be detected in more than one charge state in the same ESI experiment, and longer peptides has a tendency to carry a higher charge state, even if they do not contain more basic residues than shorter peptides.

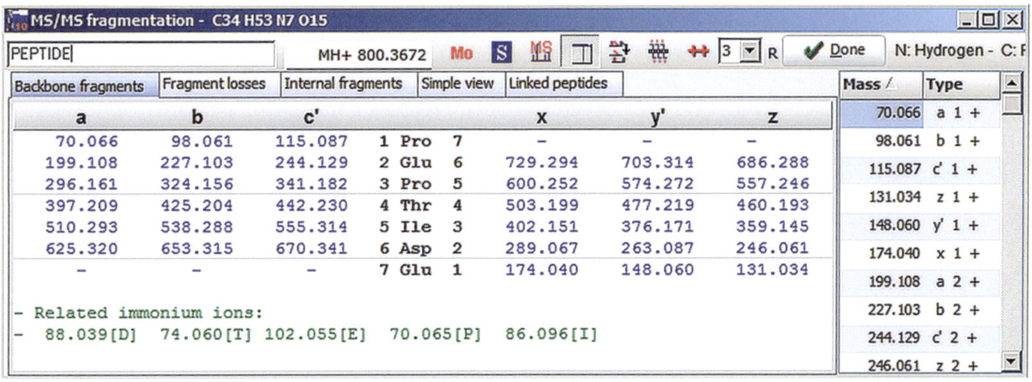

MS/MS fragmentation - C34 H53 N7 O15

PEPTIDE | MH+ 800.3672 | Mo | Done | N: Hydrogen - C: F

Backbone fragments | Fragment losses | Internal fragments | Simple view | Linked peptides

a	b	c'			x	y'	z
70.066	98.061	115.087	1 Pro	7	-	-	-
199.108	227.103	244.129	2 Glu	6	729.294	703.314	686.288
296.161	324.156	341.182	3 Pro	5	600.252	574.272	557.246
397.209	425.204	442.230	4 Thr	4	503.199	477.219	460.193
510.293	538.288	555.314	5 Ile	3	402.151	376.171	359.145
625.320	653.315	670.341	6 Asp	2	289.067	263.087	246.061
-	-	-	7 Glu	1	174.040	148.060	131.034

- Related immonium ions:
- 88.039[D] 74.060[T] 102.055[E] 70.065[P] 86.096[I]

Mass	Type
70.066	a 1 +
98.061	b 1 +
115.087	c' 1 +
131.034	z 1 +
148.060	y' 1 +
174.040	x 1 +
199.108	a 2 +
227.103	b 2 +
244.129	c' 2 +
246.061	z 2 +

Fig. 4 Calculation of the most common fragments of the peptide 'PEPTIDE' as presented by GPMAW. The values presented can be easily modified by the user (e.g., which fragments to show, number of decimals, multiple charges, average/monoisotopic mass). The right-hand panels lists all presented fragments sorted by mass, and other panels show typical fragment losses, internal fragments etc

In Fig. 5 we see a peptide identified with a significant score in a database search of ESI CID MSMS spectra from a horse bone marrow sample. However, the spectrum showed to be incorrectly assigned and is instead representing a highly modified collagen peptide with a substitution. The simple search strategy used was not able to identify the correct peptide due to all of these modifications. The peptide suggested by the search engine is a doubly charged peptide with the sequence SDPAGPP$_{ox}$GPPRRSR. The first reason to be suspicious towards the identification is that we would expect this peptide to have more than the two charges due to the three arginines. We go into more details about this example in Subheading 3.3.3.

3.3.3 Is the Intensity of the Fragment Ions as Expected?

For most of the fragment ions it is not possible to predict whether they will be of high, medium, or low intensity, or whether they will be detected at all. Sophisticated algorithms can give a hint to the relative intensities [13], but there are no simple rules, except from a few rules of thumb connected to specific residues. However, these few selectively enhanced fragmentation pathways can turn out very useful for manual evaluation as explained below. We start by describing the proline-induced fragmentation behavior of ESI spectra and then describe the enhanced fragmentation C-terminal to acidic residues seen primarily in MALDI spectra.

For ESI spectra:

A commonly known feature for ESI data is the enhanced intensity of the fragment resulting from cleavage N-terminal to proline, i.e., in the peptide bond Xaa-Pro, where Xaa can be any residue [14]. The effect of proline on the fragmentation pattern seems to be most pronounced when the residue N-terminal to proline (Xaa) is Asp, His, Val, Leu, or Ile. The fragment resulting from cleavage on the other side of the proline (C-terminal toproline) is often found to be of low intensity or even missing.

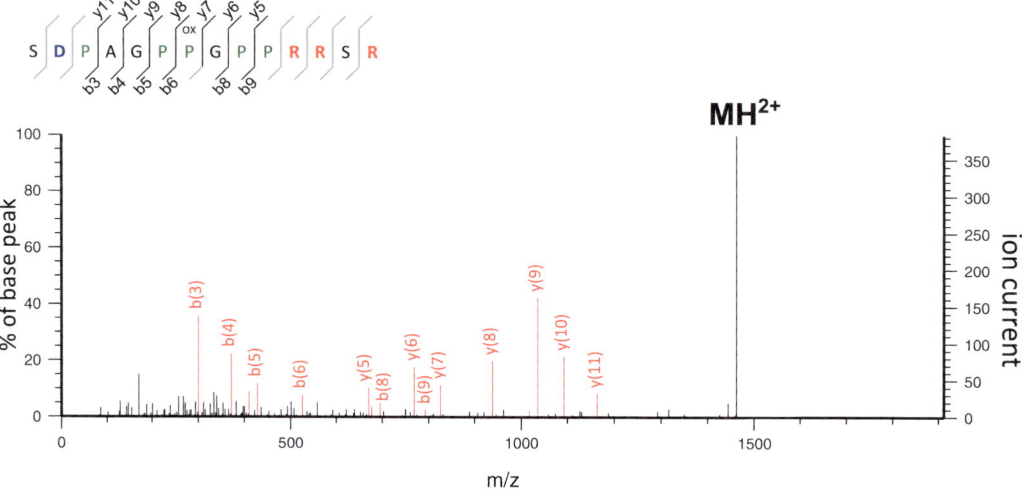

Fig. 5 MSMS spectrum of a peptide initially identified as SDPAGPP$_{ox}$GPPRRSR by the automatic search engine. Careful examination of the spectrum using the rules suggested, revealed the peptide to be a modified collagen peptide, for details please see the text

For MALDI spectra:

The proline-directed fragmentation pattern described for ESI above, also applies for MALDI data but only when the peptide does not contain an arginine. In order to understand this we have to introduce the mobile proton theory.

When a proton is "free" to move along the backbone to energetically less favored protonation sites we call it a mobile proton. *See* Fig. 6a, b. Using CID fragmentation, the proton will weaken the amide bonds along the backbone, resulting in "low cost dissociation pathways" giving rise to the dominating b- and y-ions series. For a detailed description of the fragmentation mechanisms, please read [11]. As the proton is involved in the fragmentation, this process is called "charge-directed" or "charge-induced" fragmentation. The energy required to induce the "charge-directed" fragmentation event depends on the composition of the peptide and especially on the basicity of the residues as explained below.

Arginine is the most basic residue in gas-phase reactions having the highest proton affinity. Arginine therefore effectively "immobilizes" or"sequesters" the otherwise available proton. This is the situation in Fig. 6a, c and d. In MALDI we have only a single proton, so no proton is available for migration at low energies, when an arginine is present, Fig. 6c, d. This opens up for alternative fragmentation pathways.

Such alternative fragmentation channels are called charge-remote fragmentation channels, as they do not involve the proton. The dominating charge-remote channel is selective cleavage C-terminal to acidic residues (Asp and Glu) (*see* Fig. 6d). This phenomenon is described in details in the review on fragmentation pathways of protonated peptides by Paizs and Suhai [11]. The consequence is that for arginine containing peptides, intense ions can be expected from fragmentation C-terminal to aspartic acid (D) and glutamic acid (E), with D having a higher impact than E (Asp-Xaa > Glu-Xaa).

In the spectrum of the peptide FLDGDGWTSR (*see* Fig. 3a) this phenomenon can be observed, as the two very dominating ions, y5 and y7, are the result of charge-remote fragmentation C-terminal to Asp (D). As seen in this example, the fragment containing the arginine will be the dominating ion and for tryptic peptides this will typically be the y-ion, as the arginine will be the C-terminal residue and have high proton affinity. If, instead, the arginine is located in the N-terminal part of the peptide, b-ions will be dominating. In addition to these acidic-induced intense fragments, arginine-containing peptides will often result in alternative fragments involving fragmentation in the side chain of residues as well as in the backbone (*see* **Note 3**). The effects of these charge-remote fragmentation pathways have a huge influence on the resulting MALDI MSMS spectra.

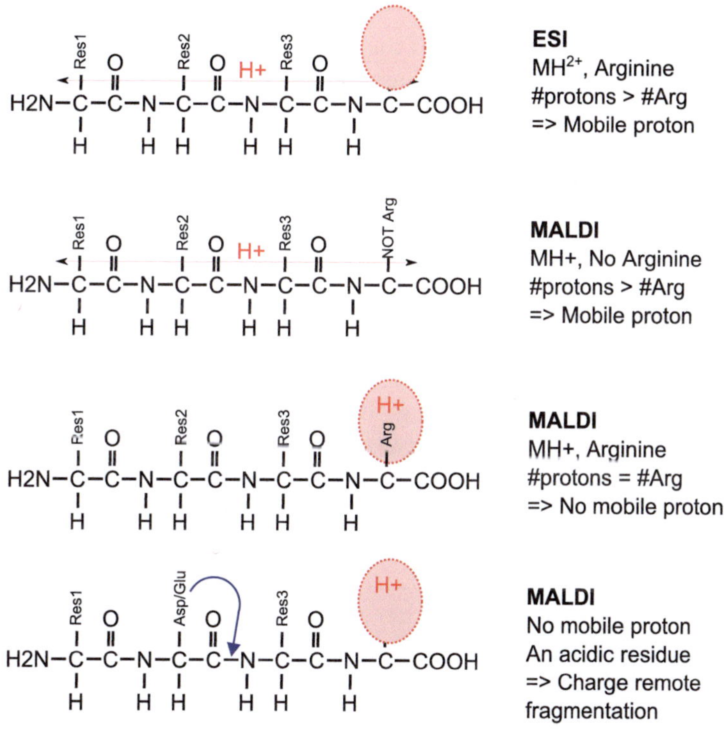

Fig. 6 Mobile and non-mobile protons. When the number of protons is higher than the number of arginine residues, the proton is mobile (a and b) otherwise it is non-mobile (c and d), and charge-remote fragmentation may take place (d)

But what about the prolines then? For arginine-containing peptides, the fragmentation N-terminal to proline is still enhanced but mainly if the previous residue is D (e.g., 15). Confused? That is understandable. But let us add one more rule of thumb before summing up.

The proton affinity of histidine is not as high as for arginine but still higher than for lysine. Histidine therefore plays a role in the relative abundance of the resulting b- and y-ions as the histidine attracts the proton, but does not sequester it to the same degree as arginine does, and the fragmentation pattern is often unpredictable. This is illustrated in the example in Fig. 6c, where b-ions are dominating the spectrum, due to the histidine being located as the third residue. Especially the b-ion resulting from cleavage C-terminal to D and N-terminal to P is found to be dominating the spectrum.

So to sum up:

- For MALDI you should distinguish between situations where you have an arginine-containing peptide and an arginine-deficient peptide.
- For arginine-containing peptides you should expect acidic residues to result in fragmentation C-terminal to these residues with D giving fragments of higher intensity than E.
- If the peptide contains the dipeptide PD you should expect the fragmentation between these two residues to be of the highest intensity.
- For arginine-deficient peptides you should not expect fragmentation C-terminal to acidic residues to dominate.
- Expect that the basic residues have a high influence on the intensity of the relevant ion series—basic residues in the C-terminal part will support the formation of y-ions whereas basic residues in the N-terminal part support formation of b-ions.

Let us go back to the example in Fig. 5, the one with the non-correctly identified peptide. We see that we have three arginines and only two protons. We know that this is not something we would expect, but for now let us follow the idea. As the arginines are holding on to the two protons we are left with absolutely no mobile protons. Therefore we should check to see whether we should expect a charge-remote fragmentation event to dominate our spectra. And indeed yes. As the second residue we have an aspartic acid and it is even followed by a proline. According to the rules described above we should expect an enhanced fragmentation event here. But should we expect the enhanced fragmentation to result in a b2 ion or a y12-ion? As the arginines are located in the C-terminal part and they have sequestered the protons we would expect a dominating y12 ion—mainly as a doubly charged ion but probably also as a singly charged ion. But we see neither! This is an example of how we can use the rules above to verify or, as in this case, reject a search result.

We can also use the guidelines to support the de novo sequencing process. Look at the spectrum in Fig. 3c. Here we miss some fragment ions, e.g., the corresponding b4/y14 ion pair. However, we see a very intense b3 ion indicating a potential proline. The difference between the b3 and the b5 corresponds to a PE dipeptide. The same is the case C-terminal to the very intense PD-directed b14 ion. Here, the difference between b14 and b16 supports the finding of a PS dipeptide. Matching our obtained sequence against a database of relevant proteins supports the resulting peptide sequence (*see* **Note 8**).

3.3.4 Is the Sequence Confirmed by the Immonium Ions/Diagnostic Ions?

Immonium ions (*see* Fig. 2) are found in the very low-mass region (below m/z 172) and some residues can give rise to more than one immonium ion (see GPMAW Tables). The immonium ions can provide a hint as to whether the correct residues are identified, but be aware that immonium ions are not always present, and they are not always proof of presence, as other ions can coexist in the same region. See GPMAW Tables for the mass values of relevant residues and their immonium ions. For the peptide FLDGDGWTSR (*see* Fig. 3a) the immonium ions for Thr, Leu/Ile, Arg, Phe, and Trp are present in the low-mass region. Immonium ions of modified residues are often used as diagnostic ions to confirm the presence of the given modification. Examples can be found in [16].

3.3.5 Are There Any Modification-Specific Ion Peaks Present?

If the identified peptide holds some modified residues, some of these may give rise to additional peaks confirming the presence of such modification.

For peptides containing an oxidized methionine, enhanced side-chain cleavage due to a neutral loss of CH_3SOH (64 Da) can be observed [17]. Such an intense ion is valuable as a diagnostic ion in order to distinguish between phenylalanine and oxidized methionine, as they have the same nominal mass. All signals, corresponding to fragments containing this oxidized methionine, will have a corresponding satellite ion with a 64 Da lower mass (the signal-to-noise ratio influence the chance of observing these ions).

Other examples of neutral losses are those of 80 or 98 Da from ions containing a phosphorylated residue, and several different neutral losses can be observed for peptides with N-linked glycans [16, 18]

4 Notes

1. Recent advances in top- and middle-down MS analysis show great promise as to quickly analyze even large proteins like immunoglobulins [19, 20]. However, these techniques still only work efficiently when the primary structure is known, and even then only part of the primary structure is verified. These problems are exaggerated when the protein is modified and for sequences like the variable region of IgG, as such modifications and variable regions cannot always be deduced from the corresponding DNA. In these cases you have to use the bottom-up approach, and even then automated sequence determination will fail to identify the sequence in many cases.
2. Traditionally, ESI MSMS instruments are coupled to LC separation systems and can be used for analysis of very complex samples resulting in thousands of MSMS spectra from a single MS experiment. In contrast, MALDI MSMS is often used for the analysis of less complex samples, e.g., characterization of peptides from a single protein.
3. Other neutral losses are due to intramolecular rearrangements involving both a backbone cleavage and the loss of a side-chain of the adjacent amino acid residues. The resulting fragments are denoted v*n*-, d*n*- and w*n*-ions [21]. These side-chain reactions require high-energy activation and an immobilized proton to occur and are therefore primarily observed in high-energy CID experiments on arginine containing peptides
4. A limitation in this strategy is that the appearance of the spectra in terms of fragments observed and the intensities of the fragments vary with the ionization method used, the type of analyzer, and the fragmentation method applied. As a consequence, only spectra obtained under similar conditions should be compared and the database should contain verified spectra of sufficient quality.
5. In the review by Allmer [22] the different types of de novo algorithms are presented and discussed. Also the benefit of combining several spectra of either the same type (e.g., grouping similar CID spectra) or different types (e.g., CID and ETD spectra of the same peptide) are described.
6. Fragment Analyzer is a small free-ware utility supplied by Lighthouse data (http, //www.gpmaw.com look in the download section). When manually annotating a spectrum during de novo sequencing, you select a peak and enter the mass into the program. This will then suggest matching peaks for the next amino acid residue in the ion series moving up or down in mass. If you locate a residue, just double-click on the residue, and the program will suggest the next residue in the chosen direction. The program is designed to be small and unobtrusive and sit on top of the mass spectrum.

7. By applying different experimental strategies involving labeling of one of the peptide terminals it is possible to get a hint as to whether two ions belong to the same or different ion series. Examples on such strategies are tryptic digestion in O18 water selectively labeling y-ions [23] and combining Lys-N digestion with differential isotopic dimethyl resulting in labeling of the N-terminal containing fragments, the b-ions [24].
8. Having obtained one or more peptide sequences by de novo sequencing it is often of interest to match these against sequences from protein databases in order to get an idea of the function of the protein based on homologous proteins. A traditional BLAST search engine is not optimized for such short sequences and it is advisable to use a specialized program like MSBlast for this purpose [25]. This program allows combined searches, where several peptide sequences are searched together in order to find proteins matching to some or all of the query sequences. The program can even deal with gaps and ambiguities in the peptide sequences.

Acknowledgements

Anne-Katrine Vestergaard is acknowledged for excellent technical assistance.

References

1. Edwards NJ (2011) Protein identification from tandem mass spectra by database searching. Methods Mol Biol 694:119–38
2. Zhang H, Cui W, Gross ML (2014) Mass spectrometry for the biophysical characterization of therapeutic monoclonal antibodies. FEBS Lett 588:308–17
3. Mann M, Højrup P, Roepstorff P (1993) Use of mass spectrometric information to identify proteins in sequence databases. Biomed Environ Mass Spectrom 22:338–345
4. Pappin DJC, Højrup P, Bleasby AJ (1993) Rapid identification of proteins by peptide-mass fingerprinting. Curr Biol 3:327–332
5. Craig R, Cortens JP, Beavis RC (2005) The use of proteotypic peptide libraries for protein identification. Rapid Commun Mass Spectrom 19:1844–1850
6. Switzar L, Giera M, Niessen WM (2013) Protein digestion, an overview of the available techniques and recent developments. J Proteome Res 12:1067–77
7. Højrup P (2009) Peptide mapping for protein characterization. In: Walker JM (ed) The protein protocols Handbook'. Humana Press, Totowa, NJ, pp 969–988
8. Whitehouse CM, Dreyer RN, Yamashita M, Fenn JB (1985) Electrospray interface for liquid chromatographs and mass spectrometers. Anal Chem 57:675–9
9. Karas M, Hillenkamp F (1988) Laser desorption ionization of proteins with molecular masses exceeding 10,000 daltons. Anal Chem 60:2299–301
10. Roepstorff P, Fohlman J (1984) Proposal for a common nomenclature for sequence ions in mass spectra of peptides. Biomed Mass Spectrom 11:601
11. Paizs B, Suhai S (2005) Fragmentation pathways of protonated peptides. Mass Spectrom Rev 24:508–48
12. Zhang J, Xin L, Shan B, Chen W, Xie M, Yuen D, Zhang W, Zhang Z, Lajoie GA, Ma B (2012) PEAKS DB: de novo sequencing assisted

database search for sensitive and accurate peptide identification. Mol Cell Proteomics 11:M111.010587

13. Degroeve S, Martens L (2013) MS2PIP, a tool for MS/MS peak intensity prediction. Bioinformatics 29:3199–203
14. Schwartz BL, Bursey MM (1992) Some proline substituent effects in the tandem mass spectrum of protonated pentaalanine. Biol Mass Spectrom 21:92–6
15. Kapp EA, Schütz F, Reid GE, Eddes JS, Moritz RL, O'Hair RA, Speed TP, Simpson RJ (2003) Mining a tandem mass spectrometry database to determine the trends and global factors influencing peptide fragmentation. Anal Chem 75:6251–64
16. Larsen MR, Trelle MB, Thingholm TE, Jensen ON (2006) Analysis of posttranslational modifications of proteins by tandem mass spectrometry. Biotechniques 40:790–8
17. Lagerwerf FM, van de Weert M, Heerma W, Haverkamp J (1996) Identification of oxidized methionine in peptides. Rapid Commun Mass Spectrom 10:1905–10
18. Bunkenborg J, Matthiesen R (2013) Inter pretation of tandem mass spectra of posttranslationally modified peptides. Methods Mol Biol 1007:139–71
19. Fornelli L, Ayoub D, Aizikov K, Beck A, Tsybin YO (2014) Middle-down analysis of monoclonal antibodies with electron transfer dissociation orbitrap fourier transform mass spectrometry. Anal Chem 86:3005–3012
20. Liu X, Dekker LJ, Wu S, Vanduijn MM, Luider TM, Tolić N, Kou Q, Dvorkin M, Alexandrova S, Vyatkina K, Paša-Tolić L, Pevzner PA (2014) De novo protein sequencing by combining top-down and bottom-up tandem mass spectra. J Proteome Res 13:3241–3248
21. Johnson RS, Martin SA, Biemann K (1988) Collision-induced fragmentation of (M + H)+ ions of peptides. Side chain specific sequence ions. Int J Mass Spectrom Ion Processes 86(29 D):137–154
22. Allmer J (2011) Algorithms for the de novo sequencing of peptides from tandem mass spectra. Expert Rev Proteomics 8:645–57
23. Takao T, Gonzalez J, Yoshidome K, Sato K, Asada T, Kammei Y, Shimonishi Y (1993) Automatic precursor-Ion switching in a 4-sector tandem mass-spectrometer and its application to acquisition of the MS/MS product ions derived from a partially O-18-labeled peptide for their facile assignments. Anal Chem 65:2394–2399
24. Hennrich ML, Mohammed S, Altelaar AFM, Heck AJR (2010) Dimethyl isotope labeling assisted de novo peptide sequencing. J Am Soc Mass Spectrom 21:1957–1965
25. Shevchenko A, Sunyaev S, Loboda A, Shevchenko A, Bork P, Ens W, Standing KG (2001) Charting the proteomes of organisms with unsequenced genomes by MALDI-quadrupole time-of-flight mass spectrometry and BLAST homology searching. Anal Chem 73:1917–26

Chapter 11

Polyclonal Peptide Antisera

Tina H. Pihl, Kristin E. Illigen, and Gunnar Houen

Abstract

Polyclonal antibodies are relatively easy to produce and may supplement monoclonal antibodies for some applications or even have some advantages.

The choice of species for production of (peptide) antisera is based on practical considerations, including availability of immunogen (vaccine) and animals. Two major factors govern the production of antisera: the nature of adaptive immune responses, which take place over days/weeks and ethical guidelines for animal welfare.

Here, simple procedures for immunization of mice, rabbits, sheep, goats, pigs, horses, and chickens are presented.

Key words Peptide antisera, Polyclonal antibodies, Mice, Rabbits, Sheep, Goats, Pigs, Horses and chickens, Immunoassays, Immunization, Bleeding, Coagulation, Serum

1 Introduction

Polyclonal antibodies (pAbs) are relatively easy to produce and may supplement monoclonal antibodies (mAbs) for some applications or even have some advantages [1–3].

Traditionally, rabbits have been the species of choice. Goats and sheep have also been much used and horses have been used for production of antitoxin sera for therapeutic applications, where large amounts of serum are needed. Chickens may be selected for the following reasons: (1) Greater evolutionary distance to mammals, (2) Ease of pAb production from egg yolk, or (3) Low reactivity with human rheumatoid factors, which are present in many human sera and may compromise immunoassays (e.g., sandwich immunoassays) [3, 4]. On the other hand, pigs may be chosen due to high similarity with humans [5].

Here, simple procedures for immunization of mice, rabbits, sheep, goats, pigs, horses, and chickens are presented. The choice of species for production of (peptide) antisera is based on practical

Gunnar Houen (ed.), *Peptide Antibodies: Methods and Protocols*, Methods in Molecular Biology, vol. 1348, DOI 10.1007/978-1-4939-2999-3_11, © Springer Science+Business Media New York 2015

considerations, including availability of immunogen (vaccine) and animals. Two major factors govern the production of antisera: the nature of adaptive immune responses, which take place over days/weeks and ethical guidelines for animal welfare.

2 Materials

Ammonium sulfate.
Polyethylene glycol (6000).
Adjuvant: Aluminum hydroxide ($Al(OH)_3$) 2 %, autoclaved (*see* **Note 1**).
Saline (0.15 M NaCl), sterile.
Sterile needles and syringes or vacutainer system and blood collection tubes with no anticoagulant.
For large animals: Sterile needles (14G), infusion tube, 1 L sterile glass bottles if a large volume of blood needs to be collected.
Autoclave.
Centrifuge.
Animals:
Mice (any strain, e.g., NMRI or Balb/c).
Rabbits (any strain, e.g., New Zealand White).
Pigs (any breed, e.g., "mini pigs").
Chickens (any breed, e.g., New Hampshire).
Sheep (any breed).
Goats (any breed).
Horses (any breed)
Animal facilities suitable for the species in question (*see* **Note 2**).

3 Methods

3.1 Vaccine Preparation

Mix peptide carrier conjugate (*see* Chapters 6 and 7) 1 mg/mL 1:1 (v:v) with autoclaved aluminum hydroxide adjuvant (2 %) (optionally incubate with end-over-end rotation overnight at 5 °C or at room temperature).

See Table 1 for recommended amounts depending on the species in question.

3.2 Immunization

All procedures involving animals must be carried out by trained and authorized personnel in approved facilities and following current ethical guidelines for animal welfare.

See Table 2 for immunization schedule. We recommend the same basic schedule for all species, basically consisting of subcutaneous (sc) immunization every second week and bleeding every second week in between.

The immunization course may be followed by testing for carrier and peptide antibodies by for example ELISA as often as required during the process (*see* Subheading 3.4).

Table 1
Immunization and bleed volumes (mL) for various species

Species	Injection route	Injection volume[a]	Bleed volume[b]
Mouse	sc	0.1	0.1–0.2
Rabbit	sc	0.5	10–20
Sheep	sc	1	50–500
Goat	sc	1	50–500
Pig	sc	1	50–500
Horse	sc	1	50–5000
Chicken	sc	0.5	10–20[c]

[a]Per animal/per site (usually 1-2 sites per animal)

[b]Approximate volumes per animal (the acceptable volumes depend on the size of the animal and the frequency of bleeding)

[c]Eggs also useful (*see* **Notes 3** and **9**)

Table 2
Immunization and bleeding schedule *see* Table 1 for volumes

Week/day	Action
0/0	Pre-bleed (optional but recommended) (*see* **Note 8**)
1/1	1′ immunization
2/8	1′ bleed
3/15	2′ immunization
4/22	2′ bleed
5/29	3′ immunization
6/36	3′ bleed
7/43	4′ immunization
8/50	4′ bleed
n/7(n − 1) + 1	n′ immunization (*see* **Note 9**)
n + 1/7n + 1	n′ bleed
–/–	Terminal bleed (*see* **Note 10**)

3.3 Bleeding

See Table 1 for approximate bleed volumes per animal for each type of animal. Maximal volumes are dependent on the weight of the individual animal [6], which can vary significantly in larger animals such as horses.

Mice are bled from the facial vein or from the heart if euthanized.

Rabbits are bled from the ear marginal vein.

Sheep are bled from the jugular vein.

Goats are bled from the jugular vein.

Pigs are bled from the cranial vena cava.

Horses are bled from the jugular vein.

Chickens are bled from a wing vein (*see* **Note 3**).

3.4 Testing

Preliminary testing is done by ELISA (*see* **Note 4**). The production of antibodies may be followed by the increase in absorbance as a function of immunization times or by determination of serum titers (*see* **Note 5**) defined as the dilution of the serum which yields half maximal response in the test system (*see* **Note 6**).

Sera should be analyzed as soon as possible after collection in a test system suitable for the intended use (e.g., immunoblotting, inhibition immunoassay, immunodiffusion, immunoprecipitation, sandwich immunoassay).

3.5 Processing

Large volumes of serum can be separated from the blood cells by letting the blood sample stand to precipitate by itself, followed by collection of the serum by aspiration and optionally filtering through a cheesecloth and/or centrifugation. Preparation of smaller volumes of serum can be made by centrifugation of the blood samples at $1000 \times g$ for 10–15 min. Blood samples should be left at room temperature for at least 30 min before centrifugation [7]. Optional: Immunoglobulins may be purified from the serum by ammonium sulfate precipitation, ion exchange chromatography, or protein A/G/L affinity chromatography (*see* **Note 7**).

3.6 Storage

Store at −20 °C, −50 °C, or −80 °C (years).

For short term storage at 5 °C: (add sodium azide to 0.02 %) (weeks–months).

4 Notes

1. We recommend the use of aluminum hydroxide ($Al(OH)_3$) as the adjuvant of choice. It is a safe adjuvant, also approved for human use, and has few side effects. $Al(OH)_3$ is known to activate the alternative complement pathway as a major mechanism of action [8]. Other adjuvants, e.g., Freunds complete/incomplete adjuvant may be used for special purposes but are not recommended.
2. If you do not have access to an animal facility, many companies offer this as a service.
3. For egg-laying hens, antibodies may also be collected and purified from egg yolk (IgY).

4. *See* [1, 2] or Chapters 12 and 21.
5. Make a dilution series starting at 1:10 and dilute further 10 times from this, etc. Further testing may be done from one of these dilutions by making twofold titrations.
6. Sometimes the endpoint titer is used, which is the dilution required to reach background signal.
7. *See* Chapter 15 and [2].
8. If a pre-immunization bleed is not available, normal serum from non-immunized animals or a serum from an animal immunized with another (irrelevant) antigen may be used as a control.
9. Polyclonal antibodies (IgY) are easily obtained from egg yolk by ammonium sulfate/PEG fractionation and/or ion exchange chromatography [9, 10] (*see* Chapter 15).
10. When the titer is sufficiently high, immunization may be discontinued but bleeding continued, if larger volumes of serum are required. The volumes and methods depend on the species and must follow local procedures and ethical guidelines.

References

1. Wild D (ed) (2013) The immunoassay handbook. Elsevier, Oxford
2. Harlow E, Lane D (1988) Antibodies—a laboratory manual. Cold Spring Harbor Laboratory Press, New York
3. Hanly C, Artwohl J, Bennett BT (1995) Review of polyclonal antibody production in mammals and poultry. ILAR J 37:93–118
4. Holm BE, Sandhu N, Tronstrøm J, Lydolph M, Trier NH, Houen G (2015) Species cross-reactivity of rheumatoid factors and implications for immunoassays. Scand J Clin Lab Invest 75:51–63
5. Meurens F, Summerfield A, Nauwynck H, Saif L, Gerdts V (2012) The pig: a model for human infectious diseases. Trends Microbiol 20:50–7
6. Diehl KH, Hull R, Morton D, Pfister R, Rabemampianina Y, Smith D, Vidal JM, van de Vorstenbosch C (2001) A good practice guide to the administration of substances and removal of blood, including routes and volumes. J Appl Toxicol 21:15–23
7. Tomlinson L, Boone LI, Ramaiah L, Penraat KA, von Beust BR, Ameri M, Poitout-Belissent FM, Weingand K, Workman HC, Aulbach AD, Meyer DJ, Brown DE, MacNeill AL, Bolliger AP, Bounous DI (2013) Best practices for veterinary toxicologic clinical pathology, with emphasis on the pharmaceutical and biotechnology industries. Vet Clin Pathol 42:252–269
8. Güven E, Duus K, Laursen I, Højrup P, Houen G (2013) Aluminum hydroxide adjuvant differentially activates the three complement pathways with major involvement of the alternative pathway. PLoS One 9;8:e71145
9. Tan SH, Mohamedali A, Kapur A, Lukjanenko L, Baker MS (2012) A novel, cost-effective and efficient chicken egg IgY purification procedure. J Immunol Methods 380:73–6
10. Polson A (1990) Isolation of IgY from the yolks of eggs by a chloroform polyethylene glycol procedure. Immunol Invest 19:253–8

Chapter 12

Production and Screening of Monoclonal Peptide Antibodies

Nicole Hartwig Trier, Anne Mortensen, Annette Schiolborg, and Tina Friis

Abstract

Hybridoma technology is a remarkable and indispensable tool for generating high-quality monoclonal antibodies. Hybridoma-derived monoclonal antibodies not only serve as powerful research and diagnostic reagents, but have also emerged as the most rapidly expanding class of therapeutic biologicals. In this chapter, an overview of hybridoma technology and the laboratory procedures used routinely for hybridoma production and antibody screening are presented, including characterization of peptide antibodies.

Key words Immunization, Fusion, Selection, Screening, Enzyme-linked immunosorbent assay, HybER, Hybridoma enhancing reagent, Isotype determination, Monoclonal antibody, Peptides

1 Introduction

For many years, antibodies to synthetic peptides have been an irreplaceable tool for many molecular immunology investigations [1–5]. In addition, monoclonal antibodies have become key components in a vast array of clinical diagnostic tests. The development of the technique for production of monoclonal antibodies has gained revolutionary influence, not only in relation to immunologic research but in biological and medical research as well.

Initial development of the basic technique for production of monoclonal antibodies is ascribed to Köhler and Milstein [6]. They showed that it was possible to combine the ability of a plasma cell to produce a monospecific antibody with the ability of a B-cell tumor to divide limitless, thus generating an immortal cell line producing monospecific antibodies. Moreover, they developed an efficient method to select for newly fused hybridomas in a mixture of hybridomas, B-cells, and non-fused tumor cells.

Monoclonal antibodies are typically raised against multiple targets, e.g., recombinant proteins, native proteins or peptides. Native or recombinant proteins were used traditionally to produce antibodies. However there are instances in which a peptide serves

Gunnar Houen (ed.), *Peptide Antibodies: Methods and Protocols*, Methods in Molecular Biology, vol. 1348, DOI 10.1007/978-1-4939-2999-3_12, © Springer Science+Business Media New York 2015

as a better choice than a protein, e.g., when raising antibodies to a specific protein isoform or a phosphorylated protein, and in cases where the protein is not available, e.g., proteins that are difficult to prepare in large amounts [7–9].

Generation of hybridoma-derived peptide antibodies is initiated by immunization of animals, and the complete process can be divided into four main stages: immunization, fusion, cloning and screening, and characterization (Fig. 1).

When generating peptide antibodies by using this approach, immunization with a small peptide itself generally do not induce an immune response with antibodies in high titers. Therefore, a

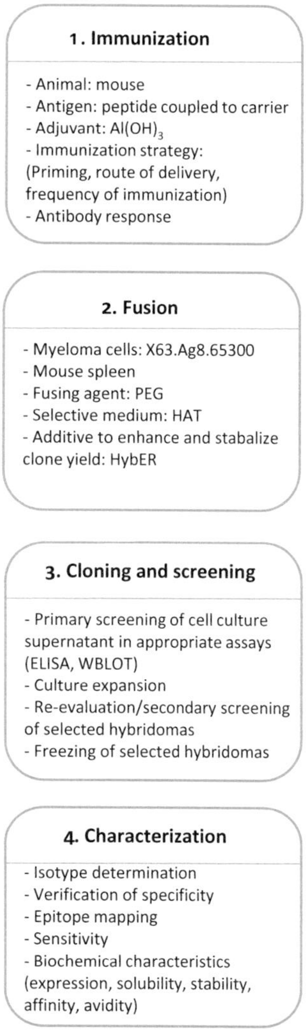

1. Immunization

- Animal: mouse
- Antigen: peptide coupled to carrier
- Adjuvant: $Al(OH)_3$
- Immunization strategy: (Priming, route of delivery, frequency of immunization)
- Antibody response

2. Fusion

- Myeloma cells: X63.Ag8.65300
- Mouse spleen
- Fusing agent: PEG
- Selective medium: HAT
- Additive to enhance and stabalize clone yield: HybER

3. Cloning and screening

- Primary screening of cell culture supernatant in appropriate assays (ELISA, WBLOT)
- Culture expansion
- Re-evaluation/secondary screening of selected hybridomas
- Freezing of selected hybridomas

4. Characterization

- Isotype determination
- Verification of specificity
- Epitope mapping
- Sensitivity
- Biochemical characteristics (expression, solubility, stability, affinity, avidity)

Fig. 1 The generation of hybridoma-derived peptide antibodies can be divided into four main stages: (**1**) Immunization, (**2**) Fusion, (**3**) Cloning and screening, and (**4**) Characterization

peptide longer than 15 amino acids or a peptide coupled to a carrier protein, e.g., keyhole limpet hemocyanin, ovalbumin, or bovine serum albumin, is used to induce an immune response in the selected animal [9–12].

Several immunization techniques can be used for production of antibodies. In general, most protocols give satisfactory results [5, 9, 12–14]. The choice of method used depends on the nature of the antigen and the type of antibody required. The method described in this chapter, concerning production of monoclonal antibodies using mice, is designed to give optimal results with minimal injury to the host animal, and has been used extensively and successfully for several years [9, 15, 16].

Following immunization, antibody clones are selected for their specific reactivity to the immunogen. Several techniques for screening of antibody specificity exist. The most basic approach employs the antigenic peptide coated to a microtiter plate, where antibody reactivity is determined using a colorimetric substrate, e.g., by enzyme-linked immunosorbent assay (ELISA) [15]. It is unpredictable whether a peptide antibody will recognize the native protein due to conformational/structural differences between synthetic peptides and peptide epitopes in the native protein and whether the antibody will recognize its target in different assay systems. One way to circumvent this is by screening with the native protein, by using peptide immunogens located within flexible surface-accessible regions or screening for reactivity in different assay systems [15, 16]. This chapter discusses potential difficulties in screening for peptide antibodies and describes a straightforward approach for primary screening of peptide antibodies.

1.1 Immunization and Hybridoma Technology

Generation of useful monoclonal antibodies with the desired antigen specificity is dependent on a number of steps in the used immunization and hybridoma technology method such as type of animal used for immunization, choice of immunogen, choice of carrier, choice of adjuvant, immunization schemes and route, source of fusion partner, B-cell immortalization procedure and selection of appropriate methods to test for antibody reactivity.

1.1.1 Choice of Animal

Mice are the most commonly used animal species for monoclonal antibody production, and BALB/c is often the strain of choice, since the majority of the murine myelomas used for cell fusion are derived from this strain, including X63.Ag8.653, Sp2/0-Ag14, FOX-NY, and NSO/1 [17]. However, we have used the outbred NMRI mouse strain that is less prone to develop myelomas as immune spleen cell donor with great success for many years. Use of a combination of different mouse strains for immunization might increase the repertoire of specific antigen epitopes of the developed antibodies.

1.1.2 Conjugation to a Carrier

Proteins and large peptides can induce an immune response, whereas small peptides up to 15 amino acids have to be coupled to a carrier in order to elicit an immune response. Due to the superior immunogenicity of large proteins such as keyhole limpet hemocyanin, ovalbumin, and bovine serum albumin, these are commonly coupled to peptides when producing peptide vaccines [18, 19]. Another highly immunogenic and effective carrier that we often use in peptide vaccines, is S3 (secreted proteins from cultures of Bacillus Calmette-Guérin (BCG)) [20]. When using S3 as carrier the mice may be primed with an intraperitoneal (I.P.) injection of BCG vaccine 3–4 weeks prior to administration of the peptide–S3 conjugate in order to enhance the immune response after the first immunization of the peptide–S3 conjugate.

1.1.3 Vaccine Adjuvants

In addition to the immunogen composed of peptide conjugated to carrier, an effective peptide vaccine formulation also has to contain an adjuvant. A vaccine adjuvant is a compound that enhances and improves the immunogenicity without having any antigenic effect itself [21, 22]. The adjuvant reduces the amount of immunogen and the number of immunizations needed to elicit the desired immune response [23]. Moreover, the adjuvant serves both as a depot of immunogen at the site of injection and as a surfactant, which promotes the availability of immunogen over a large surface area.

Despite being one of the most potent adjuvants and considered the gold standard adjuvant for use in animals, Freund's complete adjuvant (FCA) composed of water and paraffin oil emulsion with killed mycobacteria [24], is considered too toxic for human use [23]. As reviewed by Petrowsky and Aguilar [23] a large number of other types of adjuvants exits, but due to their toxicity only a few are approved for use in human vaccines. In 1926, Glenny et al. [25] described the adjuvant activity of aluminum compounds and the adjuvant property has been used in human vaccines since then. Fyfe et al [26] have used the $Al(OH)_3$ as adjuvant in a HIV vaccine for mice and reported that none of the mice showed any signs of local inflammation or lymph node swelling. In order to minimize the injury to the animal we recommend using aluminum hydroxide as adjuvant.

The understanding of the mechanism behind the action of the adjuvants is not complete, but recently is was proposed that aluminum-containing adjuvants induce and enhance activation of the adaptive immune system by acting on dendritic cells or other antigen-presenting cells [27]. Moreover, aluminum hydroxide has been demonstrated to activate the three complement pathways with major involvement of the alternative complement pathway [22].

1.1.4 Immunization Route and Time Intervals

A number of studies have shown that the route of vaccine administration strongly affects the cellular immune response, and have great impact on the vaccine-induced protection [26, 28, 29].

If the only purpose of the immunization is generation of antibody production, the route of immunogen administration seems less important, since activation of the humoral immune response with production of IgG1 antibodies with comparable titers have been obtained after intramuscular (I.M.), intraperitoneal (I.P), intravenous (I.V.), and subcutaneous (S.C) injections of different vaccines composed of different immunogens and adjuvants [26, 28, 29].

For immunization with peptide conjugated to a carrier protein, Harlow and Lane [30] suggested to choose the subcutaneous route for the primary immunization and all subsequent boosters, except for the last booster. Harlow and Lane described that S.C. injection of the last booster vaccine composed of antigen in the absence of adjuvant only resulted in a poor immune response, whereas I.P injection resulted in a fair response, and I.V. injection in the best response. We recommend to use the I.P. administration route for priming, repeated S.C. injections for immunization with antigen and adjuvant every other weeks until an antibody titer of at least 1600 is obtained, and I.P. injection for the last boosting with antigen in the absence of adjuvant, 4 days prior to fusion.

1.1.5 Fusion Partner

Many of the commonly used murine fusion partners are derived from the myeloma MOPC 21 that was developed by injection of mineral oil into the peritoneum of a Balb/c mice [30]. Myeloma cells with a deficiency mutation in the salvage pathway of purine nucleotide biosynthesis has been selected as fusion partner in order to eliminate unfused myeloma cells or myeloma cells fused with other myeloma cells, and to identify successfully fused B-cell–myeloma hybridomas. Unfused myeloma cells and myeloma–myeloma fusions will die upon addition of hypoxanthine, aminopterin, and thymidine (HAT) to the fusion medium, if the myeloma cells have a deficiency mutation in the hypoxanthine-guanine phosphoribosyl transferase gene (HGPRT), since aminopterin blocks the de novo nucleotide synthesis pathway and the defect in HGPRT prevents use of hypoxanthine in the salvage pathway. Unfused B-cells and B-cells fused to other B-cells are capable of using hypoxanthine and thymidine for the salvage pathway, and survive in the HAT medium but only for a short period due to their limited life span. Only successfully fused B-cell–myeloma hybridomas will growth unlimited in HAT medium.

Initially, the newly fused heterocaryotic hybridomas are unstable due to the increased number of chromosomes and the high risk of losing chromosomes during cell division because of errors in the segregation process during cells division, where identical sets of chromosomes are not distributed to the daughter cells. Loss of aminopterin resistance will subsequently lead to cell death, and loss of the genes coding for the immunoglobulin heavy or light chains will influence the antibody titer due to reduction and termination of antibody production [17, 30]. Use of a fusion partner from the same species as the B-cell donor promotes formation of hybridomas that are more stable than interspecies hybridomas [18]. A disadvantage of the primary developed myeloma cell lines was their production and secretion of functional antibodies. This was overcome by selection of non-secreting cell lines.

1.1.6 Fusion Method

A number of different methods can increase the frequency of cell fusion. Köhler and Milstein [6] used Sendai virus for the first hybridoma fusion, but Sendai virus has since then been replaced by other fusogenic agents, and today polyethylene glycol (PEG) is the most commonly used fusogen [30]. Electrofusion is a viral- and chemical-free method for induction of cell fusion, but the success of the method is not a trivial task, since influential parameters and factors affecting the final outcome is not yet completely known and the conclusions from different laboratories are contradictory [31].

1.1.7 Feeder Cells and Growth Media Additives

To increase clone yield and stability of the hybridomas a variety of cells including murine peritoneal macrophages can be used as feeder cells in the hybridoma culture. However, use of feeder cells also has a number of disadvantages such as use of animals, laborious production of the cells and batch to batch variations. Moreover, the feeder cells represent a risk of contamination, they metabolize nutrients, thus resulting in an increased need for change of culture medium and they may overgrow or kill the hybridomas [32]. Conditioned medium from murine macrophages and fibroblasts have successfully been used to replace feeder cells for enhancement of hybridoma growth and antibody secretion [32, 33], but use of conditioned medium is also associated with disadvantages. There is a large and unpredictable batch to batch variation of conditioned medium and addition of conditioned medium dilute the cloning medium. We overcome these disadvantages by replacing feeder cells and conditioned medium with addition of the lyophilized growth media additive, HybER (Hybridoma Enhancing Reagent) [34].

1.2 Antibody Screening

Following fusion, the B-cell–myeloma hybridomas are cultured in 96-well plates. The next step is a rapid "primary" screening process, which identify and select only hybridomas that produce peptide antibodies. The cells that produce the desired antibodies are cloned to produce many identical daughter clones. Ultimately, "primary" screening is necessary to eliminate nonspecific hybridomas at the earliest opportunity.

The most applied screening methods involve antibody capture assays, antigen capture assays and functional assays [12, 13, 16, 30]. In general, the more immunogens used for immunization, the more difficult it is to screen. As the peptide antigen used for generation of peptide antibodies often is available in relatively large amounts, it is seldom necessary to use highly sensitive immunoassays or complicated screening approaches to check for peptide antibodies. In fact, for many purposes a colorimetric antibody capture immunoassay is adequate. Here, the peptide antigen originally used for immunization is passively adsorbed to the bottom of microtiter wells, either as a free peptide or as a peptide conjugated to an irrelevant carrier protein, e.g., bovine serum albumin, as described in Subheading 3.5. Occasionally, key determinants of some peptides may become masked upon adsorption to the plate. In these cases a peptide–carrier conjugate should be used. Alternatively, biotinylated peptides coupled to a streptavidin-coated microtiter plate are used [15]. Next, antibodies in the hybridoma culture supernatant are allowed to bind to the peptides, which are detected with an appropriate second-layer reagent, typically an enzyme-linked antibody, and the assay is developed with a colorimetric substrate.

A crucial aspect of peptide antibody screening relates to profiling in different assay systems. This especially relates to antibodies, which will be used in different systems, e.g., ELISA, versus western blotting or bead-based immunoassays. This phenomenon, termed assay restriction [14, 35, 36], relates to how an antibody recognizes its target epitope in the context of the assay system used. In this case, the epitope could be masked, denatured or rendered inaccessible by the immobilization procedure adopted within a given technique. Thus, because the peptide–antibody recognizes its target in ELISA does not necessarily mean that is will recognizes its target in other immunoassays. Since peptides usually do not have the same conformation when they are free in solution, coupled to a protein carrier or adsorbed to microtiter plates, the antigenic activity of synthetic peptides/protein can vary greatly in different immunoassay formats and hence the reactivity of the produced antibody may wary [37, 38]. Thus, it is essential to test the peptide antibody in a variety of formats and most appropriate in the format of intended use before any conclusions are drawn.

In addition to screening for peptide antibodies in different assays, it can be of interest to screen for reactivity to native or denatured proteins as well [15]. One of the critical points of peptide antibodies is that it can be unpredictable, whether the antibody will recognize the native protein due to structural differences between the synthetic peptide and the peptide epitope in the native protein structure. Hence, if the intended use of the peptide antibody is to recognize for example the native protein, antibodies should indeed be screened for reactivity towards the native protein structure. In addition to this, it is essential to screen for reactivity to varying protein structures as well. When coating proteins to the surface of microtiter plates, for example through passive adsorption, a molten globule intermediate may be generated, which is structurally different from the native and a completely denatured state. As a result, the antibody may not recognize the protein or peptide in question [15].

Collectively, when conducting "primary" screening of hybridomas for further selection, it is most important to screen for antibody reactivity in the intended assay system using the intended target molecule.

Following primary screening, selected clones are expanded. Secondary screening of selected hybridomas may be conducted to verify the specificity of the antibodies and to ensure that the clones have not terminated the antibody production.

1.3 Antibody Characterization

Following selection of clones, characterization of selected peptide antibodies in terms of reactivity, specificity, and cross-reactivity can be achieved using culture supernatants or purified antibodies. The simplest approach for determination of these factors is by ELISA, where antibody reactivity to the peptide antigen and a panel of related peptides is determined [15], basically as described in Subheading 3.5. Alternatively, competition studies can be applied, where the peptide used for immunization is used as inhibitor. In relation to ELISA, this type of testing may be performed by immunoblotting or immunohistochemistry. Moreover, biochemical characteristics such as solubility, stability and binding characteristics (e.g., performance in antibody affinity chromatography) should be determined [5].

In addition to biochemical and specificity characteristics, isotype determination is a crucial step in antibody characterization. Isotype determination serves not only to define the immunoglobulin class or subclass, but also confirms the presence of a single isotype and is required for choice of appropriate isotype-matched control antibodies in different applications, e.g., immunohistochemistry. An easy and fast way to determine the isotype is by using commercially available subtyping strips. Another straightforward approach is to consider the antibody themselves as antigens and to use anti-immunoglobulin antibodies as the specific and sensitive agents of detection. Thus, the isotype of the antibody is determined by a simple antibody capture assay [39]. Alternatively, ELISA or bead-based immunoassays may be conducted, where capture antibodies that specifically recognize the heavy chain of each isotype and kappa and lambda light chains are conjugated to beads or coated into the wells of microtiter plates, whereafter the reactivity of the peptide antibodies to each isotype is determined.

It is noteworthy that although monoclonal antibodies usually are specific for a single target, this same antibody can in fact cross-react with other antigens or exhibit dual specificity [36]. This may occur when the antibody recognizes more than one antigenic determinant, because of some similarity in shape of chemical composition. Consequently, stringent evaluation of the peptide antibody and its target epitope is necessary [40], which may include epitope mapping. This particular technique allows precise determination of key amino acid residues that are important for antibody recognition and binding [41, 42]. Several approaches for epitope mapping exist. In relation to mapping of epitopes of peptide antibodies, these antibodies are often characterized using synthetic peptides [41, 43], as the epitope in many cases is easy to determine based on the limited size of the immunogenic peptide used for antibody production.

Further characterization may also include affinity measurements of peptide–antibody interactions using bio-layer interferometry, surface plasmon resonance technology, e.g., BIACore or other techniques [44–46]. Once characterized, monoclonal peptide antibodies can serve as investigative research tools, or may be applied in diagnostic assays or as therapeutic agents.

2 Materials

2.1 Immunizations, Fusion, Cell Cultivation and Cloning

1. Syringes.
2. Needles.
3. Scissors.
4. Forceps.
5. Eppendorf tubes.
6. EDTA-coated tubes.
7. Gaze.
8. Counting tubes.
9. Pasteur pipets.
10. Serological pipets.
11. 96-well culture plates with delta surface and lid.
12. CO_2 Incubator.
13. Laminar flow cabinet.
14. Homogenizer (mortar and pestle).
15. Microscope.
16. Centrifuge.
17. Counting chamber.
18. Stop watch.
19. –80 °C freezer.
20. –135 °C freezer or N_2 tank.
21. Female mice, e.g., NMRI strain or Balb/c strain.
22. BCG vaccine for priming, if S3 is used as carrier.
23. 1 mg/ml carrier (e.g., S3, keyhole limpet hemocyanin, ovalbumin, or bovine serum albumin) conjugated with peptide. Store at –20 °C.

24. Adjuvant, e.g., 10 mg/ml aluminum hydroxide. Store at 4 °C.
25. Merthiolate for conservation of the prepared ready to use vaccine. Store at 4 °C.
26. Saline, 4 °C.
27. 70 % ethanol. Store at room temperature.
28. Crystal violet in acetic acid for leucocyte staining. Store at room temperature.
29. 0.2 % nigrosin in saline. Store at room temperature.
30. Myeloma cells, e.g., X63.Ag8.653.
31. Serum-free medium:

    Dulbecco's Modified Eagle Medium (DMEM) + 2 mM L-glutamine, and 1 % penicillin–streptomycin (10,000 U/ml Pen–10,000 μg/ml Strep). Store at 4 °C and use within a week.
32. Polyethylene glycol (PEG): (PEG, 7.5 % v/v DMSO):

    14.1 ml melted PEG (47 % v/v), 2.25 ml DMSO (7.5 % v/v) and 13.65 ml DMEM. Store at 4 °C and use within a week.
33. HAT + HybER medium: DMEM + 10 % FCS, 0.038 mM hypoxanthine, 0.4 μM aminopterin, 0.1 mM thymidine, 2 mM L-glutamine, 1 % Hybridoma Enhancing Reagent (HybER), and 1 % Pen/Strep. Store at 4 °C and use within a week.

    This medium is used for cultivation of fused cells.
34. Cloning medium: DMEM + 10 % FCS, 0.038 mM hypoxanthine, 0.1 mM thymidine, 2 mM L-glutamine, 1 % HybER, and 1 % Pen/Strep. Store at 4 °C and use within a week.

    This medium is used for cultivation of the cells during the cloning process.
35. HT medium: DMEM + 10 % FCS, 0.038 mM hypoxanthine, 0.1 mM thymidine, 2 mM L-glutamine, and 1 % Pen/Strep. Store at 4 °C and use within a week.

    This medium is used for cultivation of the cells, when the cloning process is completed cells has to be expanded in order to be able to establish a cell bank composted of for example 8 vials each containing approximately $5 \times 10_5$ hybridoma cells.
36. Culture medium: DMEM + 10 % FCS, 2 mM L-glutamine, and 1 % Pen/Strep. Store at 4 °C and use within a week.

    This medium is used for cultivation of hybdridomas.
37. Freezing medium: Either DMEM + 30 % fetal calf serum (FCS), 1 % penicillin–streptomycin (Pen/strep), and 5 % dimethylsulfoxide (DMSO); or FCS + 5 % DMSO. Store at 4 °C and use within a week.

2.2 Antibody Screening

1. Coating buffer, e.g., carbonate buffer: 15 mM Na_2CO_3, 35 mM $NaHCO_3$, 0.001 % phenol red, pH 9.6. Store at 4 °C. Discard if changes in pH occur (*see* **Note 1**).
2. Peptide solution/suspension at 1 mg/ml in phosphate-buffered saline. The free peptides are supplied as a lyophilised product. Dissolve the free peptides according to the manufacturer's instructions to a concentration of 1 mg/ml (*see* **Notes 2** and **3**).
3. Cell culture supernatants/peptide antibody. Store at 4 °C.
4. 96-well microtitre plates
5. Alkaline Phosphatase (AP)-substrate buffer: 1 M diethanolamine, 0.5 mM $MgCl_2$, pH 9.8. Store at 4 °C.
6. Tris-Tween-NaCl buffer (TTN buffer): 0.05 M Tris, 0.3 M NaCl, 1 % Tween 20, pH 7.5 (*see* **Note 4**). Store at 4 °C.
7. Secondary antibody: AP-conjugated anti-mouse IgG_{1-4}/IgA/IgM antibody.
8. AP buffer: Dissolve phosphatase substrate tablets (4-nitrophenyl phosphate) in AP-substrate buffer to a final concentration of 1 mg/ml. The substrate buffer is light-sensitive and should be prepared immediately before use and kept in the dark. Remains should be discarded (*see* **Note 5**).

3 Methods

3.1 Immunization

1. In 3–5 mice, inject intraperitoneally 0.2 ml BCG vaccine per mouse (~2 human doses), 3–4 weeks before the first immunization with peptide coupled to S3.
2. Bleed the mice immediately before the first immunization with the peptide vaccine (use EDTA-containing tubes). Centrifuge the tube at 600 × *g* for 5 min and harvest plasma. This plasma sample, called Bleed 0, will serve as a baseline control to use for assay setup and during the immunization course for test of antibody reactivity.
3. Prepare a vaccine containing 20–50 μg/ml S3 conjugated with peptide and 2 mg/ml $Al(OH)_3$ by diluting S3 conjugated with peptide in saline and add it drop wise to the $Al(OH)_3$, while stirring the mixture slowly. Vaccines for all immunizations except the booster vaccine can be prepared simultaneously if 0.05 % merthiolate is added for preservation. Store at 4 °C.
4. Inject subcutaneously 0.5 ml of the peptide vaccine.
5. Repeat immunizations every other week.
6. Bled the mice 10 days after the third immunization, harvest plasma as described above and test for antigen reactivity in ELISA or the assay system for intended antibody use.

7. Repeat testing for antigen reactivity 10 days after all subsequent immunizations until the antibody titer is at least 1600. The antibody titer is defined as the reciprocal value of the dilution that gives an OD value that is half the maximal measured OD value.
8. After approximately 4–6 immunization, when the antibody titer is above 1600, inject I.P. 0.5 ml peptide–S3 conjugate in the absence of adjuvant in order to boost the number of antigen-specific B-cells in the spleen.
9. Sacrifice the mouse 4 days after the booster injection.

3.2 Preparation of Myeloma Cells for Cell Fusion

Harvest of the spleen cells from an immunized and boosted mouse results in approximately 2×10^8 antigen-specific B-lymphocytes. Myeloma cells and B-lymphocytes should be fused in a ratio of 1:5, thus requiring a minimum of 4×10^7 myeloma cells.

1. Thaw the murine myeloma fusion partner, X63.Ag8.653, 1–2 weeks prior to cell fusion.
2. Calculate the cell population doubling time by daily cell counting.
3. Three to four days before cell fusion dilute the X63.Ag8.653 cells according to the calculated population doubling time to obtain 4×10^7 exponentially growing myeloma cells with high viability on the day of fusion. For example, if the cells have a cell population doubling time of 24 h, then dilute cells to a density of 1×10^5 viable cells per ml in 50 ml culture medium.

3.3 Fusion

1. Prior to cell fusion, preheat 1 ml PEG, 5 ml serum-free medium, and 225 ml HAT + HybER medium to 37 °C and cool 100 ml serum-free medium to 4 °C.
2. Preparation of the X63.Ag8.653 myeloma cells:
   (a) Count the number of cells and determine the viability,
   (b) Transfer 4×10^7 cells to a 50 ml tube and centrifuge for 10 min at $400 \times g$ at room temperature.
   (c) Resuspend the cells in serum-free medium at a density of 1×10^6 cells/ml, and store at 37 °C, 6.5 % CO_2 until use.
3. Preparation of spleen cells:
   (a) Transfer 2–3 ml cold serum-free medium to a homogenizer.
   (b) Dip the mouse in 70 % ethanol and transfer it to a laminar flow cabinet.
   (c) Open the mouse by using sterile scissors and forceps.
   (d) Immediately, transfer the spleen to the cold serum-free medium in the homogenizer.
   (e) Open the heart and transfer heart blood to an EDTA-containing tube. Centrifuge the tube at $600 \times g$ for 5 min, harvest plasma and store it at –20 °C for later use.

(f) Grind spleen tissue with a mortar and pestle to obtain spleen cells in suspension.

(g) Filtrate the spleen cell suspension into a 50 ml tube through sterile gaze.

(h) Rinse homogenizer and gaze with 10 ml cold serum-free medium, and transfer to the spleen cell suspension.

(i) Add cold serum-free medium to a total volume of 50 ml.

(j) Centrifuge the cells for 10 min at $400 \times g$.

(k) Resuspend cell pellet in 10 ml cold serum-free medium.

(l) Use methyl violet acetic acid for leucocyte counting to count the number of viable B-lymphocytes.

4. Mixture of spleen cells and myeloma cells

(a) Add myeloma cells to the tube with spleen cells in a ratio of 1 myeloma cell to 5 spleen cells, and add cold serum-free medium to 50 ml.

(b) Centrifuge for 10 min at $400 \times g$.

(c) Remove all supernatant from the pellet.

5. Fusion

(a) Add slowly 1 ml of 37 °C warm PEG to the cell pellet, while carefully stirring with a 1 ml serological pipette.

(b) Continue the slow stirring for 2 min.

(c) During a 3 min. period, add slowly 2 ml of 37 °C warm serum-free medium while stirring.

(d) During a period of ½–1 min, add 7 ml of 37 °C warm HAT + HybER medium.

(e) Dilute to a density of approximately 1×10^6 cells/ml in HAT + HybER medium.

(f) Transfer 225 μl cell suspension per well to 96-well cell culture plates.

(g) Culture the cells at 37 °C, 6.5 % CO_2 and 90 % humidity for 7 days

(h) Replace the medium with freshly prepared HAT + HybER medium and continue cultivation for another 4 days.

(i) Harvest 150 μl culture supernatant from cell-containing wells and replace with 150 μl HT medium.

(j) Test hybridoma supernatants for antibody production in preferred assays.

(k) Select a number of wells with cells producing antigen-specific antibodies and transfer cells to T25 flasks.

(l) Follow the cell density and add fresh HT medium when the medium turns yellow or the cell density reaches $4–5 \times 10^5$ cells/ml (the maximal volume in a T25 flask is 15 ml).

(m) Test the hybridoma supernatants for antibody production again after 2 weeks of cell propagation in culture flasks.

3.4 Use of the Limiting Dilution Technique to Obtain Hybridoma Monoclonality

(a) Prepare the cloning medium.

(b) Count the number of living and dead cells (use for example 0.2 % nigrosin staining).

(c) Dilute the cells in 5–10 ml cloning medium to a density of 4.4 cells/ml, 2.2 cells/ml, and 1.1 cell/ml.

(d) Seed 225 μl of cell suspension into 20 wells per cell density (use the 60 center wells) in 96-well-culture plates.

(e) Incubate the cells at 37 °C, 6.5 % CO_2, and 90 % humidity for up to 14 days or until clones appear in the wells.

(f) Test culture supernatants from for example 25 cell-containing wells in selected assays.

(g) Expand antibody-producing cells from selected wells, and repeat the cloning step of antibody-producing cells until positive antibody-specific reaction is obtained in culture supernatant from all tested wells. Then the first cloning step is considered finished.

(h) Subclone the cells as described above except that the cells in the lowest density should be seeded in 60 wells and cells in two other densities should each be seeded in 30 wells.

(i) Test culture supernatant from for example 60 cell-containing wells in selected test assays.

(j) Expand the monoclonal hybridoma cells in HT medium in culture flasks in order to create a cell bank composed of for example 8 vials each containing 5×10^6 cells with a viability of at least 80 %.

(k) Count the number of viable cells by using 0.2 % nigrosin.

- Centrifuge the cells for 10 min at $400 \times g$.

(l) Resuspend the cells at a density of 5×10^6 cells/ml in freezing medium and immediately transfer the cells to a −80 °C freezer in a specialized freezing container to secure a controlled freezing process of −1 °C per min.

(m) Transfer the cells to a −135 °C freezer or liquid N_2 for storage.

3.5 Enzyme-Linked Immunosorbent Assay for Deter mination of Peptide Antibodies

1. Dilute free peptides or peptides coupled to another carrier than the one used for immunization to a final concentration of 1 μg/ml in coating buffer (*see* **Notes 1–3**).
2. Coat microtiter plates with 100 μl of the peptide solution in each well (*see* **Note 6**). Incubate overnight at 4 °C (*see* **Note 7**).

3. Remove any non-adsorbed peptides and wash the plates three times with TTN buffer (250 μl/well).
4. Add 250 μl TTN as blocking buffer to each well to block free binding sites and incubate at room temperature for 20 min (*see* **Note 4**).
5. Empty the wells and add 100 μl of plasma or cell culture supernatants (*see* **Note 8**).
6. Incubate for 1 h at room temperature on a platform shaker.
7. Wash the wells three times as described in **step 3**.
8. Add 100 μl of AP-conjugated secondary antibody reagent diluted in TTN to a final concentration of 1 μg/ml (*see* **Note 5**).
9. Incubate the microtitre plate on a platform shaker for 1 h at room temperature.
10. Following incubation with secondary antibody repeat washing steps described in **step 3**.
11. Detect the presence of bound antibodies by adding 100 μl of freshly prepared *p*-NPP substrate in AP buffer solution to each well. Place the plates on a platform shaker and read the plate when the solution within the wells turns yellow (*see* **Note 5**).
12. The absorbance is measured at 405 nm, with background subtraction at 650 nm on a microtitre plate reader or on an equivalent instrument measuring the wavelength of 405 nm and a reference wavelength of 650 nm (*see* **Note 9**).

4 Notes

1. Various coating buffers may be applied, such as carbonate buffer, PBS, and tris buffer.
2. A common issue with synthetic peptides, especially those containing hydrophobic amino acid residues, is insolubility in aqueous solutions. Other solvents recommended for peptide solvation include dimethylformamide (DMF), dimethylsulfoxide (DMSO) or different mixtures of DMF and water or DMSO and water. However, note that DMSO may oxide SH groups to disulfides. Some peptides may also be soluble in acetonitrile–water mixtures.
3. After lyophilisation, peptides retain significant amounts of water. Peptides are oxidized over time at −20 °C and slowly degrade. Thus, the peptide stock solution should be stored in small aliquots upon arrival to prevent degradation caused by repeated freezing and thawing.
4. Alternatively, TTN buffer can be replaced with PBS supplemented with Tween 20.

5. Alternatively, other color reagents may be applied for detection of antibody reaction, e.g., appropriate peroxidase-conjugated secondary antibody in combination with e.g., O-phenylediamine (OPD) or Tetramethylbenzidine (TMP) substrate and stop solution.
6. Coat with the peptide originally used for immunization (but without conjugation to the carrier used for immunization) or alternatively the peptide conjugated to an irrelevant carrier protein or the whole protein, possibly in a denatured version depending on the original location of the immunogen in the protein structure.
7. Alternatively, coat the plates with antigens for 2 h at room temperature.
8. Suitable starting dilutions are 1:100 for plasma and 1:10 for hybridoma culture supernatants.
9. Measure the absorbance at an appropriate wavelength according to the selected color reaction detection system.

References

1. Akhidova EV, Volkova TD, Koroev DO, Kim I, Filatova MP, Vladimirova NM, Karmakova TA, Zavalishina LE, Andreeva I, Vol'pina OM (2010) Antibodies to synthetic peptides for the detection of survivin in tumor tissues. Bioorg Khim 36:178–186
2. Armstrong A, Hildreth JE, Amzel LM (2013) Structural and thermodynamic insights into the recognition of native proteins by anti-peptide antibodies. J Mol Biol 425:2027–2038
3. Nakagawa M, Ohmido N, Ishikawa K, Uchiyama S, Fukui K, Azuma T (2008) Anti-peptide antibodies for examining the conformation, molecular assembly and localization of an intracellular protein, ribosomal protein S6, in vivo. J Biochem 143:325–332
4. Schulz S, Rocken C, Schulz S (2006) Immunocytochemical localisation of plasma membrane GHRH receptors in human tumours using a novel anti-peptide antibody. Eur J Cancer 42:2390–2396
5. Trier NH, Hansen PR, Houen G (2012) Production and characterization of peptide antibodies. Methods 56:136–144
6. Kohler G, Milstein C (1975) Continuous cultures of fused cells secreting antibody of predefined specificity. Nature 256:495–497
7. Kao DJ, Hodges RS (2009) Advantages of a synthetic peptide immunogen over a protein immunogen in the development of an anti-pilus vaccine for Pseudomonas aeruginosa. Chem Biol Drug Des 74:33–42
8. Moisa AA, Kolesanova EF (2011) Synthetic peptide vaccines. Biomed Khim 57:14–30
9. Skovbjerg H, Koch C, Anthonsen D, Sjostrom H (2004) Deamidation and cross-linking of gliadin peptides by transglutaminases and the relation to celiac disease. Biochim Biophys Acta 1690:220–230
10. Hancock DC, O'Reilly NJ (2005) Synthetic peptides as antigens for antibody production. Methods Mol Biol 295:13–26
11. Lateef SS, Gupta S, Jayathilaka LP, Krishnanchettiar S, Huang JS, Lee BS (2007) An improved protocol for coupling synthetic peptides to carrier proteins for antibody production using DMF to solubilize peptides. J Biomol Tech 18:173–176
12. Lee BS, Huang JS, Jayathilaka GD, Lateef SS, Gupta S (2010) Production of antipeptide antibodies. Methods Mol Biol 657:93–108
13. Hancock DC, Evan GI (1998) Production and characterization of antibodies against synthetic peptides. Methods Mol Biol 80:15–22
14. Maleki LA, Majidi J, Baradaran B, Abdolalizadeh J, Akbari AM (2013) Production and characterization of murine monoclonal antibody against synthetic peptide of CD34. Hum Antibodies 22:1–8
15. Holm BE, Bergmann AC, Hansen PR, Koch C, Houen G, Trier NH (2015) Antibodies with specificity for native and denatured forms of ovalbumin differ in reactivity between enzyme-linked immunosorbent assays. APMIS 123(2):136–145. doi:10.1111/apm.12329

16. Koch C, Jensen SS, Oster A, Houen G (1996) A comparison of the immunogenicity of the native and denatured forms of a protein. APMIS 104:115–125
17. Koch C (1993) Monoklonale antistoffer. In: Kielberg V, Brünner N, Briand P (eds) Celledyrkning—En praktisk håndbog i dyrkning af mammale celler'. Foreningen af Danske Lægestuderendes Forlag, København, pp 201–210
18. Grimaldi CM, French DL (1995) Monoclonal Antibodies by Somatic Cell Fusion. ILAR J 37:125–132
19. Helling F, Shang A, Calves M, Zhang S, Ren S, Yu RK, Oettgen HF, Livingston PO (1994) GD3 vaccines for melanoma: superior immunogenicity of keyhole limpet hemocyanin conjugate vaccines. Cancer Res 54:197–203
20. Khalil IF, Alifrangis M, Recke C, Hoegberg LC, Ronn A, Bygbjerg IC, Koch C (2011) Development of ELISA-based methods to measure the anti-malarial drug chloroquine in plasma and in pharmaceutical formulations. Malar J 10:249
21. Degen WG, Jansen T, Schijns VE (2003) Vaccine adjuvant technology: from mechanistic concepts to practical applications. Expert Rev Vaccines 2:327–335
22. Guven E, Duus K, Laursen I, Hojrup P, Houen G (2013) Aluminum hydroxide adjuvant differentially activates the three complement pathways with major involvement of the alternative pathway. PLoS One 8, e74445. doi:10.1371/journal.pone.0074445
23. Petrovsky N, Aguilar JC (2004) Vaccine adjuvants: current state and future trends. Immunol Cell Biol 82:488–496
24. Freund J, Casals J, Hosmer EP (1937) Sen sitization and antibody formation after injection of tubercle bacili and paraffin oil. Proc Soc Exp Biol Med 37:509–513
25. Glenny AT, Pope CG, Waddington H, Wallance U (1926) The antigenic value of toxoid precipitated by potassium alum. J Pathol Becteriol 29:38–39
26. Fyfe L, Maingay J, Robinson AC, Howie SE (1991) Murine immune response to HIV-1 p24 core protein following subcutaneous, intraperitoneal and intravenous immunization. Immunology 74:467–472
27. Ghimire TR, Benson RA, Garside P, Brewer JM (2012) Alum increases antigen uptake, reduces antigen degradation and sustains antigen presentation by DCs in vitro. Immunol Lett 147:55–62
28. Budimir N, de Haan A, Meijerhof T, Gostick E, Price DA, Huckriede A, Wilschut J (2013) Heterosubtypic cross-protection induced by whole inactivated influenza virus vaccine in mice: influence of the route of vaccine administration. Influenza Other Respir Viruses 7:1202–1209
29. Mohanan D, Slutter B, Henriksen-Lacey M, Jiskoot W, Bouwstra JA, Perrie Y, Kundig TM, Gander B, Johansen P (2010) Administration routes affect the quality of immune responses: A cross-sectional evaluation of particulate antigen-delivery systems. J Control Release 147:342–349
30. Harlow E, Lane D (1988) Monoclonal antibodies. In: Antibodies: a laboratory manual. Cold Spring Harbor Laboratory, New York, pp 139–244
31. Kanduser M, Usaj M (2014) Cell electrofusion: past and future perspectives for antibody production and cancer cell vaccines. Expert Opin Drug Deliv 11:1885–1898
32. Sugasawara RJ, Cahoon BE, Karu AE (1985) The influence of murine macrophage-conditioned medium on cloning efficiency, antibody synthesis, and growth rate of hybridomas. J Immunol Methods 79:263–275
33. Walker KZ, Gibson J, Axiak SM, Prentice RL (1986) Potentiation of hybridoma production by the use of mouse fibroblast conditioned media. J Immunol Methods 88:75–81
34. Schwelberger HG, Feurle J, Houen G (2013) New tools for studying old questions: antibodies for human diamine oxidase. J Neural Transm 120:1019–1026
35. Jefferis R, Reimer CB, Skvaril F, de Lange G, Ling NR, Lowe J, Walker MR, Phillips DJ, Aloisio CH, Wells TW (1985) Evaluation of monoclonal antibodies having specificity for human IgG sub-classes: results of an IUIS/WHO collaborative study. Immunol Lett 10:223–252
36. Nelson PN, Fletcher SM, MacDonald D, Goodall DM, Jefferis R (1991) Assay restriction profiles of three monoclonal antibodies recognizing the G3m(u) allotype. Development of an allotype specific assay. J Immunol Methods 138:57–64
37. Muller S, Plaue S, Couppez M, Van Regenmortel MH (1986) Comparison of different methods for localizing antigenic regions in histone H2A. Mol Immunol 23:593–601
38. Van Regenmortel MH (1987) Protein structure and antigenicity. Int J Rad Appl Instrum B 14(4):277–280
39. Hornbeck P, Fleisher TA, Papadopoulos NM (2001) Isotype determination of antibodies. Curr Protoc Immunol Chapter 2, Unit. 2.2
40. Bull H, Choy M, Manyonda I, Brown CA, Waldron EE, Holmes SD, Booth JC, Nelson PN (1999) Reactivity and assay restriction

profiles of monoclonal and polyclonal antibodies to acid phosphatases: a preliminary study. Immu nol Lett 70:143–149

41. Amrutkar SD, Trier NH, Hansen PR, Houen G (2012) Fine mapping of a monoclonal antibody to the N-Methyl D-aspartate receptor reveals a short linear epitope. Biopolymers 98:567–575
42. Welner S, Trier NH, Houen G, Hansen PR (2013) Identification and mapping of a linear epitope of centromere protein F using monoclonal antibodies. J Pept Sci 19:95–101
43. Petersen NH, Hansen PR, Houen G (2011) Fast and efficient characterization of an anti-gliadin monoclonal antibody epitope related to celiac disease using resin-bound peptides. J Immunol Methods 365:174–182
44. Gibbs E, Oger J (2008) A biosensor-based characterization of the affinity maturation of the immune response against interferon-beta and correlations with neutralizing antibodies in treated multiple sclerosis patients. J Interferon Cytokine Res 28:713–723
45. Stubenrauch K, Wessels U, Vogel R, Schleypen J (2009) Evaluation of a biosensor immunoassay for simultaneous characterization of isotype and binding region of human anti-tocilizumab antibodies with control by surrogate standards. Anal Biochem 390:189–196
46. Wegner GJ, Lee HJ, Corn RM (2002) Characterization and optimization of peptide arrays for the study of epitope-antibody interactions using surface plasmon resonance imaging. Anal Chem 74:5161–5168

Chapter 13

Production of Epitope-Specific Antibodies by Immunization with Synthetic Epitope Peptide Formulated with CpG-DNA-Liposome Complex Without Carriers

Dongbum Kim, Younghee Lee, and Hyung-Joo Kwon

Abstract

Antibody production using synthetic peptides has been investigated extensively to develop therapeutic antibodies and prophylactic vaccines. Previously, we reported that a complex of CpG-DNA and synthetic peptides corresponding to B cell epitopes, encapsulated in a phosphatidyl-β-oleoyl-γ-palmitoyl ethanolamine (DOPE):cholesterol hemisuccinate (CHEMS) complex, significantly enhanced the synthetic peptide-specific IgG production. Here, we describe synthetic peptide-based epitope screening and antibody production without conventional carriers.

Key words Epitope, Peptide, Antibody production, CpG-DNA, CpG-DNA-liposome complex

1 Introduction

The synthetic peptide-based antibody production against B cell epitopes has gained attention because it has the potential to be used in developing therapeutic antibodies and vaccines for cancers and infectious diseases [1, 2]. The antibody production using synthetic peptides as an antigen has several advantages; it is easy and inexpensive to prepare antigens for immunization and the specificity of obtained antibodies is restricted to defined epitopes. However, there are some disadvantages such as weak immunogenicity and the need for carrier protein. To improve the efficacy of peptide-based antibody production, researchers used liposomes as a vehicle for vaccine delivery [3–5]; they also formulated adjuvants such as flagella [6] and CpG-DNA [7] to enhance the magnitude of immune responses.

CpG-DNA, which contains synthetic oligodeoxynucleotides (ODNs) and bacterial DNA including unmethylated CpG dinucleotides flanked by specific base sequences, has been examined as an immune adjuvant for vaccine development [8, 9]. Previously, we

Gunnar Houen (ed.), *Peptide Antibodies: Methods and Protocols*, Methods in Molecular Biology, vol. 1348, DOI 10.1007/978-1-4939-2999-3_13,

screened and identified natural phosphodiester bond CpG-DNA (PO-ODN) with immunostimulatory activity from *Mycobacterium bovis* chromosomal DNA. The PO-ODN, namely MB-ODN 4531(O), contains three CpG motifs and has powerful adjuvant effects enhancing the induction of Ag-driven Th1 responses without severe side effects [10]. Furthermore, we found that the CpG-DNA was most potent when encapsulated in a phosphatidyl-β-oleoyl-γ-palmitoyl ethanolamine (DOPE):cholesterol hemisuccinate (CHEMS) complex (Lipoplex(O)). We extended the research to the selection of B cell epitopes and revealed that complexes of several peptides and Lipoplex(O) without carriers significantly enhanced each peptide-specific IgG production involving TLR9 [11, 12]. As a typical example, we successfully screened the B cell epitope of the transmembrane 4 superfamily member 5 protein (TM4SF5) which has been implicated in human hepatocarcinoma (Fig. 1). We also revealed the potent production of epitope-specific antibodies in mice immunized with a complex of human TM4SF5 peptide (hTM4SF5R2-3) and Lipoplex(O) (Figs. 2 and 3) [11, 13]. We then produced a monoclonal antibody by immunization with a complex of antigenic peptide (hTM4SF5R2-3) and Lipoplex(O). We currently reported that the anti-hTM4SF5R2-3 peptide-specific antibody can reduce a hepatocarcinoma mass in a xenograft mouse model after Huh-7 cells have been injected into nude mice [14]. Through the same strategy, we also confirmed production of epitope-specific antibodies against influenza virus HA protein, hepatitis C virus E protein, or human respiratory syncytial virus G protein [12, 15].

It is likely that the enhanced delivery through the DOPE:CHEMS complex and the adjuvant effect of CpG-DNA facilitate the antibody production using peptides as an antigen without conventional carriers. As our novel strategy enables rapid selection of functional B cell epitopes, it may be promptly used for the development of specific epitope-targeted peptide vaccines and production of therapeutic antibodies in case of emergency. Here, we provide a detailed method for the synthetic peptide-based epitope screening and antibody production.

2 Materials

Prepare all solutions using cell culture grade distilled water and analytical grade reagents. Prepare and store all reagents at room temperature (unless indicated otherwise).

2.1 Mice

1. Obtain 4-week-old female BALB/c (H-2^b) mice from an animal service company.
2. Zoletil 50: Dissolve the powder (250 mg/vial) in 5 mL of solvent which the supplier provides.
3. Rompun: 10 mL solution/vial.

a

```
1..197 amino acids
MCTGKCARCV GLSLITLCLV CIVANALLLV RNGETSWTNT NHLSLQVWLM
GGFIGGGLMV LCPGIAAVRA GGKGCCGAGC CGNRCRMLRS VFSSAFGVLG
AIYCLSVSGA GLRNGPRCLM NGEWGYHFED TAGAYLLNRT LWDRCEAPPR
VVPWNVTLFS LLVAASCLEI VLCGIQLVNA TIGVFCGDCR KKQDTPH
```

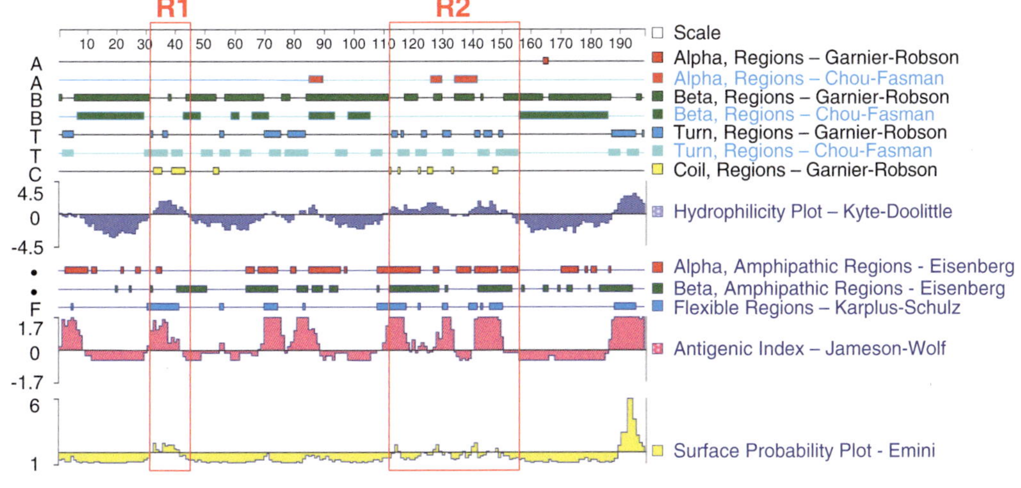

b

Protein	Peptides	Sequences	Location	Length
	hTM4SF5R1	NGETSWTNTNHLSL	32-45	14
	hTM4SF5R2-1	RNGPRCLMNGEWGY	113-126	14
hTM4SF5	hTM4SF5R2-2	GEWGYHFEDTAGAY	122-135	14
	hTM4SF5R2-3	NRTLWDRCEAPPRV	138-151	14
	hTM4SF5R2-4	WDRCEAPPRVVPWN	142-155	14
	hTM4SF5R2-5	GAYLLNRTLWDRCEA	133-147	15

Fig. 1 Analysis of the TM4SF5 amino acid sequence and prediction of the putative B cell epitopes. (**a**) The amino acid sequence of TM4SF5 was analyzed using parameters described in Subheading 3 and two potential putative epitope regions R1 and R2 were identified. (**b**) Six candidate peptides were selected and synthesized [11]. This information was reproduced from [BMC Immunology 2011] under the terms of the Creative Commons Attribution License (http://creativecommons.org/licenses/by/2.0), which permits unrestricted use, distribution, and reproduction in any medium, provided the original work is properly cited

4. Prepare Zoletil 50 + Rompun anesthesia for sacrifice. Add 0.6 mL of Zoletil 50 and 0.4 mL of Rompun to 9 mL of distilled water.

2.2 Lipoplex(O) Components

1. Synthesize CpG-DNA (MB-ODN 4531(O)). MB-ODN 4531(O) is composed of 20 bases containing three CpG motifs (underlined): AGCAG<u>CG</u>TT<u>CG</u>TGT<u>CG</u>GCCT. The purity of the synthesized MB-ODN 4531(O) has to be confirmed (over 92 %) based on HPLC and LC-MS. Dissolve MB-ODN 4531(O) (10 mg/mL) with distilled water (GIBCO, Carlsbad,

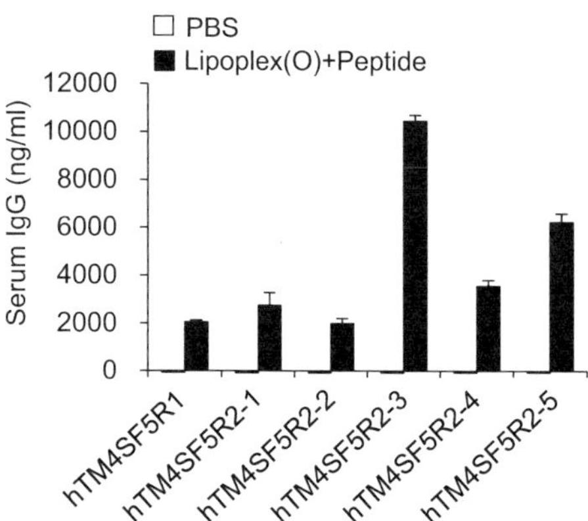

Fig. 2 Identification of a potent B cell epitope by immunization and ELISA experiments. BALB/c mice were immunized with each peptide and Lipoplex(O) complex according to the procedure described here. The amounts of anti-each peptide-specific total IgG was measured using ELISA. In the case of TM4SF5 protein, hTM4SF5R2-3 was confirmed to be a potent B cell epitope [11]. This figure was reproduced from BMC Immunology (2011) under the terms of the Creative Commons Attribution License

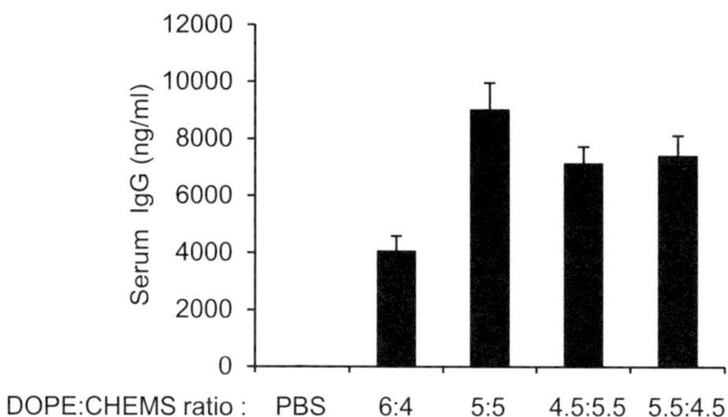

Fig. 3 Optimal DOPE:CHEMS ratio. The CpG-DNA((MB-ODN 4531(O))-peptide (TM4SF5R2-3)-DOPE:CHEMS complex was prepared varying the lipid composition. After injection into the mice with the complex three times, sera were prepared and analyzed for the amount of total IgG against TM4SF5R2-3 peptide. The most optimal ratio molar of DOPE:CHEMS was 5:5, which was effective more than twofold compared to the ratio of 6:4

CA, USA). Confirm that the endotoxin content of the MB-ODN 4531(O) is less than 1 ng/mg of ODN when measured by *Limulus amebocyte* assay. Store at −70 °C (*see* **Note 1**).

2. Peptide synthesis: Peptides can be synthesized by the Fmoc solid-phase method using an automated peptide synthesizer. After deprotection of the synthesized peptides from the resin, purify and analyze the peptides with reverse-phase HPLC and a C8 analytical RP column (purity greater than 90 %). The peptides can be

identified by means of a mass spectrometer. Dissolve the peptides (10 mg/mL) in distilled water and store at −70 °C (*see* **Note 2**).

3. Liposomes: Obtain the liposomes Cholesteryl hemisuccinate (CHEMS) as a powder form and 1,2-dioleoyl-*sn*-glycero-3-phosphoethanolamine (DOPE) dissolved in chloroform at the concentration of 10 mg/mL.
4. 10 mL glass tubes.
5. Liquid nitrogen.
6. Nitrogen gas.
7. Vortex mixer.
8. Magnetic stirrer.
9. Sonicator.
10. 0.22 μm cellulose acetate filters.

2.3 ELISA Components

1. Immunoplates.
2. Coating buffer: 0.1 M sodium carbonate, pH 9.6.
3. PBST: PBS containing 0.05 % Tween 20.
4. Sealing tapes.
5. Multichannel pipettes.
6. Buffer reservoir.
7. Blocking solution: PBS containing 1 % BSA. Store at 4 °C.
8. HRP-conjugated donkey anti-mouse IgG.
9. HRP-conjugated rat anti-mouse IgG1 and anti-mouse IgG2a antibody.
10. Tetramethylbenzidine (TMB) peroxidase substrate solution.
11. TMB Stop solution (KPL).
12. Microplate ELISA reader.

3 Methods

3.1 Epitope Prediction

1. Retrieve the amino acid sequence of target antigen from NCBI database.
2. Analyze the amino acid sequence based on their hydrophilicity values from Parker et al. [16], surface accessibility values from Emini et al. [17], β-turn region values from Chou and Fasman [18], and antigenicity index from Kolaskar and Tongaonkar [19] (http://tools.immuneepitpoe.org/main/index.htmL).
3. Average the parameters over six amino acid residues and the regions above the threshold value 0.5–1.0 were chosen for each prediction factor (*see* **Note 3**).
4. Select the putative B cell epitopes (Fig. 1).

3.2 Peptide Synthesis

1. Prepare 15 doses of a complex comprising synthetic peptide and Lipoplex(O) for mice injection.
   50 μg of CpG-DNA and 50 μg of peptide in 200 μL liposome per 1 dose.
2. At least 2 mg of peptide has to be synthesized for one experiment. 50 μg of peptide per each mouse (20 g body weight), three mice per each experiment, four times injection and 5 μg/well for ELISA (480 μg/96-well plate).

3.3 Preparation of Complexes Comprising Synthetic Peptide and Lipoplex(O)

1. Mix the liposomes DOPE (10 mg/mL in chloroform, 1.5 mL) and CHEMS (12 μg) by vortexing in the glass tube, which makes the molar ratio of DOPE:CHEMS 1:1 (*see* **Note 4**).
2. Evaporate chloroform to make a solvent-free lipid film on the bottom of glass tube by blowing with nitrogen gas.
3. Add ethanol (0.3 mL) to the film.
4. Dilute 80 μL of MB-ODN 4531(O) stock solution (10 mg/mL) with distilled water to make the final volume of 1.44 mL.
5. Dilute 80 μL of peptide stock solution (10 mg/mL) with distilled water to make the final volume of 1.44 mL.
6. Mix diluted MB-ODN 4531(O) and peptide solution (total 2.88 mL).
7. Take 2.7 mL of the mixture containing MB-ODN 4531(O) (750 μg) and peptide (750 μg) and add it to the 0.3 mL ethanol-treated film (**step 3**) (*see* **Note 5**).
8. Mix vigorously with vortex mixer for 30 s.
9. Insert magnetic stirring bar into the glass tube and stir at room temperature for 30 min, at 400 rpm with magnetic stirrer.
10. Adjust the pH of aqueous solutions to 7.0 with 0.1 N HCl or 0.5 N NaOH.
11. Mildly sonicate the hydrated lipid solutions containing CpG-DNA and peptide (Lipoplex(O)) for 30 s (pulse mode; output 1, 80 % duty cycle).
12. Filter Lipoplex(O) with a 0.22 μm cellulose acetate filter.
13. Freeze-thaw Lipoplex(O) solution three times for 1 h interval with liquid nitrogen to achieve a unilamellar structure of liposome.
14. Store at −70 °C.

3.4 Mice Immunization and Serum Collection

1. All procedures involving animal studies have to be approved by the Institutional Animal Care and Use Committee.
2. BALB/c mice have to be maintained under specific-pathogen-free conditions in a controlled environment (20–25 °C, 32–37 % humidity).

3. Inject the mice intraperitoneally with 200 μL of the complex solution comprising synthetic peptide and Lipoplex(O) (50 μg of peptides supplemented 50 μg of MB-ODN 4531(O) encapsulated with DOPE/CHEMS liposomes) on three occasions at 10 days intervals.
4. Ten days after final injection, sacrifice the mice under Zoletil 50 + Rompun anesthesia by intraperitoneal injection (0.5 mL/mouse). All efforts have to be made to minimize suffering of mice.

3.5 Serum ELISA

1. To measure the amounts of total IgG against peptide epitope, coat 96-well immunoplates with 5 μg/mL of the peptide in 100 μL of ELISA coating buffer overnight at 4 °C.
2. Wash with 100 μL of blocking solution and incubate in 100 μL of blocking solution for 1 h at room temperature.
3. Wash the plates with PBST.
4. Dilute the sera to 1:50 with PBS, add the diluents to the wells of each plate, and incubate for 2 h.
5. Wash the plates with PBST three times.
6. Incubate the plates with HRP-conjugated donkey anti-mouse IgG antibody (ratio 1:5000) for 1 h.
7. Wash the plates with PBST three times and PBS one time.
8. Apply 100 μL of TMB substrate solution and stop the reaction with 100 μL of TMB Stop solution.
9. Analyze the plates at 450 nm using a micro-plate ELISA reader.
10. To measure IgG1 and IgG2a levels, perform the procedure as described above except using HRP-conjugated rat anti-mouse IgG1 or IgG2a rather than anti-mouse IgG at the **step 6** (*see* Fig. 2).
11. Perform general hybridoma cell fusion to obtain clones expressing peptide-specific monoclonal antibodies [20].

4 Notes

1. Impurities are removed to prevent degradation of phosphodiester bond MB-ODN 4531(O).
2. Water soluble peptide should be prepared.
3. Consider the length and structure of the peptide. Although we have an experience to produce epitope-specific antibodies even with 7 amino acid peptide as an antigen, you had better select longer peptides. Depending on your purpose, you can select amino acid sequences located on the surface of the protein and consider protein-protein interaction.

4. We found that the 1:1 molar ratio of DOPE and CHEMS was the best (Fig. 3).
5. Use the same preparation method in the case when you use other phosphorothioate-backbone CpG-DNA (PS-ODN) instead of MB-ODN 4531(O). To our experience, liposome complexes including synthetic peptide and PS-ODN can also induce antibody production in a manner similar with Lipoplex(O) we used here. However, there are backbone-related side effects such as transient splenomegaly and anti-DNA antibody production [21]. Furthermore, it is much expensive (about 20-fold) to synthesize PS-ODN.

Acknowledgement

This work was supported by grants from the National Research Foundation (2012R1A2A2A01009887, 2013M3A9A9050126, 2013R1A2A2A03067981) funded by the Ministry of Science, ICT & Future Planning in the Republic of Korea.

References

1. Ben-Yedidia T, Arnon R (1997) Design of peptide and polypeptide vaccines. Curr Opin Bio technol 8:442–448
2. Bijker MS, Melief CJ, Offringa R et al (2007) Design and development of synthetic peptide vaccines: past, present and future. Expert Rev Vaccines 6:591–603
3. Simoes S, Moreira JN, Fonseca C et al (2004) On the formulation of pH-sensitive liposomes with long circulation times. Adv Drug Deliv Rev 56:947–965
4. Chikh G, Schutze-Redelmeier MP (2002) Liposomal delivery of CTL epitopes to dendritic cells. Biosci Rep 22:339–353
5. Felnerova D, Viret JF, Gluck R et al (2004) Liposomes and virosomes as delivery system for antigens, nucleic acids and drugs. Curr Opin Biotechnol 15:518–529
6. Ben-Yedidia T, Marcus H, Reisner Y et al (1999) Intranasal administration of peptide vaccine protects human/mouse radiation chimera from influenza infection. Int Immunol 11:1043–1051
7. Li WM, Dragowska WH, Bally MB et al (2003) Effective induction of CD8+ T-cell response using CpG oligodeoxynucleotides and HER-2/neu-derived peptide co-encapsulation in liposomes. Vaccine 21:3319–3329
8. Kreig AM (2002) CpG motifs in bacterial DNA and their immune effects. Annu Rev Immunol 20:709–760
9. Krieg AM (2006) Therapeutic potential of Toll-like receptor 9 activation. Nat Rev Drug Discov 5:471–484
10. Lee KW, Jung J, Lee Y et al (2006) Immuno stimulatory oligodeoxynucleotide isolated from genome wide screening of Mycobacterium bovis chromosomal DNA. Mol Immunol 43:2107–2118
11. Kim D, Kwon S, Rhee JW et al (2011) Production of antibodies with peptide-CpG-DNA-liposome complex without carriers. BMC Immunol 12:29
12. Kim D, Kwon HJ, Lee Y (2011) Activation of Toll-like receptor 9 and production of epitope specific antibody by liposome-encapsulated CpG-DNA. BMB Rep 44:607–612
13. Kwon S, Kim D, Park BK et al (2012) Prevention and therapy of hepatocellular carcinoma by vaccination with TM4SF5 epitope-CpG-liposome complex without carriers. PLoS One 7:e33121
14. Kwon S, Choi KC, Kim YE et al (2014) Monoclonal antibody targeting of the cell surface molecule TM4SF5 inhibits the growth of

hepatocellular carcinoma. Cancer Res 74:3844–3856

15. Rhee JW, Kim D, Park BK et al (2012) Immunization with a hemagglutinin-derived synthetic peptide formulated with a CpG-DNA-liposome complex induced protection against lethal influenza virus infection in mice. PLoS One 7:e48750
16. Parker JM, Hodges PC (1986) New hydrophilicity scale derived from high-performance liquid chromatography peptide retention data: correlation of predicted surface residues with antigenicity and x-ray-derived accessible sites. Biochemistry 25:5425–5432
17. Emini EA, Hughes JV, Perlow DS et al (1985) Induction of hepatitis A virus-neutralizing antibody by a virus-specific synthetic peptide. J Virol 55:836–839
18. Chou PY, Fasman GD (1978) Prediction of the secondary structure of proteins from their amino acid sequence. Adv Enzymol Relat Areas Mol Biol 47:45–148
19. Kolaskar AS, Tongaonkar PC (1990) A semi-empirical method for prediction of antigenic determinants on protein antigens. FEBS Lett 276:172–174
20. Yokoyama WM, Christensen M, Santos GD et al (2006) Production of monoclonal antibodies. In: Coligan JE, Bierer B (eds) Current protocols in immunology. Wiley, New York, Chapter 2, Unit 2 5
21. Kim D, Rhee JW, Kwon S et al (2009) Immunostimulation and anti-DNA antibody production by backbone modified CpG-DNA. Biochem Biophys Res Commun 379: 362–367

Chapter 14

Thioredoxin-Displayed Multipeptide Immunogens

Angelo Bolchi, Elena Canali, Andrea Santoni, Gloria Spagnoli, Daniele Viarisio, Rosita Accardi, Massimo Tommasino, Martin Müller, and Simone Ottonello

Abstract

Fusion to carrier proteins is an effective strategy for stabilizing and providing immunogenicity to peptide epitopes. This is commonly achieved by cross-linking of chemically synthesized peptides to carrier proteins. An alternative approach is internal grafting of selected peptide epitopes to a scaffold protein via double stranded-oligonucleotide insertion or gene synthesis, followed by recombinant expression of the resulting chimeric polypeptide. The scaffold protein should confer immunogenicity to the stabilized and structurally constrained peptide, but also afford easy production of the antigen in recombinant form. A macromolecular scaffold that meets the above criteria is the redox protein thioredoxin, especially bacterial thioredoxin. Here we describe our current methodology for internal grafting of selected peptide epitopes to thioredoxin as tandemly arranged multipeptide repeats ("Thioredoxin Displayed Multipeptide Immunogens"), bacterial expression and purification of the recombinant thioredoxin–multipeptide fusion proteins and their use as antigens for the production of anti-peptide antibodies for prophylactic vaccine as well as diagnostic purposes.

Key words Thioredoxin, Display site, Peptide epitope, Internal grafting, Recombinant peptide antigens, Thioredoxin-displayed multipeptide immunogens, Thermal purification, Anti-peptide antibodies

1 Introduction

Anti-peptide antibodies are valuable reagents for a number of applications ranging from target antigen detection in various experimental settings (e.g., immunoblotting, immunoprecipitation, immunohistochemistry, and in vivo imaging) to vaccinology. They have become even more popular in recent years due to the explosion of genomics and the steadily increasing availability of sequence information for peptide epitope prediction.

The most common way to exploit this information for the construction of peptide-based immunogens relies on the (largely random) covalent cross-linking of chemically synthesized peptides

Gunnar Houen (ed.), *Peptide Antibodies: Methods and Protocols*, Methods in Molecular Biology, vol. 1348,
DOI 10.1007/978-1-4939-2999-3_14,

to carrier proteins such as ovalbumin and hemocyanin [1–3]. While quite straightforward and effective, this approach requires a substantial chemical peptide synthesis/coupling expertise, may suffer from the intrinsic variability of cross-linking conjugation, which can lead to modification of the antigenic determinant and not always provides optimal immunogenicity, especially in the case of weak epitopes (reviewed in [4]). Possible ways to overcome these limitations may rely on chemically defined conjugation to synthetic backbones as in the Multiple Antigenic Peptide (MAP) [5–7] and lipopeptide [8] approaches.

In a different experimental setup, selected peptide epitopes are recombinantly fused to specific sites of carrier proteins, including self-assembling polypeptides of viral [9–11] or nonviral (e.g., [12–14]) origin, in either an end-to-end (e.g., [15]) or an internal (e.g., [16]) fusion configuration to produce well-defined chimeric antigens. The internal grafting configuration, although less commonly utilized, has some potential advantages over the terminal fusion setup. For example, a higher resistance to proteolysis [17–19], a superior scaffold-induced constraining capacity [20] and the fact that internal grafting more closely resembles the flanking sequence context of natural, non-terminally located peptide epitopes. A general advantage of scaffold protein–peptide fusions over cross-linked conjugates or MAPs is their ease of derivatization and conversion to alternative antigenic formats. For example, incorporation into higher-order self-assembling structures such as virus-like particles and capsomers [9, 10, 21] to increase immunogenicity and to extend plasma half-life, addition of specific immunogenicity-enhancing modules (e.g., [12, 22–24]) as well as adaptation to different expression hosts (e.g., yeast and insect cells, in addition to *E. coli*) and modes of delivery, including viral vectorization and DNA vaccination (*see* **Note 6**).

Internally grafted scaffold protein–peptide assemblies are being mainly used for the construction of various antibody-mimicking chimeric proteins ("binders"), in which the peptide moiety is either grafted to internal secondary structure elements or to exposed surface loops [25–27]. We transferred this approach to recombinant peptide immunogen construction, using the well characterized *Escherichia coli* thioredoxin A (EcTrx) protein as scaffold [28–30]. EcTrx is a small (109 amino acid residues), soluble and nontoxic protein containing a surface-exposed loop stabilized at the base by a disulfide bond (polypeptide segment 33–36; Fig. 1a). It has previously been shown to be suitable for peptide aptamer display [31–33], but has also been successfully employed as an immune peptide scaffold in an end-to-end, rather than internal grafting configuration [15]. At the nucleotide sequence level, this loop corresponds to a unique *Cpo*I restriction site that can be conveniently used for directional in-frame cloning of peptide-encoding double-stranded (ds) oligonucleotides, thus allowing the facile incorporation of multiple DNA inserts in a head-to-tail arrangement.

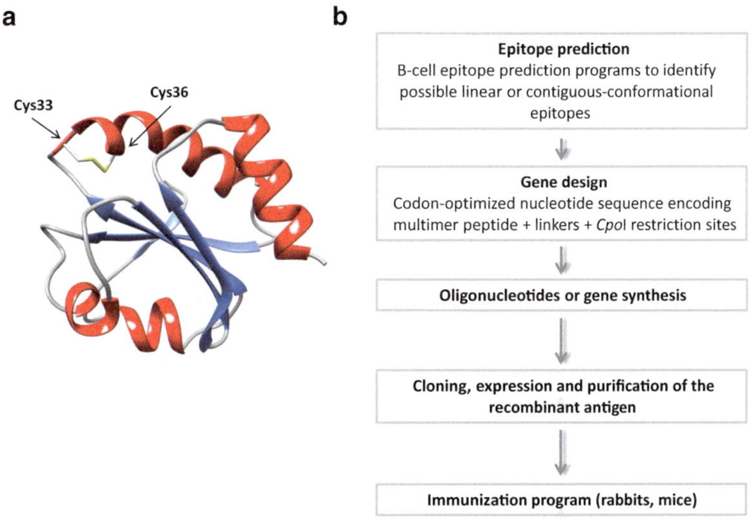

Fig. 1 Thioredoxin structure and TDMI production flowchart. (**a**) 3D structure of *E. coli* thioredoxin (PDB accession code: 2trx) with the two cysteines (Cys33 and Cys36, shown as disulfide bonded residues) located at the peptide display site marked with *arrows* and represented in a *stick format*. (**b**) Schematic representation of the main steps involved in the production of anti-peptide antibodies using TDMIs as antigens

EcTrx, and other thioredoxins as well, contain intrinsic putative T-cell epitopes, whose functionality is experimentally supported by the fairly strong immune responses elicited by Trx-displayed peptide antigens in various (e.g., Balb/C mice, rabbits, and guinea pigs; [28–30] and our unpublished results) but not all (e.g., C57BL/6 mice; [34] and our unpublished results) animal models.

We have thus exploited thioredoxin as a scaffold protein capable of conferring immunogenicity to short, tandemly repeated peptides in an approach called Thioredoxin-Displayed Multipeptide Immunogens (TDMI) [28, 29] (Fig. 1b). The choice of using multiple peptide repeats instead of single peptide epitope insertions rests on a number of previous (largely empirical) observations indicating that multi-epitope antigens are more effectively recognized by antigen presenting cells (APC) [15, 35–38] (*see* also **Notes 3** and **4**). Although the exact peptide housing capacity of Trx has not yet been determined, we have been able to produce, and immunologically test, Trx-peptide fusions carrying from one up to 15 copies of tandemly repeated peptide epitopes (ranging in size from 15 to 19 amino acids each) with an interposed three-amino acids spacer (e.g., GlyGlyPro) aimed at minimizing junctional epitope formation [28, 29]. In at least one experimental setup (a 19 aa. peptide from human papillomavirus minor capsid protein L2), immunogenicity was significantly enhanced when peptide copy number was augmented from one to three copies, but remained essentially constant upon further increase (up to 15 copies) [29]. A similar trend, with sometimes a drop of immunogenicity upon insertion of peptide repeats with a total length higher than 300 aa., was observed with other peptide epitopes (A.B. and S.O., unpublished), suggesting that this length (and corresponding peptide multiplicity) may correspond to the maximal housing (and perhaps constraining) capacity of Trx (*see* **Note 4**). Indeed, purified EcTrx harboring three copies of a 19 aa HPV16-L2 peptide was found to be almost completely oxidized, indicating that this insert size (a total of 68 aa including the GGP spacers) is fully compatible with intramolecular disulphide bond formation between Trx active site cysteine residues (40 and unpublished results). Based on these observations, threefold tandem repeats of peptides ranging in the size from 7 to 25 amino acids/peptide unit were subsequently used for most TDMI applications. A further increase in overall peptide multiplicity can be achieved by linking together individual Trx-(peptide)3 units end-to-end, but in at least one case this was found not to appreciably enhance immunogenicity [39]. We also observed the spontaneous formation, especially under non-reducing conditions, of higher molecular weight, disulfide-bonded Trx-(peptide)3 intermolecular aggregates, whose immunogenicity was again not appreciably different from that of the fully monomeric Trx-(peptide)3 antigen (40 and unpublished results).

Another advantage of thioredoxin compared to other scaffold proteins is its widespread occurrence in organisms ranging from (archae)bacteria to humans. This allows for the easy retrieval, by cDNA cloning or gene synthesis, of more or less divergent Trx homologs, some of which may exhibit more favorable (general or application-specific) properties compared to EcTrx (*see* **Notes 1** and **5**). For example, we have recently shown that a far-divergent thioredoxin from the hyperthermophilic archaebacterium *Pyrococcus furiosus* (PfTrx) is superior to EcTrx with regard to thermal stability, which allows for long-term storage of PfTrx-based antigens at temperatures higher than 50 °C, proteolytic stability and multipeptide solubilization/constraining capacity [40]. Another favorable feature of PfTrx is the lack of anti-PfTrx scaffold antibody reactivity against host (e.g., mouse, rabbit, and human) thioredoxins [40] (*see* also **Note 1**). This is an important prerequisite for immunization, and especially vaccination studies as it prevents "carrier-induced epitopic suppression" (i.e., a dampening of the immune response due to preexisting anti-scaffold antibodies; [41]) as well as unwanted and potentially harmful immune reactions between anti-scaffold antibodies and scaffold-homologous, cross-reactive host proteins. The same property can also be exploited to eliminate anti-scaffold antibody reactivity in post-immunization screenings through the use of two different (non-cross-reactive) thioredoxins (e.g., EcTrx and PfTrx) as primary immunogen and capture antigen, respectively (*see* **Note 5**).

The main applications of TDMI, so far, have been centered on the development of prophylactic recombinant peptide vaccines

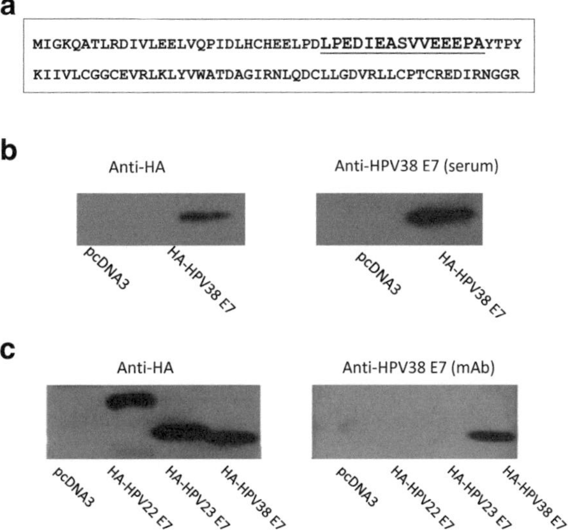

Fig. 2 Immunoblot analysis with antibodies generated against a TDMI. Amino acid sequence of the targeted HPV38 E7 antigen, with the 15 aa. peptide selected as immune epitope shown in bold and underlined (**a**). Human embryonic kidney cells (HEK293) transfected with a hemagglutinin (HA)-tagged version of the HPV38 E7 gene cloned in the pcDNA3 vector (**b**) or the same cell line transfected with HA-tagged E7 genes from HPV22, 23 and 38 (**c**) were used as sources of cellular extracts. Cells were lysed in IP buffer (20 mM Tris–HCl [pH 8], 200 mM NaCl, 0.5 % Nonidet P-40, 1 mM EDTA, 10 mM NaF, 0.1 mM Na_3VO_4, 1 mM phenylmethylsulfonyl fluoride, and 1 μg/ml leupeptin and aprotinin) and the lysates were fractionated on SDS-polyacrylamide gels (40 μg total protein/lane), which were then electrotransferred (130 mA, 1 h 30 min) to PVDF-membranes. Unfractionated immune-serum from rabbits immunized with an EcTrx-based HPV 38 E7-TDMI (1:5000 dilution; see panel (**a**)) and a mouse monoclonal antibody (unpurified supernatant from three times subcloned monoclonal hybridomas) obtained from a Balb/c mouse immunized with the same TDMI were used as primary antibodies in (**b**) and (**c**), respectively. Peroxidase-conjugated goat anti-rabbit IgGs (panel (**b**), *right*) and goat anti-mouse IgGs (panel (**c**), *right*) were used as secondary antibodies; a peroxidase-conjugated anti-HA mAb was used as control antibody for the blots shown on the left-side in panels (**b**) and (**c**)

against Alzheimer's disease, using the amyloid-β 1–15 peptide as epitope [28], and human papillomavirus (HPV) infection, using a peptide epitope derived from the N-terminal portion (aa 20–38) of the HPV16 minor capsid protein L2 [29, 30, 40]. In the latter application we also successfully exploited various Trx-L2 antigens for the production of HPV (cross)neutralizing monoclonal antibodies [42]. Recently, we extended the use of TDMIs to the production of anti-peptide diagnostic antibodies.

As an example, we show in Fig. 2 the immunoblot performance of TDMI-generated polyclonal and monoclonal antibodies targeting HPV38 E7. Together with HPV38 E6, the E7 oncoprotein is involved in human keratinocyte immortalization by cutaneous HPV types [43] and increases susceptibility to chemical or UV-induced skin carcinogenesis in mice [44]. It thus represents an important diagnostic and immune prophylaxis target.

Although TDMIs have originally been developed with EcTrx, we recently switched to PfTrx as a superior and more convenient scaffold [40]. In the sections below, we thus refer to both scaffold proteins.

2 Materials

2.1 Oligonucleotides and Chimeric Gene Constructs

1. Codon-optimized phosphorylated oligonucleotides and chimeric (Trx-multipeptide) gene constructs coding for the selected peptide epitopes (*see* Subheading 3.1, **step 1**) are chemically synthesized by various companies.

2.2 cDNA Cloning and TDMI Construct Assembly Reagents

1. Phosphorylated single-stranded oligonucleotides.
2. *Cpo*I restriction enzyme.
3. Annealing buffer: 10 mM Tris–HCl pH 8, 50 mM NaCl, 1 mM EDTA.
4. 1 % agarose solution.
5. DNA gel extraction kit.
6. pEcTrx [29] or pPfTrx [40].
7. Calf intestinal alkaline phosphatase (CIAP).
8. T4 DNA ligase.
9. Taq polymerase.
10. T7 promoter and T7 terminator universal primers.
11. Deoxynucleotide (dNTP) mix solution.

2.3 Subcloning of the Synthetic TDMI Gene into pET26

1. Chemically synthesized, vector-cloned Trx-multipeptide genes (e.g., pEX-A).
2. pET26 plasmid.
3. *Nde*I restriction enzyme.
4. 1 % agarose solution.
5. DNA gel extraction kit.
6. Calf Intestinal Alkaline Phosphatase.
7. DNA ligation kit (e.g., Mighty mix).

2.4 Bacterial Transformation

1. Electrocompetent *E. coli* BL21 (DE3) codon plus cells or other BL21 (DE3) *E. coli* strains (e.g., SoluBL21 or BL21 Gen-X) as appropriate.
2. Electroporation cuvettes (*see* also **Note 2**).

3. LB medium: dissolve 10 g tryptone, 5 g yeast extract, and 10 g NaCl in 950 ml deionized water; adjust the pH to 7.0 with 1 N NaOH and bring volume to 1 l; sterilize by autoclaving.
4. LB agar medium: prepare LB medium as above and add 15 g/l agar before autoclaving.
5. Kanamycin (10 μg/ml) and chloramphenicol (34 μg/ml) stock solutions.

2.5 TDMI Expression

1. LB medium plus kanamycin/chloramphenicol (*see* Subheading 2.4 above).
2. 1 M isopropyl β-D-thiogalactopyranoside (IPTG) stock solution.
3. Lysis buffer: 25 mM Tris–HCl pH 8, 0.3 M NaCl plus protease inhibitors mix (0.5 mM PMSF, 0.5 mM benzamidine, 1 μM leupeptin, 1 μM pepstatin).
4. Acrylamide–bis-acrylamide (29:1) 30 % stock solution for the preparation of 11 % polyacrylamide-SDS gels.

2.6 TDMI Purification

Metal-Affinity Purification

1. Metal (Ni/Co)-affinity purification resin (e.g., TALON® Metal Affinity Resin).
2. Equilibration buffer: 50 mM sodium-phosphate buffer pH 8, 0.3 M NaCl.
3. Wash buffer: 50 mM sodium-phosphate buffer pH 8, 0.3 M NaCl plus 10 mM imidazole.
4. Elution buffer: sodium-phosphate buffer pH 8, 0.3 M NaCl plus 200 mM imidazole.
5. Phosphate buffered saline (PBS) for antigen storage/delivery: 135 mM NaCl, 25 mM KCl, 10 mM disodium hydrogen phosphate buffer pH 7.4.

Thermal Purification

1. 62.5 % polyethylene glycol (PEG) 1000 stock solution.
2. 5 M NaCl.
3. PBS for antigen storage/delivery as above.

3 Methods

3.1 Epitope Prediction and Synthesis of Peptide-Encoding DNA

1. Analyze the amino acid sequence of the protein of interest with the following B-cell epitope prediction programs in order to identify putative linear or continuous-conformational (but not structural-discontinuous) epitopes: BCPred [45], AAP [46], FBCPred [47], ABCPRED [48], BcePred [49], BepiPred [50], Best [51].

2. Use a "majority vote" criterion to select the "best" epitope (i.e., the epitope supported by most of the above prediction programs).
3. Reverse translate the amino acid sequence of the selected epitope into the corresponding nucleotide sequence taking into account *E. coli* codon usage [52]. For multipeptide epitopes, insert a nucleotide sequence coding for the GGP tripeptide spacer between individual peptide units.
4. Add a *Cpo*I restriction site to both ends of the peptide-encoding DNA sequence.
5. Produce the desired constructs using either the ds-oligonucleotide or the synthetic gene approach (*see* Subheadings 3.2 and 3.3, below).

3.2 TDMI Expression Plasmid Assembly from ds-Oli gonucleotides

1. Chemical synthesis of two complementary oligonucleotides that, after annealing (see below), generate an epitope-coding ds-DNA cassette with *Cpo*I-compatible sticky ends (e.g., sense oligo: GTCCG NNN NNN NNN GGCG; antisense oligo GACCGCC N′N′N′ N′N′N′ N′N′N′ CG). Make sure that the epitope-coding dsDNA is in frame with the thioredoxin coding sequence.
2. Dissolve the two complementary oligonucleotides in annealing buffer at the same molar concentration; for the annealing reaction use a thermal cycler set with a program that includes a 2 °C decrease in temperature every minute from 80 to 20 °C.
3. Digest the pTrx vector (pEcTrx or pPfTrx) with *Cpo*I.
4. Electrophoretically fractionate the restriction digest on a 1 % agarose gel and cut the appropriate (linearized vector) band.
5. Extract the band using a gel extraction kit.
6. Dephosphorylate the digested pTrx vector by CIAP treatment.
7. Ligate the annealed ds-oligonucleotides to the linearized and dephosphorylated pTrx vector (at a 1:3, vector–insert molar ratio) by overnight incubation at 15 °C with T4 DNA ligase (*see* Fig. 3, panel a).
8. Transform bacterial cells by electroporation and plate on LB-agar containing 10 μg/ml kanamycin and 34 μg/ml chloramphenicol.
9. Identify Trx-peptide constructs bearing a single peptide repeat by colony-PCR, using T7 promoter and T7 terminator primers.
10. After sequence verification, digest the Trx-peptide plasmid with *Cpo*I, electrophoretically fractionate the restriction digest on a 1 % agarose gel and cut the appropriate bands as above.
11. Ligate the extracted DNA insert fragment to the linearized and dephosphorylated pTrx vector at a 1:50, vector–insert molar ratio, as above.

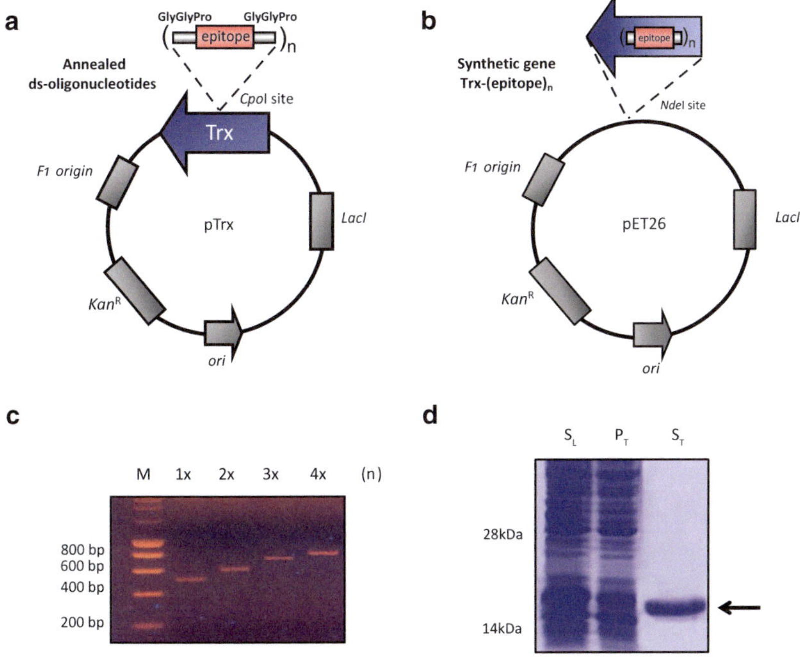

Fig. 3 Plasmid maps, gene construct verification and production of a TDMI. Maps of plasmids utilized for TDMI construction with the ds-oligonucleotide or the synthetic gene approach are shown in (**a**) and (**b**), respectively. (**c**) TDMI DNA constructs, harboring different numbers of repeats of the E7-HPV38 peptide epitope (from 1 to 4 copies as indicated), size-fractionated on a 1.5 % agarose gel. (**d**) SDS-PAGE analysis of a heat purified E7-HPV38 TDMI (*lane ST*, marked with an *arrow*) bearing three copies of the selected peptide epitope (*see* Subheading 3.1, **step 1**); *SL*, total proteins recovered in the supernatant after bacterial cell lysis; *PT*, total proteins recovered in the pellet after heat-treatment; *ST*, purified TDMI recovered in the supernatant after heat-treatment

12. Transform bacterial cells by electroporation and plate on LB agar plus antibiotics as above.
13. Identify Trx-multipeptide constructs of the desired multiplicity (usually, three repeats) by colony-PCR as above (*see* Fig. 3c for an example).
14. Following sequence verification, use the transformed bacterial cells for antigen production.

3.3 TDMI Expression Plasmid Construction from a Preassembled Synthetic Gene

1. Direct chemical synthesis of a TDMI cDNA containing *Nde*I restriction sites at both ends (*see* Subheading 2.1) and (*see* **Note 2**).
2. Digest pET26 and the pEX-A vector (or other source vector carrying the synthetic Trx-multipeptide gene) with *Nde*I.
3. Electrophorese the restriction digest on a 1 % agarose gel, cut the appropriate bands and gel-extract them as above.

4. Ligate the gel-purified Trx-multipeptide insert to the dephosphorylated (CIAP-treated) pET26 vector (1:3, vector–insert molar ratio; DNA ligation kit (e.g., Mighty Mix), overnight incubation at 15 °C as above) (*see* Fig. 3b).
5. Transform the ligation mixture into electrocompetent BL21(DE3) codon plus *E. coli* cells, or similar, by electroporation and plate cells on LB agar containing 10 μg/ml kanamycin and 34 μg/ml chloramphenicol (*see* **Note 2**).
6. Identify recombinant clones and check insert orientation by colony-PCR.
7. After sequence verification, directly use Trx-multipeptide transformants for antigen expression.

3.4 Antigen Expression and Purification

1. To induce recombinant antigen expression, add 1 mM IPTG to Trx-multipeptide transformant pre-cultures (OD600 = 0.6) and culture induced cells at 37 °C for 3 h. In case of poor solubility, decrease culture temperature to 20 °C with an induction time of 16 h. Special BL21(DE3) *E. coli* strains (e.g., SoluBL21) can be used to increase solubility.
2. After centrifugation (15 min, 7000 × *g* at 4 °C) and resuspension in lysis buffer (freshly supplemented with protease inhibitors), bacterial cells are lysed by sonication (20 min total sonication time performed with 2 s bursts alternated with 6 s resting on ice) carried out at constant power (4 W/cm^2; Sonicator 3000). Clarify the bacterial lysate by centrifugation for 30 min at 15,000 × *g* (4 °C) to recover the cleared supernatant containing soluble proteins.
3. TDMI purification by affinity chromatography or thermal purification.
   (a) For purification by metal-affinity chromatography, load the soluble lysate (supernatant containing the His-tagged Trx-multipeptide protein) on a metal-affinity resin (e.g., Talon); wash the column and elute the His-tagged antigen with imidazole-containing buffer as per manufacturer's instructions (see also [29]).
   (b) For thermal purification, add 12.5 % (w/v) PEG 1000 and 0.25 M NaCl (NaCl only in the case of PfTrx) to the lysate supernatant, incubate at 70 °C for 20 min, cool on ice for 10 min and centrifuge at 12,000 × *g* for 10 min to recover the heat-stable (and soluble) Trx-multipeptide antigen, usually at a relative purity ≥80 % (*see* Fig. 3d for a representative example).
4. Following purification, the antigen is dialyzed against PBS and concentrated (usually to 1 mg/ml) by ultrafiltration (e.g., cellulose membranes). Protein concentration and purity are

determined by both SDS-PAGE (11 % polyacrylamide gel) analysis, Coomassie Blue G-250 staining (Protein Assay with bovine serum albumin as standard) and UV absorption spectroscopy using predicted A280 extinction coefficients calculated for each Trx-multipeptide [53, 54].

5. The purified recombinant antigen is then sterilized by filtration on 0.22 μm syringe filters. If required, especially for vaccination studies involving the use of non-Freund immune adjuvants, the final antigen preparation is detoxified by adding 1 % (w/v) Triton X-114, mixing, incubating for 5 min on ice and 5 min at 37 °C, followed by centrifugation at 18,000 ×*g* for 5 min at 37 °C [55].

4 Notes

1. Although TDMIs, including the HPV38-E7 example, have previously been based on *E. coli* thioredoxin (EcTrx), we have recently moved to the far-divergent thioredoxin from *Pyrococcus furiosus* (PfTrx) as preferred scaffold. As mentioned in the Introduction, specific practical advantages of PfTrx are a higher thermal stability, which allows for an easier and more consistent thermal purification not requiring PEG/NaCl supplementation as well as long-term storage at temperatures >50 °C, a higher solubilization and peptide insert constraining capacity, and the lack of immune reactivity of anti-PfTrx antibodies with endogenous/host thioredoxin and related (Trx fold-containing) proteins. The nucleotide sequence of PfTrx, which includes the *Cpo*I restriction site artificially inserted at the display site of thioredoxin, is reported in [40]; the pPfTrx expression vector is available on request.
2. Peptide epitopes containing multiple (>3) tandem repeats of certain amino acids (e.g., proline) may be difficult to express in *E. coli*. This problem can in part be alleviated by: (a) the use of commercially available (e.g., Codon-plus) or home-made modified *E. coli* strains overexpressing tRNAs for particular codons [56]; (b) properly managing “codon usage” (i.e., utilize, when possible, high-usage codons for the same amino acid read by different tRNA isoacceptors).
3. Cyclic synthetic peptides bearing two terminal, disulfide bridge-forming Cys residues, are often used as carrier protein-conjugated immunogens. If required, a similar setup can be reproduced with TDMI by inserting flanking Cys residues (in addition to the two terminal cysteines associated to the thioredoxin display site) within the spacer sequence interposed between individual peptide units.

4. Tandemly repeated (3×) peptide epitopes, which are required to increase immunogenicity in the context of vaccination studies relying on relatively weak, human use-approved immune adjuvants, can perhaps be replaced by a single peptide epitope in the case of immunization studies aimed at producing diagnostic antipeptide antibodies with the use of the fairly strong Freund adjuvant. The latter setup may attenuate junctional epitope formation between individual peptide-spacer-peptide units, thus reducing background reactivity and leading to an overall improved antibody quality.
5. In addition to standard affinity purification methods [57], which require the availability of the full-length antigen, immune sera produced by TDMI can be depleted of background anti-Trx antibodies by batch or column treatment with the empty thioredoxin scaffold immobilized on agarose/sepharose beads. To eliminate anti-scaffold antibody reactivity in post-immunization tests (e.g., ELISA) and get a more reliable estimate of target-specific antibodies, one could also take advantage of the lack of immune cross-reactivity between PfTrx and EcTrx by using for such tests a Trx antigen (e.g., EcTrx-multipeptide) different from that utilized for immunization (e.g., PfTrx-multipeptide).
6. For special needs, e.g., large-scale production of particular antigens in eukaryotic cells, PfTrx-multipeptide constructs can be easily transferred and expressed as secreted polypeptides in the metylotrophic yeast *Pichia pastoris* with the use of available vector/secretion expression systems [58].

Acknowledgments

This work was supported by grants from the Italian Association for Cancer Research (AIRC, grant IG 12956) and from the Regione Emilia Romagna, Ricerca Regione-Università 2010–2012, Strategic Programme "A tailored approach to the immune-monitoring and clinical management of viral and autoimmune diseases" to S.O. We thank Arturo Roberto Viscomi (GSK, San Polo di Torrile, Parma, Italy) for precious and highly competent help at an early stage of this work. E.C. was supported by a postdoctoral fellowship from AIRC.

References

1. Edwards RJ, Singleton AM, Murray BP, Davies DS, Boobis AR (1995) Short synthetic peptides exploited for reliable and specific targeting of antibodies to the C-termini of cytochrome P450 enzymes. Biochem Pharmacol 49: 39–47
2. Angeletti RH (1999) Design of useful peptide antigens. J Biomol Tech 10:2–10

3. Trier NH, Hansen PR, Vedeler CA, Somnier FE, Houen G (2012) Identification of continuous epitopes of HuD antibodies related to paraneoplastic diseases/small cell lung cancer. J Neuroimmunol 243:25–33
4. Olive C, Toth I, Jackson D (2001) Tech nological advances in antigen delivery and synthetic peptide vaccine developmental strategies. Mini Rev Med Chem 1:429–438
5. Tam JP (1988) Synthetic peptide vaccine design: synthesis and properties of a high-density multiple antigenic peptide system. Proc Natl Acad Sci U S A 85:5409–5413
6. Posnett DN, McGrath H, Tam JP (1988) A novel method for producing anti-peptide antibodies. Production of site-specific antibodies to the T cell antigen receptor beta-chain. J Biol Chem 263:1719–1725
7. Niederhafner P, Sebestik J, Jezek J (2005) Peptide dendrimers. J Pept Sci 11:757–788
8. Zaman M, Toth I (2013) Immunostimulation by synthetic lipopeptide-based vaccine candidates: structure-activity relationships. Front Immunol 4:318
9. Chuan YP, Wibowo N, Connors NK, Wu Y, Hughes FK, Batzloff MR, Lua LH, Middelberg AP (2014) Microbially synthesized modular virus-like particles and capsomeres displaying group A streptococcus hypervariable antigenic determinants. Biotechnol Bioeng 111: 1062–1070
10. Peabody DS, Manifold-Wheeler B, Medford A, Jordan SK, do Carmo Caldeira J, Chackerian B (2008) Immunogenic display of diverse peptides on virus-like particles of RNA phage MS2. J Mol Biol 380:252–263
11. Branco LM, Grove JN, Geske FJ, Boisen ML, Muncy IJ, Magliato SA, Henderson LA, Schoepp RJ, Cashman KA, Hensley LE, Garry RF (2010) Lassa virus-like particles displaying all major immunological determinants as a vaccine candidate for Lassa hemorrhagic fever. Virol J 7:279
12. Ogun SA, Dumon-Seignovert L, Marchand JB, Holder AA, Hill F (2008) The oligomerization domain of C4-binding protein (C4bp) acts as an adjuvant, and the fusion protein comprised of the 19-kilodalton merozoite surface protein 1 fused with the murine C4bp domain protects mice against malaria. Infect Immun 76:3817–3823
13. Rudra JS, Tian YF, Jung JP, Collier JH (2010) A self-assembling peptide acting as an immune adjuvant. Proc Natl Acad Sci U S A 107: 622–627
14. Liu J, Liu J, Xu H, Zhang Y, Chu L, Liu Q, Song N, Yang C (2014) Novel tumor-targeting, self-assembling peptide nanofiber as a carrier for effective curcumin delivery. Int J Nanomed 9:197–207
15. Kovacs-Nolan J, Mine Y (2006) Tandem copies of a human rotavirus VP8 epitope can induce specific neutralizing antibodies in BALB/c mice. Biochim Biophys Acta 1760: 1884–1893
16. Krause S, Schmoldt HU, Wentzel A, Ballmaier M, Friedrich K, Kolmar H (2007) Grafting of thrombopoietin-mimetic peptides into cystine knot miniproteins yields high-affinity thrombopoietin antagonists and agonists. FEBS J 274:86–95
17. Backstrom M, Lebens M, Schodel F, Holmgren J (1994) Insertion of a HIV-1-neutralizing epitope in a surface-exposed internal region of the cholera toxin B-subunit. Gene 149: 211–217
18. Guo L, Wang J, Qian S, Yan X, Chen R, Meng G (2000) Construction and structural modeling of a single-chain Fv-asparaginase fusion protein resistant to proteolysis. Biotechnol Bioeng 70:456–463
19. Jacquet A, Daminet V, Haumont M, Garcia L, Chaudoir S, Bollen A, Biemans R (1999) Expression of a recombinant Toxoplasma gondii ROP2 fragment as a fusion protein in bacteria circumvents insolubility and proteolytic degradation. Protein Expr Purif 17:392–400
20. Ofek G, Guenaga FJ, Schief WR, Skinner J, Baker D, Wyatt R, Kwong PD (2010) Elicitation of structure-specific antibodies by epitope scaffolds. Proc Natl Acad Sci U S A 107:17880–17887
21. Van Braeckel-Budimir N, Haijema BJ, Leenhouts K (2013) Bacterium-like particles for efficient immune stimulation of existing vaccines and new subunit vaccines in mucosal applications. Front Immunol 4:282
22. del Guercio MF, Alexander J, Kubo RT, Arrhenius T, Maewal A, Appella E, Hoffman SL, Jones T, Valmori D, Sakaguchi K, Grey HM, Sette A (1997) Potent immunogenic short linear peptide constructs composed of B cell epitopes and Pan DR T helper epitopes (PADRE) for antibody responses in vivo. Vaccine 15:441–448
23. Dempsey PW, Allison ME, Akkaraju S, Goodnow CC, Fearon DT (1996) C3d of complement as a molecular adjuvant: bridging innate and acquired immunity. Science 271: 348–350
24. Carayanniotis G, Barber BH (1987) Adjuvant-free IgG responses induced with antigen coupled to antibodies against class II MHC. Nature 327:59–61

25. Hosse RJ, Rothe A, Power BE (2006) A new generation of protein display scaffolds for molecular recognition. Protein Sci 15:14–27
26. Binz HK, Amstutz P, Pluckthun A (2005) Engineering novel binding proteins from non-immunoglobulin domains. Nat Biotechnol 23: 1257–1268
27. Weidle UH, Auer J, Brinkmann U, Georges G, Tiefenthaler G (2013) The emerging role of new protein scaffold-based agents for treatment of cancer. Cancer Genomics Proteomics 10:155–168
28. Moretto N, Bolchi A, Rivetti C, Imbimbo BP, Villetti G, Pietrini V, Polonelli L, Del Signore S, Smith KM, Ferrante RJ, Ottonello S (2007) Conformation-sensitive antibodies against alzheimer amyloid-beta by immunization with a thioredoxin-constrained B-cell epitope peptide. J Biol Chem 282:11436–11445
29. Rubio I, Bolchi A, Moretto N, Canali E, Gissmann L, Tommasino M, Muller M, Ottonello S (2009) Potent anti-HPV immune responses induced by tandem repeats of the HPV16 L2 (20–38) peptide displayed on bacterial thioredoxin. Vaccine 27:1949–1956
30. Seitz H, Canali E, Ribeiro-Muller L, Palfi A, Bolchi A, Tommasino M, Ottonello S, Muller M (2014) A three component mix of thioredoxin-L2 antigens elicits broadly neutralizing responses against oncogenic human papillomaviruses. Vaccine 32:2610–2617
31. LaVallie ER, DiBlasio EA, Kovacic S, Grant KL, Schendel PF, McCoy JM (1993) A thioredoxin gene fusion expression system that circumvents inclusion body formation in the E coli cytoplasm. Biotechnology 11:187–193
32. Colas P, Cohen B, Jessen T, Grishina I, McCoy J, Brent R (1996) Genetic selection of peptide aptamers that recognize and inhibit cyclin-dependent kinase 2. Nature 380:548–550
33. Klevenz B, Butz K, Hoppe-Seyler F (2002) Peptide aptamers: exchange of the thioredoxin-A scaffold by alternative platform proteins and its influence on target protein binding. Cell Mol Life Sci 59:1993–1998
34. Nieto K, Weghofer M, Sehr P, Ritter M, Sedlmeier S, Karanam B, Seitz H, Muller M, Kellner M, Horer M, Michaelis U, Roden RB, Gissmann L, Kleinschmidt JA (2012) Development of AAVLP(HPV16/31 L2) particles as broadly protective HPV vaccine candidate. PLoS One 7, e39741
35. Broekhuijsen MP, van Rijn JM, Blom AJ, Pouwels PH, Enger-Valk BE, Brown F, Francis MJ (1987) Fusion proteins with multiple copies of the major antigenic determinant of foot-and-mouth disease virus protect both the natural host and laboratory animals. J Gen Virol 68(Pt 12):3137–3143
36. Enea V, Arnot D, Schmidt EC, Cochrane A, Gwadz R, Nussenzweig RS (1984) Circumsporozoite gene of plasmodium cynomolgi (Gombak):cDNA cloning and expression of the repetitive circumsporozoite epitope. Proc Natl Acad Sci U S A 81:7520–7524
37. Schuman JT, Grinstead JS, Apostolopoulos V, Campbell AP (2005) Structural and dynamic consequences of increasing repeats in a MUC1 peptide tumor antigen. Biopolymers 77: 107–120
38. Liu W, Chen YH (2005) High epitope density in a single protein molecule significantly enhances antigenicity as well as immunogenicity: a novel strategy for modern vaccine development and a preliminary investigation about B cell discrimination of monomeric proteins. Eur J Immunol 35:505–514
39. Seitz H, Dantheny T, Burkart F, Ottonello S, Muller M (2013) Influence of oxidation and multimerization on the immunogenicity of a thioredoxin-l2 prophylactic papillomavirus vaccine. Clin Vaccine Immunol 20:1061–1069
40. Canali E, Bolchi A, Spagnoli G, Seitz H, Rubio I, Pertinhez TA, Muller M, Ottonello S (2014) A high-performance thioredoxin-based scaffold for peptide immunogen construction: proof-of-concept testing with a human papillomavirus epitope. Sci Rep 4:4729
41. Jegerlehner A, Wiesel M, Dietmeier K, Zabel F, Gatto D, Saudan P, Bachmann MF. Carrier induced epitopic suppression of antibody responses induced by virus-like particles is a dynamic phenomenon caused by carrier-specific antibodies. Vaccine 28:5503–5512
42. Rubio I, Seitz H, Canali E, Sehr P, Bolchi A, Tommasino M, Ottonello S, Muller M (2011) The N-terminal region of the human papillomavirus L2 protein contains overlapping binding sites for neutralizing, cross-neutralizing and non-neutralizing antibodies. Virology 409:348–359
43. Caldeira S, Zehbe I, Accardi R, Malanchi I, Dong W, Giarre M, de Villiers EM, Filotico R, Boukamp P, Tommasino M (2003) The E6 and E7 proteins of the cutaneous human papillomavirus type 38 display transforming properties. J Virol 77:2195–2206
44. Viarisio D, Mueller-Decker K, Kloz U, Aengeneyndt B, Kopp-Schneider A, Grone HJ, Gheit T, Flechtenmacher C, Gissmann L, Tommasino M. E6 and E7 from beta HPV38 cooperate with ultraviolet light in the development of actinic keratosis-like lesions and

squamous cell carcinoma in mice. PLoS Pathog 7:e1002125

45. El-Manzalawy Y, Dobbs D, Honavar V (2008) Predicting linear B-cell epitopes using string kernels. J Mol Recognit 21:243–255
46. Chen J, Liu H, Yang J, Chou KC (2007) Prediction of linear B-cell epitopes using amino acid pair antigenicity scale. Amino Acids 33:423–428
47. El-Manzalawy Y, Dobbs D, Honavar V (2008) Predicting flexible length linear B-cell epitopes. Comput Syst Bioinformatics Conf 7:121–132
48. Saha S, Raghava GP (2006) Prediction of continuous B-cell epitopes in an antigen using recurrent neural network. Proteins 65: 40–48
49. Saha S, Raghava G (2004) BcePred: prediction of continuous B-cell epitopes in antigenic sequences using physico-chemical properties. Lect Notes Comput Sci 3239:197–204
50. Larsen JE, Lund O, Nielsen M (2006) Improved method for predicting linear B-cell epitopes. Immunome Res 2:2
51. Gao J, Faraggi E, Zhou Y, Ruan J, Kurgan L (2012) BEST: improved prediction of B-cell epitopes from antigen sequences. PLoS One 7, e40104
52. Puigbo P, Guzman E, Romeu A, Garcia-Vallve S (2007) OPTIMIZER: a web server for optimizing the codon usage of DNA sequences. Nucleic Acids Res 35:W126–W131
53. Gill SC, von Hippel PH (1989) Calculation of protein extinction coefficients from amino acid sequence data. Anal Biochem 182:319–326
54. Pace CN, Vajdos F, Fee L, Grimsley G, Gray T (1995) How to measure and predict the molar absorption coefficient of a protein. Protein Sci 4:2411–2423
55. Liu S, Tobias R, McClure S, Styba G, Shi Q, Jackowski G (1997) Removal of endotoxin from recombinant protein preparations. Clin Biochem 30:455–463
56. Dieci G, Bottarelli L, Ballabeni A, Ottonello S (2000) tRNA-assisted overproduction of eukaryotic ribosomal proteins. Protein Expr Purif 18:346–354
57. Ayyar BV, Arora S, Murphy C, O'Kennedy R (2012) Affinity chromatography as a tool for antibody purification. Methods 56:116–129
58. Uchima CA, Tokuda G, Watanabe H, Kitamoto K, Arioka M (2012) Heterologous expression in Pichia pastoris and characterization of an endogenous thermostable and high-glucose-tolerant beta-glucosidase from the termite Nasutitermes takasagoensis. Appl Environ Microbiol 78:4288–4293

Chapter 15

The Purification of Natural and Recombinant Peptide Antibodies by Affinity Chromatographic Strategies

Hui Ma and Richard O'Kennedy

Abstract

The purification of peptide antibodies (e.g., IgG, IgY, scFv, and Fab) are described in this chapter. Affinity chromatographic purification, a very convenient and effective antibody purification strategy, is used to isolate peptide antibodies based on specific binding, i.e., binding of the antibody to a column on which its specific ligand is immobilized with subsequent elution of the purified antibody. In addition, the application of purification methods based on the use of proteins A, G, and L, each of which bind to specific domains on an antibody/fragment, or the use of specific tags (e.g., histidine and biotin) attached to antibodies or antigens are also described.

Key words IgG, IgY, ScFv, Fab, Affinity chromatographic purification, Antibody–antigen binding

1 Affinity Chromatographic Purification of Peptide Antibodies

Affinity chromatographic purification is a method used to separate mixtures based on the specific and reversible interaction between the immobilized ligand and its target. This can occur with receptor-ligand and antigen–antibody binding [1]. For antibody purification, the antibody mixture is initially dissolved in a buffer called the mobile phase, then the mobile phase carries the mixture through the stationary phase (e.g., protein A/G/L or antigen-Immobilized and nickel/copper/zinc/cobalt-charged agarose resin) which is usually packed into a column. The specific antibodies will be bound onto the resin and the nonspecific antibodies and other impurities will be removed by the wash buffer running repeatedly through the column. Finally, the specific antibodies will be eluted by application of an elution buffer, often of a lower pH or with a high salt content, to the column (*see* Fig. 1). If the antibodies are eluted using a low pH elution buffer, neutralization buffer is always added to preserve reactivity of purified antibodies, as monoreactive antibodies may gain the ability to bind to multiple antigens after exposure to low pH [2].

Gunnar Houen (ed.), *Peptide Antibodies: Methods and Protocols*, Methods in Molecular Biology, vol. 1348, DOI 10.1007/978-1-4939-2999-3_15, © Springer Science+Business Media New York 2015

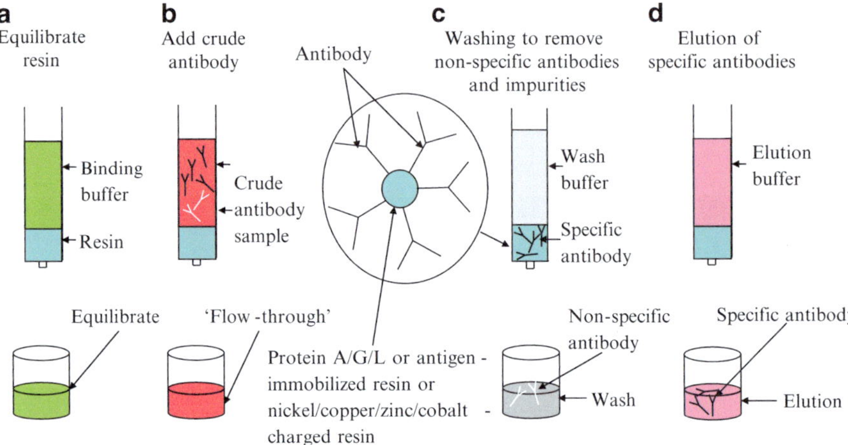

Fig. 1 Illustration of affinity chromatographic purification on a column. A typical procedure for affinity chromatographic purification of antibodies, including equilibration (**a**), binding (**b**), washing (**c**) to remove impurities and elution of specific antibodies (**d**)

Mammals and birds are the main hosts used for peptide antibody generation. Since IgG is the main antibody isotype found in mammalian blood and IgY is the major antibody in birds, and scFv and Fab are the most popular recombinant antibody formats, only their affinity purification is described in this chapter.

1.1 IgG Purification Using Protein A, G, and L

IgG isolation has been widely performed by using protein A, G, and L, which are the most popular immunoglobulin-binding proteins through their binding to the fragment crystallizable (Fc) or Fab region [3].

Protein A (~56 kDa) was firstly isolated from the cell wall of the bacterium *Staphylococcus aureus*. It binds immunoglobulins primarily through the Fc region, as well as the Fab region of the human V_H3 family [4]. Protein G is expressed by the cell wall of group C (~58 kDa, C40 protein G) and G (~65 kDa, G148 protein G) *Streptococcal species* [5]. Protein G was found to bind immunoglobulins through Fab and Fc regions, with a higher affinity than Protein A [6]. Protein L (~76 kDa) was originally derived from the surface of *Peptostreptococcus magnus*. Unlike protein A and G, protein L predominantly binds immunoglobulins through kappa light chain interactions [7].

The IgG binding domains of protein A and G and the kappa light chain binding region of protein L have been expressed by *Escherichia coli* for commercial use. For example, Thermo Fisher Scientific Incorporation produces and commercializes recombinant protein A (~47 kDa) consisting of four Fc-binding domains, recombinant protein G (~22 kDa) containing two Fc-binding domains and recombinant protein L (~36 kDa) containing four immunoglobulin-binding domains, which enable the binding of kappa I, III, and IV in human and kappa I in mouse.

1.2 ScFv and Fab Purification Using Metal-Charged Affinity Resins

Immobilized metal ion affinity chromatography (IMAC) is one of the most popular purification techniques for recombinant antibody fragments (i.e., scFv and Fab) [8].

IMAC is an affinity-separation technique based on the coordinate covalent bonding of specific groups (e.g., histidine) to metals (e.g., nickel, copper, zinc, and cobalt ions). For instance, histidine tags are often used for affinity purification of recombinant antibodies because they have a high affinity for nickel ions. The histidine tag is usually located on the C-terminal end of the scFv or Fab to prevent purification of truncated proteins. After incubating scFv or Fab-containing samples with the nickel-charged resin, the histidine-tagged recombinant antibodies will be bound onto the resin, while the non-tagged antibodies and impurities are then washed away with washing buffer, and, finally, the histidine-tagged recombinant antibodies are eluted using a low pH elution buffer (neutralization buffer is always added to preserve reactivity of purified antibodies).

1.3 IgG, IgY, scFv, and Fab Purification Using Peptide-Antibody Binding-Based Affinity Chromatographic Strategies

The peptide-specific affinity purification is performed via peptide-antibody binding. The peptide can be immobilized to a solid support surface through a unique functional group (e.g., sulfhydryl on a single terminal cysteine or C-terminal carboxyl amino acids in a peptide) or through streptavidin-biotin interactions (one of the strongest non-covalent interactions in nature) [9]. After passing the antibody sample over the peptide surface, the peptide-specific antibody will be bound onto the solid support surface, while the non-specific antibodies and impurities are then washed away with washing buffer, and finally the peptide-specific antibodies are eluted.

To achieve the ideal peptide-specific purification, peptide antibodies should first be isolated using general affinity purification strategies (IgG purification using protein A, G, and L or scFv and Fab purification using IMAC; *see* Subheadings 1.1, 1.2, 3.1 and 3.2), followed by antibody-peptide binding-based affinity purification. For chicken IgY, since IgG affinity resins (i.e., protein A, protein G, and protein L) do not bind chicken IgY, the popular IgY purification strategy (*see* Subheading 3.3), based on the techniques of delipidation-and-precipitation, is used before specific affinity purification [10].

2 Materials for Peptide Antibody Affinity Purification

2.1 Equipment and Consumables

Centrifuge tube, 50 mL.

Micro tube, 1.5 and 2 mL.

10,000 MWCO Viva spin.

Filter, 0.2 μL.

Pasteur pipette.

Gravity flow column, 20 mL.

Pierce™ BCA Protein Assay Kit.

Molecular grade water.

Sodium azide (Warning: Sodium azide has acute toxicity (oral, dermal, and inhalation), is hazardous to the aquatic environment (acute hazard—category 1 and chronic hazard—categories 1,2)).

2.2 Buffers and Commercial Kits

2.2.1 Buffers for IgG Purification Using Protein A, G, or L

Protein A-Agarose.

Protein G-Agarose.

Protein L-Agarose.

Phosphate buffered saline (PBS): 150 mM NaCl, 2.5 mM KCl, 10 mM Na_2HPO_4, 18 mM KH_2PO_4, pH 7.4.

Binding/Washing buffer for Protein A-Agarose, Protein L-Agarose: PBS, pH 7.4.

Binding buffer for Protein G-Agarose: 20 mM sodium phosphate buffer, pH 7.0. Weigh 3.27 g of $Na_2HPO_4{\cdot}7H_2O$ and 0.94 g of NaH_2PO_4, then adjust pH to 7.0 and bring volume to 1 L with distilled water.

Elution buffer: 100 mM glycine–HCl, pH 2.5–3.0.

Neutralization buffer: 1 M Tris–HCl, pH 8.0–9.0.

2.2.2 Buffers for scFv and Fab Purification Using Metal-Charged Affinity Resins

Ni^{+}-NTA agarose resin.

Cobalt, copper or zinc chelating resin.

Copper chelating resin.

Zinc chelating resin.

PBS (10×): 1.5 M NaCl, 25 mM KCl, 100 mM Na_2HPO_4, 180 mM KH_2PO_4.

PBS (1×): 150 mM NaCl, 2.5 mM KCl, 10 mM Na_2HPO_4, 18 mM KH_2PO_4, pH 7.4.

Binding buffer: 0.5 M NaCl, 20 mM imidazole in PBS (*see* **Note 1**).

Running buffer: 0.5 M NaCl, 20 mM imidazole in PBS and 0.1 % (v/v) Tween 20 (*see* **Notes 1** and **2**).

Elution buffer: 100 mM NaAc, pH 4.4.

Neutralization buffer: 1:1 (v/v) 100 mM NaOH and 10× PBS.

2.2.3 Commercial Kits and Buffers for General IgY Purification

Eggcellent™ IgY purification kit.

PBS: 150 mM NaCl, 2.5 mM KCl, 10 mM Na_2HPO_4, 18 mM KH_2PO_4, pH 7.4.

2.2.4 Buffers and Commercial Kits for Antibody Purification Using Peptide-Antibody Binding-Based Affinity Chromatographic Strategies

Buffers and Commercial Kits for Antibody Purification Using Streptavidin-Agarose

EZ-Link® Micro sulfo-NHS-LC-biotinylation kit.

High capacity streptavidin-agarose resin.

Binding/Wash Buffer: PBS (150 mM NaCl, 2.5 mM KCl, 10 mM Na_2HPO_4, 18 mM KH_2PO_4), pH 7.4.

Elution buffer: 100 mM glycine–HCl buffer, pH 2.5.

Neutralization buffer: 2 M Tris–HCl, pH 8.5.

Commercial Kits and Buffers for Antibody Purification Using a Sulfo-Linked Resin

SulfoLink® Immobilization Kit.

Binding/Wash Buffer: PBS (150 mM NaCl, 2.5 mM KCl, 10 mM Na_2HPO_4, 18 mM KH_2PO_4), pH 7.4.

Elution Buffer: 100 mM glycine–HCl buffer, pH 2.5.

Neutralization Buffer: 1 M Tris–HCl, pH 9.

Commercial Kits and Buffers for Antibody Purification Using a Carboxy-Linked Resin

CarboxyLink™ Immobilization Kit.

Coupling Buffer: Dissolve contents of the BupH MES Buffered Saline Pack in 500 μL of distilled water.

Binding/Wash Buffer: PBS (150 mM NaCl, 2.5 mM KCl, 10 mM Na_2HPO_4, 18 mM KH_2PO_4), pH 7.4.

Elution Buffer: 100 mM glycine–HCl buffer, pH 2.5.

Neutralization Buffer: 1 M Tris–HCl, pH 9.

3 Methods

Caution: Always ensure that the column does not dry out during purification procedures.

3.1 IgG Purification from Serum Using Protein A, G, or L

1. Add 2 mL of protein A, protein G, or protein L agarose to a 20 mL column (0.5 × 20 cm).
2. Equilibrate the column with 10 mL of PBS.
3. Place a cap on the base of the column.
4. Remove debris in the serum by centrifugation at 3220 × *g* for 10 min at 4 °C.
5. Add 5 mL of serum supernatant to the equilibrated protein A, protein G, or protein L column from **step 3** and mix gently on a roller mixer overnight at 4 °C.
6. Allow the column to drain into a 50 mL tube. Collect the "flow-through" and reapply it to the column. Repeat this process three times.
7. Wash the column five times with 10 mL of PBS.

8. Elute the antibody with 10 mL of elution buffer and collect into a 50 mL tube containing 1 mL of neutralization buffer.
9. The neutralized IgG is then thoroughly buffer exchanged three times against 5 mL of filtered PBS using a 10,000 MWCO Viva spin column by centrifuging at 3220 × *g* at 4 °C.
10. The IgG concentration is determined by the Bicinchoninic acid (BCA) assay (*see* **Note 3**).
11. The purified IgG antibody is then dispensed into clean 1.5 mL micro tubes and stored at −20 °C or, alternately, at −80 °C.

3.2 Purification of Histidine Tag-Labeled scFv and Fab Using Metal-Charged Affinity Resin

Carry out all procedures at 4 °C unless otherwise specified.

1. Add 3 mL of nickel, cobalt, copper, or zinc chelating resin to a 20 mL column (0.5 × 20 cm).
2. Equilibrate the column with 20 mL of running buffer.
3. Apply the filtered (using 0.2 μL filter) lysate supernatant (centrifuged at 20,000 × *g* for 20 min at 4 °C; *see* **Note 4**) to the equilibrated column from **step 2** and mix gently on a roller mixer for 1 h at 4 °C.
4. Allow the column to drain into a 50 mL tube. Collect the "flow-through" and reapply it to the column. Repeat this process three times.
5. Wash the column with 30 mL of running buffer to remove any loosely bound non-specific proteins.
6. Elute the specific scFv or Fab fraction with 8 mL of elution buffer and collect into a 50 mL tube containing 4 mL of filtered neutralization buffer.
7. The neutralized scFv or Fab is then thoroughly buffer exchanged three times against 5 mL of filtered PBS using a 10,000 MWCO Viva spin column by centrifuging at 3220 × *g* at 4 °C.
8. The scFv or Fab concentration is determined by the BCA assay (*see* **Note 3**).
9. The purified antibody is then dispensed into clean 1.5 mL micro tubes and stored at −20 °C or, alternately, at −80 °C.

3.3 Purification of IgY Antibody from Chicken Egg Yolk Using Eggcellent™ IgY Purification Kit

1. Separate the cold egg yolk (pre-stored at 4 °C; one kit contains sufficient reagents to purify IgY from 5 egg yolks) from the egg white using the Eggcellent™ Egg Separator and rinse the yolk with distilled water.
2. Remove adhering egg white carefully by rolling the egg sac onto a clean, dry paper towel.
3. Puncture the egg sac using a pasteur pipette and collect egg yolk in a clean beaker.
4. Record the volume of the egg yolk in the beaker.

5. Slowly add five volumes of cold Eggcellent™ delipidation reagent to the egg yolk and mix well by gentle stirring.
6. Cover the beaker and incubate the egg yolk with Eggcellent™ delipidation reagent at 4 °C for 2–24 h, followed by centrifugation for 15 min at 10,000 × *g* at 4 °C.
7. Measure and record the volume of supernatant, then transfer the supernatant to a clean beaker.
8. Add an equal volume of cold Eggcellent™ IgY precipitation reagent into the supernatant and mix gently for at least 2 min.
9. Cover the beaker and incubate the suspension with Eggcellent™ IgY precipitation reagent at 4 °C for 1 h to overnight, followed by centrifugation for 15 min at 10,000 × *g* at 4 °C.
10. Discard the supernatant and add a volume of PBS equal to the original volume of the egg yolk to the pellet.
11. Dissolve the IgY pellet completely by gentle mixing.
12. The IgY concentration is determined by the BCA assay (*see* **Note 3**).
13. The purified IgY is then dispensed into clean 1.5 mL micro tubes and stored at −20 °C or, alternately, at −80 °C.

3.4 Specific Purification of Antibodies (IgG, IgY, scFv, and Fab) Using Peptide-Antibody Binding-Based Affinity Chromatographic Strategies

For the isolation of anti-peptide antibodies, the peptides may need to be immobilized on a resin prior to use for affinity chromatographic isolation of the associated antibodies. Peptides are synthesized with a biotin moiety or may be biotinylated on the $-NH_2$ groups of lysine or alpha-amino groups using the protocol described below.

3.4.1 Specific Purification of IgG, IgY, scFv, and Fab Using Streptavidin-Agarose

Biotinylation of Peptide Using an EZ-Link® Micro Sulfo-NHS-LC-Biotinylation Kit

1. Prepare peptide in 0.5–2 mL of PBS to a final concentration of 1 mg/mL.
2. Equilibrate one vial of sulfo-NHS-LC-biotin (1 mg, from kit), which is stored at −20 °C, to room temperature for 5–10 min.
3. Add ultrapure water (180 μL) to a vial of equilibrated sulfo-NHS-LC-biotin (from **step 1**) to give a 10 mM biotin reagent solution.
4. Add an appropriate volume (calculated based on the use of following formulae from the EZ-Link® Micro sulfo-NHS-LC-biotinylation kit protocol for volume estimation) of 10 mM biotin reagent solution to the peptide solution from **step 1**:

(a) Calculation of millimoles of sulfo-NHS-LC-biotin to add to the reaction to give a 20-fold molar excess:

$$\mathrm{mL}\left(\mathrm{peptide}\right)\frac{\mathrm{mg}\left(\mathrm{peptide}\right)}{\mathrm{mL}\left(\mathrm{peptide}\right)}\times\frac{\mathrm{mmol}\left(\mathrm{peptide}\right)}{\mathrm{mg}\left(\mathrm{peptide}\right)}\times\frac{20\,\mathrm{mmol}\left(\mathrm{biotin}\right)}{\mathrm{mmol}\left(\mathrm{peptide}\right)}=\mathrm{mmol\ biotin}$$

20 = Recommended molar fold excess of biotin per 1–10 mg/mL protein sample

(b) Calculation of microliters of 10 mM sulfo-NHS-LC-biotin to add to the reaction:

$$\mathrm{mmol}\ (\mathrm{biotin})\times\frac{1{,}000{,}000\ \mu\mathrm{L}}{\mathrm{L}}\times\frac{\mathrm{L}}{10\ \mathrm{mmol}}=\mu\mathrm{L}\ (\mathrm{biotin})$$

Notes the molecular weight of sulfo-NHS-LC-biotin is 577 kDa and the volume of water in which 2.2 mg of sulfo-NHS-LC-biotin is dissolved to make a 10 mM solution is 400 μL.

5. Place the biotin reagent and peptide mixture from **step 4** on ice and incubate for 2 h.
6. The biotinylated peptide is then thoroughly buffer exchanged three times against 5 mL of filtered PBS using a suitable MWCO Viva spin column by centrifugation at 3220 × *g* at 4 °C.
7. The peptide concentration is determined by the BCA assay (*see* **Note 3**).
8. The biotinylated peptide sample is then dispensed into clean 1.5 mL micro tubes and store at −20 °C or, alternately −80 °C for future use.

Specific Purification of IgG, IgY, scFv, and Fab Anti-peptide Antibodies Using Biotin-Streptavidin Interaction: Utilization of Biotin-Conjugated Peptide (Antigen)

Carry out all procedures at 4 °C unless otherwise specified.

1. For sufficient antibody and peptide (antigen) binding, mix two parts of biotinylated peptide with one part of antibody (1–50 mM) in a volume of 500 μL on a roller mixer overnight.
2. Add 2 mL of high capacity streptavidin-agarose resin to a 20 mL column (0.5 × 20 cm).
3. Wash the resin with 10 mL of PBS.
4. Immediately apply the biotinylated peptide/antibody mixture (which was incubated overnight) to the resin before it is dry. Then transfer the resin and peptide/antibody mixture into a clean 2 mL tube and mix gently on a roller mixer for 1 h.
5. Allow the column to drain into a 50 mL tube. Collect the "flow-through" and reapply it to the column. Repeat this process three times.
6. Wash the column with 30 mL of wash buffer to remove any loosely bound non-specific antibody.

7. Elute the specific antibody with 10 mL of elution buffer and collect into a 50 mL tube containing 1 mL of filtered neutralization buffer.
8. The neutralized antibody is then thoroughly buffer exchanged three times against 5 mL of filtered PBS using a 10,000 MWCO Viva spin column by centrifugation at 3220 × *g* at 4 °C.
9. The antibody concentration is determined by the BCA assay (*see* **Note 3**).
10. Dispense the purified antibody into clean 1.5 mL micro tubes and store at −20 °C or, alternately −80 °C.

3.4.2 Specific Purification of IgG, IgY, scFv, and Fab Using a SulfoLink® Immobilization Kit

Covalent Immobilization of Sulfhydryl-Containing Peptides to a Sulfo-Linked Column

1. Column preparation: equilibrate the sulfo-linked column from the SulfoLink® immobilization kit (stored at 4 °C) to room temperature and wash the resin with 2 mL of coupling buffer three times by centrifugation at 1000 × *g* for 1 min.
2. Sample preparation: dissolve or dilute 0.1–1 mg sulfhydryl-containing peptide in 2 mL of coupling buffer. If peptide is oxidized perform tris(2-carboxyethyl)phosphine (TCEP) reduction (add 0.1 mL of 25 mM TCEP to the 2 mL of peptide and incubate mixture at room temperature for 30 min). Save 0.1 mL preparation of peptide for coupling efficiency analysis.
3. Apply 2–3 mL of the peptide in coupling buffer from **step 1** to the prepared column from **step 1** and mix on a roller mixer at room temperature for 15 min.
4. The non-bound peptide is then be removed, by centrifugation at 1000 × *g* for 1 min.
5. Save the "flow-through" from **step 4**. The coupling efficiency can be determined by comparing the peptide concentrations of the "flow-through" to the starting preparation of peptide (*see* **Note 5**).
6. Wash the column with 2 mL of wash solution by centrifugation at 1000 × *g* for 1 min. Repeat this step three times.
7. Wash the column with 2 mL of coupling buffer by centrifugation at 1000 × *g* for 1 min. Repeat this step once.
8. Blocking of the nonspecific binding sites: add 15.8 mg l-cysteine HCl to 2 mL of coupling buffer to give a final concentration of 50 mM cysteine. Apply the cysteine solution to the column and mix for 15 min at room temperature, followed by incubation of the reaction without mixing for 30 min and draining the column.
9. For storage, wash the column three times using 2 mL of degassed PBS buffer with 0.05 % (w/v) sodium azide by centrifugation at 1000 × *g* for 1 min. Then add 2 mL of degassed PBS buffer with 0.05 % (w/v) sodium azide for long-term storage at 4 °C.

Specific Purification of IgG, IgY, scFv, and Fab Using Sulfhydryl-Containing Peptides

Carry out all procedures at 4 °C unless otherwise specified.

1. Remove the storage solution (PBS or PBS with 0.05 % (w/v) sodium azide) by centrifugation at 1000×*g* for 1 min and equilibrate the sulfo-linked column (preparation described in Subheading 3.4.2.1) by washing the column three times using 2 mL of PBS.
2. Add antibody sample (≤2 mL) in PBS to the column and mix on a roller mixer at room temperature for 1 h.
3. Collect the "flow-through" in a 2 mL tube by centrifugation at 1000×*g* for 1 min and reapply it to the column. Repeat this process three times. Save the entire "flow-through" to evaluate binding efficiency and capacity (*see* **Note 6**).
4. To wash the resin, add 2 mL of PBS and centrifuge at 1000×*g* for 1 min. Repeat this step two to four times.
5. Elute the antibody with 2 mL of elution buffer and collect into a 50 mL tube containing 100 μL of neutralization buffer and centrifuge at 1000×*g* for 1 min. Repeat this step two to three times.
6. The neutralized antibody is then thoroughly buffer exchanged three times against 5 mL of filtered PBS using a 10,000 MWCO Viva spin column by centrifuging at 3220×*g* at 4 °C.
7. The antibody concentration is determined by the BCA assay (*see* **Note 3**).
8. Dispense the purified antibody into clean 1.5 mL tubes and stored at −20 °C or, alternately, at −80 °C.

3.4.3 Specific Purification of IgG, IgY, scFv, and Fab Using a CarboxyLink™ Immobilization Kit

Covalent Immobilization of Carboxyl-Containing Peptides to an Amino-Linked Column

1. Column preparation: equilibrate the carboxy-linked column from the CarboxyLinked™ immobilization kit (stored at 4 °C) to room temperature and wash the resin three times with 2 mL of coupling buffer by centrifugation at 1000×*g* for 1 min.
2. Sample preparation: dissolve 1–10 mg of peptide in 2 mL of coupling buffer.
3. Apply 2–3 mL of the peptide sample from **step 2** to the prepared column from **step 1** and mix on a roller mixer at room temperature for 10 min.
4. Add 0.5 mL of the coupling buffer to one vial of EDC (previously equilibrated to room temperature), and immediately add the EDC solution to the sample and resin mixture from **step 3**.
5. Mix the reaction gently on a roller mixer at room temperature for 3 h.
6. Place the column from **step 1** upright and drain the reaction solution from **step 5** into a clean collection tube.
7. Without changing collection tubes, gently add 2 mL of wash buffer to the column and collect all the "flow-through."

The coupling efficiency can be determined by comparing the peptide concentrations of the "flow-through" to the starting preparation of peptide (*see* **Note 7**).

8. Wash the column with 8 mL of wash solution to remove the non-bound peptide.
9. Equilibrate the column by passing 6 mL of PBS buffer through the column.
10. For immediate use of the column, equilibrate with a buffer appropriate for the binding interaction. For storage of the column, equilibrate with PBS containing 0.05 % (w/v) sodium azide, and store the capped column upright at 4 °C.

Specific Purification of IgG, IgY, scFv, and Fab Using Carboxyl-Containing Peptides

Carry out all procedures at 4 °C unless otherwise specified.

1. Remove the storage solution (PBS or PBS containing 0.05 % (w/v) sodium azide) by centrifugation at 1000×g for 1 min and equilibrate the carboxy-linked column (prepared in Subheading 3.4.3.1) by washing the column with 2 mL of PBS three times.
2. Add antibody sample (≤2 mL) in PBS to the column from **step 1** and mix on a roller mixer at room temperature for 1 h.
3. Collect the "flow-through" in a 2 mL tube by centrifugation at 1000×g for 1 min and reapply it to the column. Repeat this process three times. Save the entire "flow-through" to evaluate binding efficiency and capacity (*see* **Note 6**).
4. To wash the resin, add 2 mL of PBS and centrifuge at 1000×g for 1 min. Repeat this step four times.
5. Elute the antibody with 8 mL of elution buffer and collect into a 50 mL tube containing 400 μL of neutralization buffer.
6. The neutralized antibody is then thoroughly buffer exchanged three times against 5 mL of filtered PBS using a 10,000 MWCO Viva spin column by centrifuging at 3220×g at 4 °C.
7. The antibody concentration is determined by the BCA assay (*see* **Note 3**).
8. Dipense the purified antibody into clean 1.5 mL micro tubes and stored at −20 °C or, alternately, at −80 °C.

4 Notes

1. Optimization of imidazole concentration for scFv and Fab purification.

   In order to obtain the ideal purification, the concentration of imidazole needs to be optimized. If a non-specific band is detected, it is suggested that a higher concentration of imidazole (e.g., 25, 30, and 40 mM) should be used; if a weak but pure antibody band is detected, it is suggested a lower concentration of imidazole (e.g., 15 mM) should be used.

2. Recipe for binding and running buffer for scFv and Fab purification.

   Binding buffer: weigh 14.71 g NaCl and 0.68 g imidazole and make it up to 500 mL with PBS.

   Running buffer: add 200 μL of Tween 20 (as Tween 20 is very sticky, cut end of blue tip to aspirate Tween 20 easily) and dissolve in 200 mL of binding buffer.

3. Determination of antibody concentration using the bicinchoninic acid (BCA) protein assay.

   A range of BSA concentrations (0–2000 μg/μL) are prepared, based on the BCA kit protocol, for the protein standard curve. Four dilutions (1/5, 1/10, 1/20, 1/40) are made for unknown samples. Twenty five microliters of BSA standards and unknown sample dilutions (in duplicate) are added into ELISA plate wells. Two hundred microliters of Reagent A and Reagent B mixture (50:1) from the BCA kit are added into the wells and mixed gently on a plate shaker for 30 s. The plate is covered with tinfoil and incubated at 37 °C for 30 min. The plate is placed on a bench to bring it to room temperature. The absorbances of the wells are measured at 562 nm using a Safire 2 plate reader. Concentrations of unknown antibodies are then calculated from the linear range of the standard curve.

4. Preparation for scFv and Fab purification.

   Normally the "freeze and thaw" procedure and sonication are applied to release recombinant antibody from bacterial cells. The temperature for freezing is −80 °C, and for thawing is usually 37 °C. However, a lower temperature (e.g., 16 °C) for thawing is preferred as it prevents antibody degeneration and aggregation. The conditions for sonication also can be optimized, e.g., the amplitude, sonication time, "pulse on/off" time and concentration of imidazole in the sonication buffer.

5. Determination of peptide coupling efficiency of a sulfo-linked column.

   The collected sample (~2 mL) contains the non-bound peptide from the sulfo-linked column preparation. To measure coupling efficiency, compare the absorbance at 280 nm of this solution to the starting peptide sample (~2 mL).

6. Determination of antibody binding efficiency.

   The collected "flow-through" (~2 mL) contains the non-bound antibody from the sulfo-linked or carboxy-linked column. To measure binding efficiency and capacity, compare the absorbance at 280 nm of this solution to the starting antibody sample (~2 mL).

7. Determination of peptide coupling efficiency of a carboxy-linked column.

   The collected sample (~4 mL) contains the non-bound peptide from the carboxy-linked column preparation. To measure coupling efficiency, compare the absorbance at 280 nm of this solution to the starting peptide sample, which is about 2 mL, accounting for the twofold dilution effect.

Acknowledgements

This work is supported by Science Foundation Ireland under CSET Grant No. 05/CE3/B754 and 10/CE/B1821.

References

1. Arora S, Ayyar BV, O'Kennedy R (2014) Affinity chromatography for antibody purification. Methods Mol Biol 1129:497–516
2. McMahon MJ, O'Kennedy R (2000) Polyreactivity as an acquired artefact, rather than a physiologic property, of antibodies: evidence that monoreactive antibodies may gain the ability to bind to multiple antigens after exposure to low pH. J Immunol Methods 241:1–10
3. Darcy E, Leonard P, Fitzgerald J, Danaher M, O'Kennedy R (2011) Purification of antibodies using affinity chromatography. Methods Mol Biol 681:369–382
4. Lund LN, Gustavsson PE, Michael R, Lindgren J, Nørskov-Lauritsen L, Lund M, Houen G, Staby A, St Hilaire PM (2012) Novel peptide ligand with high binding capacity for antibody purification. J Chromatogr A 1225:158–167
5. Sauer-Eriksson AE, Kleywegt GJ, Uhlén M, Jones TA (1995) Crystal structure of the C2 fragment of streptococcal protein G in complex with the Fc domain of human IgG. Structure 3:265–278
6. Grodzki AC, Berenstein E (2010) Antibody purification: affinity chromatography – protein A and protein G Sepharose. Methods Mol Biol 588:33–41
7. Nilson BH, Lögdberg L, Kastern W, Björck L, Akerström B (1993) Purification of antibodies using protein L-binding framework structures in the light chain variable domain. J Immunol Methods 164:33–40
8. Fitzgerald J, Leonard P, Darcy E, O'Kennedy R (2011) Immunoaffinity chromatography. Methods Mol Biol 681:35–59
9. Ma H, Meng J, Wang J, Hearty S, Dolly JO and O'Kennedy R (2014) Targeted delivery of a SNARE protease to sensory neurons using a single chain antibody (scFv) against the extracellular domain of P2X3 inhibits the release of a pain mediator. Biochem J 462:247–256
10. Edupuganti SR, Edupuganti OP, O'Kennedy R, Defrancq E, Boullanger S (2013) Use of T-2 toxin-immobilized amine-activated beads as an efficient affinity purification matrix for the isolation of specific IgY. J Chromatogr B Analyt Technol Biomed Life Sci 923–924:98–101

Chapter 16

Isolation of Camelid Single-Domain Antibodies Against Native Proteins Using Recombinant Multivalent Peptide Ligands

Norah A. Alturki, Kevin A. Henry, C. Roger MacKenzie, and Mehdi Arbabi-Ghahroudi

Abstract

Generation of antibodies against desired epitopes on folded proteins may be hampered by various characteristics of the target protein, including antigenic and immunogenic dominance of irrelevant epitopes and/or steric occlusion of the desired epitope. In such cases, peptides encompassing linear epitopes of the native protein represent attractive alternative reagents for immunization and screening. Peptide antigens are typically prepared by fusing or conjugating the peptide of interest to a carrier protein. The utility of such antigens depends on many factors including the peptide's amino acid sequence, display valency, display format (synthetic conjugate vs. recombinant fusion) and characteristics of the carrier. Here we provide detailed protocols for: (1) preparation of DNA constructs encoding peptides fused to verotoxin (VT) multimerization domain; (2) expression, purification, and characterization of the multivalent peptide-VT ligands; (3) concurrent panning of a non-immune phage-displayed camelid V_HH library against the peptide-VT ligands and native protein; and (4) identification of V_HHs enriched via panning using next-generation sequencing techniques. These methods are simple, rapid and can be easily adapted to yield custom peptide-VT ligands that appear to maintain the antigenic structures of the peptide. However, we caution that peptide sequences should be chosen with great care, taking into account structural, immunological, and biophysical information on the protein of interest.

Key words Single-domain antibody, V_HH, Camelid, Peptide, Phage display, NGS

1 Introduction

Single-domain antibodies (sdAbs or V_HHs), the monomeric antigen-binding variable domains of camelid heavy-chain antibodies, have many desirable therapeutic properties including small size, stability and modularity (reviewed in refs. 1–3). Like monoclonal antibodies, sdAbs can be generated against proteins and other biomolecules using in vitro display technologies to mine the repertoires of naïve and/or immunized animals [4]. However, many targets of therapeutic interest are transmembrane receptors

Gunnar Houen (ed.), *Peptide Antibodies: Methods and Protocols*, Methods in Molecular Biology, vol. 1348,
DOI 10.1007/978-1-4939-2999-3_16, © Springer Science+Business Media New York 2015

that are intrinsically difficult to purify and manipulate in their native conformations [5]. Even for soluble folded proteins, isolating sdAbs against preselected epitopes can be challenging, for instance if the protein is poorly expressed or cytotoxic or if the epitope of interest is sterically occluded [6], antigenically or immunologically subdominant [7], or conformationally labile [6].

In some such cases, peptides have been used successfully as alternatives to conventional recombinant protein antigens for immunization and screening [8]. Unfortunately, peptide antigens are generally weakly immunogenic and often fail to adequately recapitulate the antigenic structures of folded proteins [9], especially in solid-phase binding and panning experiments in which the peptide is nonspecifically adsorbed to a plastic surface. Many factors contribute to a peptide's ability to form B-cell epitopes and select binding sdAbs from phage-displayed libraries, such as: (1) its primary amino acid sequence [10]; (2) its method of synthesis and level of purity (for synthetic peptides; [11]); (3) context-dependent structuring effects of chemical cross-linkers (for synthetic peptide conjugates) or surrounding amino acid sequences (for recombinant fusions; [12]); (4) display valency [13]; and (5) its phase, especially the immobilization strategy in solid phase [14]. Recombinant display of peptide antigens fused to a carrier protein has several important advantages over chemical conjugation of synthetic peptides, including low cost, ease of production and uniformity of peptide valency and structure, and also circumvents solubility issues of some synthetic peptide sequences. Even peptides whose three-dimensional structures are indistinguishable from the corresponding epitope on native proteins, however, have the potential to select topologically restricted sdAbs that do not cross-react with native protein [15]. Thus, isolation of sdAbs that bind to both a given peptide epitope and the cognate native protein can be very technically challenging.

In this chapter, we describe a set of protocols for isolating sdAbs from phage-displayed libraries against desired epitopes on a folded protein using biotinylated pentavalent peptide antigens produced in *Escherichia coli*. We include detailed methods for: (1) peptide selection, (2) preparation and characterization of the recombinant peptide antigens, (3) construction and panning of phage-displayed sdAb libraries, and (4) next-generation sequencing (NGS) approaches to detect subtle co-enrichment of sdAb-bearing phage after panning on peptides and whole protein. These protocols are presented as a case study of a model antigen, IGFBP7, but should be amenable to modification for other peptides and proteins. While the methods described in this chapter were developed primarily for screening sdAb libraries, we expect that the peptide reagents described here may be useful for immunization as well.

2 Materials

Prepare all solutions using ultrapure water (prepared by purifying deionized water to attain a sensitivity of 18 MΩ cm at 25 °C) and analytical grade reagents. Prepare and store all reagents at room temperature (unless indicated otherwise). Diligently follow all waste disposal regulations when disposing waste materials. All oligonucleotides listed in this protocol were purchased commercially desalted but with no additional purification unless otherwise indicated.

2.1 Peptide Selection

See Subheading 3.1.

2.2 Preparation of VT-Peptide DNA Constructs

1. *E. coli* TG1 electrocompetent cells.
2. pVT2-HH expression vector (a variant of pVT2 [16] modified to encode a human hinge region).
3. *E. coli* AVB101 cells harboring the pACYC184 plasmid.
4. QIAprep® spin miniprep kit.
5. Primer BioT5R (5′-GATCTGGGCAGCTCGGGTCTTAAT GATATTTTTGAAGCTCAGAAGATTGAATGGCA TGAAGGAGGTGGGTCCGAAAATCTGTA| TTTTCAGGGCCATCACCATCACCATCACTAGT GAA-3′), 5′ phosphorylated.
6. Primer BioT6F (5′-AGCTTTCACTAGTGATGGTGATGG TGATGGCCCTGAAAATACAGATTTTCGGACCC ACCTCCTTCATGCCATTCAATCTTCTGAGCTTC AAAAATATCATTAAGACCCGAGCTG CCCA-3′), 5′ phosphorylated.
7. Primer PBio9RP (5′-CAGGCCGCGCTGCATGAAATTCC GGTGAAAAAAGGCGAAGGCGCGGAACTGGGGCC-3′), 5′ phosphorylated.
8. Primer PBio10FP (5′-CCAGTTCCGCGCCTTCGCCTTT TTTCACCGGAATTTCATGCAGCGCGG-3′).
9. Primer PBio11RP (5′-CAGGCCCAGGCGAGCGCGAGC GCGAAAATTACCGTGGTGGATGCGCTGCAT GAAATTCCGGTGAAAGGGCC-3′), 5′ phosphorylated.
10. Primer PBio12FP (5′-CTTTCACCGGAATTTCATGCAGCG CATCCACCACGGTAATTTTCGCGCTCGCGC TCGCCTGGG-3′), 5′ phosphorylated.
11. Sterile ultrapure Milli-Q H_2O.
12. LigaFast™ Rapid DNA Ligation System.
13. Fermentas FastDigest® restriction enzymes *Apa*I, *Bbs*I, *Bgl*II, and *Hin*dIII.
14. QIAquick® gel extraction kit.

15. Agarose gel electrophoresis equipment and power supply.
16. Heating block.
17. 0.5 M ethylenediaminetetraacetic acid (EDTA) stock solution (per L): 186.1 g EDTA-$Na_2 \cdot 2H_2O$ in ultrapure H_2O. Adjust pH to 8.0 with NaOH and store at room temperature.
18. 50× TAE buffer (per L): 242 g (2 M) Tris base, 57.1 mL glacial acetic acid, 100 mL 0.5 M EDTA solution in ultrapure H_2O. Store at room temperature. Dilute 1:50 in H_2O to prepare 1× TAE buffer.
19. 1 % (w/v) agarose gel, prepared in 1× TAE buffer.
20. Disposable electroporation cuvettes, 0.2 cm gap width.
21. MicroPulser™ electroporator or similar instrument.
22. 2 M glucose (per L): 360 g glucose dissolved in ultrapure H_2O. Sterilize by passing through a 0.22 μm filter and store at room temperature.
23. SOC broth (per L): 20 g tryptone, 5 g yeast extract (BD), 0.58 g NaCl (ACP), 0.19 g KCl (ACP), 0.95 g $MgCl_2$ (ACP), 1.2 g $MgSO_4$ (EMD) dissolved in ultrapure H_2O. Sterilize by autoclaving, cool to ~50 °C then add 10 mL 2 M glucose solution.
24. 100 mg/mL ampicillin disodium salt (Amp) stock solution prepared in ultrapure water. Sterilize by passing through a 0.22 μm filter and store at −20 °C.
25. 10 mg/mL chloramphenicol (Cam) stock solution prepared in ethanol. Store at −20 °C in the dark (*see* **Note 1**).
26. LB broth or agar (per L): 10 g tryptone, 5 g yeast extract, 10 g NaCl dissolved in ultrapure H_2O. For agar plates, add 15 g agar (BD). Sterilize by autoclaving, cool to ~50 °C then add filter-sterilized ampicillin and/or chloramphenicol to a final concentration of 100 μg/mL and 34 μg/mL, respectively.
27. 37 °C incubator with shaker.
28. ND-1000 spectrophotometer or similar instrument.
29. GeneAmp® PCR System 9700 thermal cycler or similar instrument.
30. DNA Sequencer, or access to a commercial DNA sequencing service.
31. Primer M13RP (5′-CAGGAAACAGCTATGAC-3′).
32. Primer M13FP (5′-GTAAAACGACGGCCAGT-3′).
33. dNTP mix, 10 mM each.
34. Taq DNA polymerase and 10× PCR reaction buffer containing $MgCl_2$.
35. 0.22 μm Steriflip-GP and Stericup-GP Express® PLUS membrane filters.

36. T4 DNA ligase and ligase buffer.
37. 70 % (v/v) ethanol.
38. Kimwipes.
39. Polypropylene 50 mL Oakridge centrifuge tubes.
40. Pasteur pipettes.
41. Whatman filter paper.

2.3 Expression and Purification of VT-Peptide Fusion Proteins

1. LB agar (*see* Subheading 2.2, **item 26**).
2. 2 M glucose (*see* Subheading 2.2, **item 22**).
3. 100 mg/mL Amp (*see* Subheading 2.2, **item 24**).
4. 1 M $MgSO_4$ solution (per L): 120.4 g $MgSO_4$ dissolved in ultrapure H_2O. Sterilize by passing through a 0.22 μm filter and store at room temperature.
5. 1.5 M KH_2PO_4 solution (per L): 204.1 g KH_2PO_4 dissolved in ultrapure H_2O. Adjust pH to 7.2 with HCl, sterilize by passing through a 0.22 μm filter and store at room temperature.
6. 2× YT broth (per liter): 16 g tryptone, 10 g yeast extract, 5 g NaCl dissolved in ultrapure H_2O. Sterilize by autoclaving, cool to ~50 °C then supplement with 100 μg/mL Amp and/or 34 μg/mL Cam as appropriate.
7. 2× YT + broth (per liter): 16 g tryptone, 10 g yeast extract, 5 g NaCl dissolved in ultrapure H_2O. Sterilize by autoclaving, cool to ~50 °C then supplement with 5 mM glucose, 50 mM KH_2PO_4 and 5 mM $MgSO_4$. Supplement with 100 μg/mL Amp and/or 34 μg/mL Cam as appropriate.
8. 1 M L-arabinose solution (per L): 150.1 g L-arabinose dissolved in ultrapure H_2O. Sterilize by passing through a 0.22 μm filter and store at room temperature.
9. 5 mM D-biotin solution (per L): 0.12 g D-biotin dissolved in warm 1 M bicine buffer solution, pH 8.3. Sterilize by passing through a 0.22 μm filter and store at 4 °C.
10. 37 °C incubator with shaker.
11. J2-21 M/E high-speed centrifuge or similar instrument.
12. GENESYS™ 20 visible spectrophotometer or similar instrument.
13. 1 M isopropyl-β-D-thio-galactopyranoside (IPTG) solution (per liter): 238.3 g IPTG dissolved in ultrapure H_2O. Sterilize by passing through a 0.22 μm filter and store at −20 °C.
14. 2 mg/mL chicken egg lysozyme stock solution prepared in ultrapure water. Store at −20 °C.
15. Sterile ultrapure Milli-Q H_2O.

16. Lysis buffer (per L): 6.1 g (50 mM) Tris base, 1.46 g (25 mM) NaCl dissolved in ultrapure H_2O. Adjust pH to 8.0 with HCl and store at room temperature. Supplement with one complete EDTA-free protease inhibitor tablet per 100 mL buffer immediately prior to use.
17. DNase I, resuspended at 15 U/μL in 10 mM Tris–HCl pH 7.6 containing 2.5 mM $MgCl_2$.
18. Sorvall RT6000B refrigerated high-speed centrifuge with swinging bucket rotor or similar instrument.
19. 0.22 μm Steriflip-GP and Stericup-GP Express® PLUS membrane filters.
20. 5 mL HiTrap™ chelating HP column.
21. 5 mg/mL $NiCl_2$ solution, prepared in ultrapure H_2O and 0.22 μm sterile filtered.
22. Buffer A (per L): 2.38 g (10 mM) 4-(2-hydroxyethyl)-1-piperazineethanesulfonic acid (HEPES), 0.68 g (10 mM) imidazole, 29.2 g (500 mM) NaCl dissolved in ultrapure H_2O. Adjust pH to 7.5 with NaOH, sterilize by passing through a 0.22 μm filter and store at room temperature
23. Buffer B (per L): 2.38 g (10 mM) HEPES, 34 g (500 mM) imidazole, 29.2 g (500 mM) NaCl dissolved in ultrapure H_2O. Adjust pH to 7.5 with NaOH, sterilize by passing through a 0.22 μm filter and store at room temperature.
24. ÄKTAxpress fast protein liquid chromatography purification system (FPLC) or similar instrument.
25. 3.5 kDa MWCO dialysis tubing.
26. ND-1000 spectrophotometer or similar instrument.
27. Phosphate-buffered saline (PBS; per L): 8 g (137 mM) NaCl, 0.2 g (2.37 mM) KCl, 1.44 g (10 mM) Na_2HPO_4, 0.24 g (1.8 mM) KH_2PO_4 dissolved in ultrapure H_2O. Adjust pH to 7.4 with HCl, sterilize by autoclaving and store at room temperature (*see* **Note 2**).
28. 500 mL polypropylene centrifuge bottles.

2.4 Characterization of VT-Peptide Fusion Proteins

1. 30 % (w/v) acrylamide–bis-acrylamide (29:1) solution.
2. 10 % (w/v) sodium dodecyl sulfate (SDS) solution prepared in ultrapure H_2O. Store at room temperature.
3. 10 % (w/v) ammonium persulfate (APS) solution prepared in ultrapure H_2O. Store at 4 °C.
4. Sterile ultrapure Milli-Q H_2O.
5. UltraPure™ *N*,*N*,*N*′,*N*′-tetramethylethylenediamine (TEMED).
6. 1.5 M Tris–HCl buffer (per L): 181.7 g Tris base dissolved in ultrapure H_2O. Adjust pH to 8.8 with HCl and store at room temperature.

7. 0.5 M Tris–HCl buffer (per L): 60.6 g Tris base dissolved in ultrapure H_2O. Adjust pH to 8.8 with HCl and store at room temperature.
8. 10× sodium dodecyl sulfate–polyacrylamide gel electrophoresis (SDS-PAGE) running buffer (per L): 30.4 g (250 mM) Tris base, 141.8 g (1.9 M) glycine, 10 g (35 mM) SDS dissolved in ultrapure H_2O. Dilute to 1× with H_2O before use.
9. Isopropanol.
10. Precision Plus Protein™ prestained molecular weight standards.
11. 4× protein sample loading buffer (per 10 mL): 4 mL glycerol, 0.4 g SDS, 4 mL 1.5 M Tris–HCl pH 8.8, 0.8 mL 0.1 % (w/v) bromophenol blue solution. Adjust final volume to 10 mL with ultrapure H_2O and add 50 μL β-mercaptoethanol per 1 mL buffer immediately prior to use.
12. SDS-PAGE equipment and power supply.
13. Heating block or water bath.
14. Coomassie blue stain (per 200 mL): 50 mL isopropanol, 20 mL glacial acetic acid, 50 mg Coomassie brilliant blue. Adjust final volume to 200 mL with ultrapure H_2O.
15. Destaining buffer (per 200 mL): 20 mL glacial acetic acid, 60 mL methanol, 120 mL ultrapure H_2O.
16. Immobilon-P® polyvinylidene difluoride (PVDF) membrane.
17. 10× transfer buffer (per L): 30.2 g (250 mM) Tris base, 72 g (0.96 M) glycine, 20 % (v/v) methanol dissolved in ultrapure H_2O.
18. Trans-Blot® semi-dry electrophoretic transfer cell, or similar instrument.
19. PBS-T: PBS (*see* Subheading 2.3, **item 27**) supplemented with 0.1 % (v/v) Tween-20.
20. PBS-M: PBS (*see* Subheading 2.3, **item 27**) supplemented with 2 % (w/v) skim milk powder.
21. Mouse anti-6×His tag® antibody [HIS-1], alkaline-phosphatase (AP) conjugated.
22. AP-conjugated streptavidin.
23. HiLoad™ 16/600 Superdex™ 200 column.
24. ÄKTA FPLC system.
25. HEPES-buffered saline (HBS)-EP buffer (per L): 2.4 g (10 mM HEPES), 8.8 g (150 mM) NaCl, 6.9 mL 0.5 M EDTA stock solution (3.44 mM), 0.005 % (v/v) P20 surfactant in ultrapure H_2O. Adjust pH to 7.4 with NaOH, sterilize by passing through a 0.22 μm filter and store at room temperature.
26. Methanol.

2.5 Preparation and Panning of Phage-Displayed sdAb Library

1. LAC-M naïve sdAb library [17–19] or other naïve or immune sdAb library in an appropriate phage or phagemid vector. The library phage should be polyethylene glycol (PEG)-purified and at a titer of ~10^{12} cfu/mL.
2. NUNC MaxiSorp™ plates.
3. Parafilm.
4. Streptavidin.
5. Paper towels.
6. StartingBlock (PBS) blocking buffer.
7. Purified VT-peptide ligands and native IGFBP7 protein.
8. Purified VT-irrelevant protein fusion protein as a negative panning control.
9. PBS (*see* Subheading 2.3, **item 27**).
10. PBS-T (*see* Subheading 2.4, **item 18**).
11. 0.22 μm syringe-driven filters.
12. 10 mL syringes.
13. PEG solution (per L): 200 g (20 % w/v) PEG (average molecular weight 6000 or 8000 Da), 146.1 g (2.5 M) NaCl dissolved in warm ultrapure H_2O. Sterilize by autoclaving and store at room temperature (*see* **Note 3**).
14. 50 mL Falcon tubes.
15. 1.5 mL Eppendorf tubes.
16. 100 mM triethylamine (TEA) solution (per 100 mL): 3.6 mL TEA in ultrapure H_2O. Prepare fresh each day and store at room temperature.
17. 1 M Tris–HCl buffer (per L): 121.1 g Tris base dissolved in ultrapure H_2O. Adjust pH to 7.4 with HCl, sterilize by autoclaving and store at room temperature.
18. 12.5 μg/mL tetracycline (Tet) hydrochloride stock solution prepared in 95 % ethanol. Store the solution in a light-tight container at −20 °C.
19. 2× YT-Tet broth and agar (per liter): 16 g tryptone, 10 g yeast extract, 5 g NaCl dissolved in ultrapure H_2O. For agar plates, add 15 g agar per L of media. Sterilize by autoclaving, cool to ~50 °C then supplement with 5 mM glucose and 12.5 μg/mL Tet.
20. Large and small diameter Petri plates.
21. *E. coli* TG1 cells.
22. Vortex.
23. 37 °C and 32 °C incubator with shaker.
24. Refrigerated 4 °C microcentrifuge.
25. U-bottom 1.5 mL Eppendorf tubes.
26. LB broth (*see* Subheading 2.2, **item 26**).

2.6 Next-Generation DNA Sequencing

1. Primer NGS-MJ7 (5′-CGCTCTTCCGATCTCTGNNNNN GCCCAGCCGGCCATGGCC-3′).
2. Primer NGS-MJ8 (5′-TGCTCTTCCGATCTGACNNNNN NTGAGGAGACGGTGACCTGG-3′)
3. Primer P5-seqF (5′-AATGATACGGCGACCACCGAGATCT ACACTCTTTCCCTACACGACGCTCTTCCGATCT CTG-3′), HPLC-purified.
4. Primer P7-index1-seqR (5′-CAAGCAGAAGACGGCATACG AGATCGTGATGTGACTGGAGTTCAGACGTGT GCTCTTCCGATCTGAC-3′), HPLC-purified.
5. Primer P7-index2-seqR (5′-CAAGCAGAAGACGGCATAC GAGATACATCGGTGACTGGAGTTCAGACGT GTGCTCTTCCGATCTGAC), HPLC-purified.
6. Primer P7-index3-seqR (5′-CAAGCAGAAGACGGCATACG AGATGCCTAAGTGACTGGAGTTCAGACGTG TGCTCTTCCGATCTGAC-3′), HPLC-purified.
7. Primer P7-index4-seqR (5′-CAAGCAGAAGACGGCATACG AGATTGGTCAGTGACTGGAGTTCAGACGTGT GCTCTTCCGATCTGAC-3′), HPLC-purified.
8. AmpliTaq Gold DNA polymerase with buffer II.
9. dNTP mix, 10 mM each.
10. GeneAmp® PCR System 9700 thermal cycler or similar instrument.
11. 0.2 mL PCR tubes, strips or plates.
12. Agarose gel electrophoresis equipment and power supply.
13. 50× TAE buffer: *see* Subheading 2.2, **item 18**.
14. 1 % (w/v) agarose gel, prepared in 1× TAE buffer.
15. PureLink® PCR purification kit.
16. 1.5 mL Eppendorf tubes.
17. Phusion® high-fidelity DNA polymerase.
18. QIAquick® gel extraction kit.
19. Dark Reader® transilluminator or similar instrument.
20. GelGreen™ nucleic acid gel stain.
21. AMPure XP beads.
22. 70 % (v/v) ethanol.
23. Sterile ultrapure Milli-Q H_2O.
24. ND-1000 spectrophotometer or similar instrument.
25. MiSeq Sequencing System.
26. 500-cycle Reagent Kit V2.

3 Methods

3.1 Peptide Selection

We recommend choosing peptide sequences based on the following general criteria: (1) length between 10 and 30 amino acids; (2) an overall hydrophilic character as predicted using any of a number of freely available online tools [20]; (3) ≤50 % hydrophobic residue (Leu, Val, Ile, Met, Phe, Trp) content; (4) surface accessibility and for transmembrane receptors, location within extracellular loops or domains, as shown either by structural data (where available) or by computationally predicted secondary structure, solvent accessibility and tertiary structure by homology modeling; and (5) where epitope mapping data are available, peptides can be chosen to contain known linear B-cell epitopes of interest.

3.2 Preparation of VT-Peptide DNA Constructs

To produce the pVT2-HH-Bio vector (Fig. 1a), the parental vector pVT2-HH must be modified to introduce a biotin acceptor peptide (BAP; GLNDIFETQKIEWHE; [21]) and an AcTEV protease cleavage site between the *Bgl*II and *Hind*III restriction sites, immediately downstream of the VT multimerization domain.

1. Digest pVT2-HH with *Bgl*II and *Hind*III for 20 min at 37 °C using the following conditions: 1 μg pVT2-HH DNA, 1 μL each enzyme, 2 μL 10× FastDigest® buffer, 20 μL final volume. Purify the digested DNA using a QIAquick® gel extraction kit and elute in 30 μL ultrapure H_2O. Verify complete digestion by agarose gel electrophoresis.

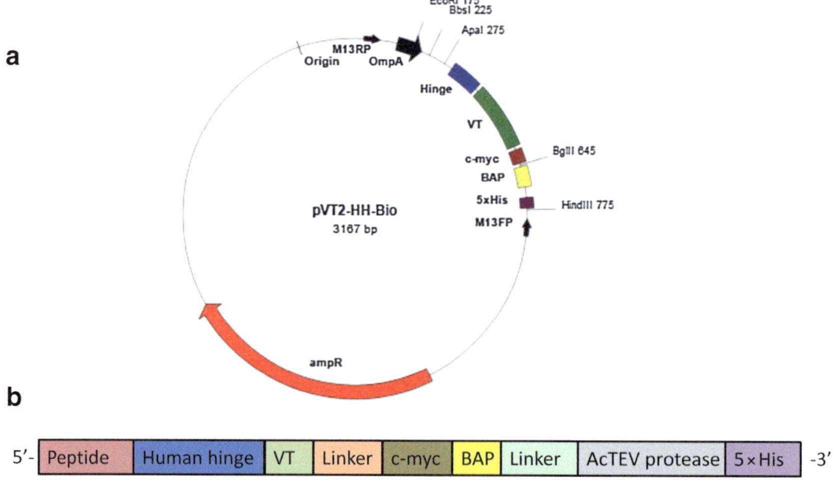

Fig. 1 Overview of pVT2-HH-Bio vector construction. (**a**) Map of pVT2-HH-Bio expression vector. (**b**) Schematic representation of DNA constructs encoding VT-peptide ligands, located between the *Bbs*I and *Hind*III sites of pVT2-HH-Bio

2. Mix 10 pmol of each phosphorylated oligonucleotide BioT5R and BioT6F in 80 μL ultrapure H_2O, boil at 100 °C for 10 min then cool to room temperature to allow annealing of the complementary DNA strands.
3. Ligate 1 μL of the annealed biotin acceptor peptide DNA with 90 ng of digested pVT2-HH vector for 1 h at room temperature in a 10-μL ligation reaction containing 1 U T4 DNA ligase and 1× ligase buffer.
4. Transform 80-μL aliquots of electrocompetent *E. coli* TG1 cells with 3 μL of the ligation mixture. Briefly, thaw cells and add the ligation mixture, then place in prechilled electroporation cuvettes and electroporate at 1800 V (25 μF and 200 Ω) using a MicroPulser™ or similar instrument. The pulse time should be ~ 5 ms. Allow cells to recover at 37 °C in 1 mL SOC for 1 h with gentle shaking. Plate serial dilutions of transformed cells on LB-amp plates and incubate at 37 °C overnight.
5. Screen single bacterial colonies for the presence of inserts by colony PCR in 20-μL reactions containing 2 μL 10× buffer containing $MgCl_2$, 0.4 μL 10 mM dNTP mix, 0.2 μL each of M13RP and M13FP primers (10 pmol/μL), 0.2 μL of Taq DNA polymerase. Touch a pipette tip to each colony, then drop the tip into the PCR reaction and leave for ~2 min, then cycle as follows: 94 °C for 5 min; 30 cycles of (94 °C for 30 s, 55 °C for 30 s, 72 °C for 1 min); 72 °C for 7 min.
6. Electrophorese 5 μl of the amplicons generated in **step 5** in a 1 % (w/v) agarose gel in TAE buffer. The size of the amplicon should be approximately 654 bp for positive clones and 530 bp for the parental vector.
7. Isolate plasmid DNA from 2 mL LB-Amp overnight cultures of positive clones using a QIAprep® spin miniprep kit and verify the sequences of positive clones using both M13RP and M13FP primers (sense and antisense strands, respectively).

Next, oligonucleotides encoding peptides of interest can be cloned between the *Bbs*I and *Apa*I restriction sites of pVT2-HH-Bio, immediately upstream of the human hinge preceding the VT multimerization domain. A diagram of the final secreted peptide-VT fusion protein is shown in Fig. 1b.

1. Digest pVT2-HH-Bio with *Bbs*I and *Apa*I for 20 min at 37 °C using the following conditions: 1 μg pVT2-HH-Bio DNA, 1 μL each enzyme, 2 μL 10× FastDigest® buffer, 20 μL final volume. Purify the digested DNA using a QIAquick® gel extraction kit and elute in 30 μL ultrapure H_2O. Verify complete digestion by agarose gel electrophoresis.
2. Boil 10 pmol of each phosphorylated oligonucleotide pair encoding IGFBP7 peptides (PBio9RP and PBio10FP;

PBio11RP and PBio12FP) in 80 μL ultrapure H_2O at 100 °C for 10 min then cool to room temperature to anneal.

3. Ligate 1 μL of the annealed IGFBP7 peptide DNA with 90 ng of pVT2-HH-Bio vector for 1 h at room temperature in a 10-μL ligation reaction containing 1 U of T4 DNA ligase and 1× ligase buffer.
4. Transform 80-μL aliquots of electrocompetent *E. coli* TG1 cells with 3 μL of the ligation mixture. Briefly, thaw cells and add the ligation mixture, then place in prechilled electroporation cuvettes and electroporate 1800 V (25 μF and 200 Ω) using a MicroPulser™ or similar instrument. The pulse time should be ~5 ms. Allow cells to recover at 37 °C in 1 mL SOC for 1 h with gentle shaking. Plate serial dilutions of transformed cells on LB-amp plates and incubate at 37 °C overnight.
5. Screen single bacterial colonies for the presence of inserts by colony PCR in 20 μL reactions containing 2 μL 10× buffer containing $MgCl_2$, 0.4 μL 10 mM dNTP mix, 0.2 μL each of M13RP and M13FP primers (10 pmol/μL), 0.2 μL of Taq DNA polymerase. Touch a pipette tip to each colony, then drop the tip into the PCR reaction and leave for ~2 min, then cycle as follows: 94 °C for 5 min; 30 cycles of (94 °C for 30 s, 55 °C for 30 s, 72 °C for 1 min); 72 °C for 7 min.
6. Electrophorese 5 μL of the amplicons generated in **step 5** in a 1 % (w/v) agarose gel in TAE buffer. The size of the amplicon should be approximately 714 bp for positive clones and 654 bp for the parental vector.
7. Isolate plasmid DNA from 2 mL LB-Amp overnight cultures of positive clones using a QIAprep® spin miniprep kit and verify the sequences of positive clones using both M13RP and M13FP primers (sense and antisense strands, respectively). The final constructs are henceforth referred to as VT-peptide 3 and VT-peptide 4.
8. Isolate pACYC184 plasmid DNA from *E. coli* AVB101 cells using a QIAprep® spin miniprep kit and elute in 30 μL EB buffer.
9. Co-transform 50-μL aliquots of electrocompetent *E. coli* TG1 cells simultaneously with both recombinant pVT2-HH-Bio and pACYC184 plasmids and plate on LB-Amp-Cam. Briefly, thaw cells and add 1–10 ng each of VT-peptide-HH-Bio and pACYC184 plasmid, then place in prechilled electroporation cuvettes and electroporate at 1800 V (with 25 μF and 200 Ω) using a MicroPulser™ or similar instrument. The pulse time should be ~ 5 ms. Allow cells to recover at 37 °C in 1 mL SOC for 1 h with gentle shaking. Plate serial dilutions of transformed cells on LB-Amp-Cam plates and incubate at 37 °C overnight.

3.3 Expression and Purification of VT-Peptide Fusion Proteins

His-tagged VT-peptide fusion proteins can now be inducibly expressed, in vivo biotinylated and purified from bacterial cytoplasmic using standard immobilized metal affinity chromatography (IMAC) techniques.

1. Inoculate 10 mL 2× YT-Amp-Cam starter cultures with single colonies harboring both pVT2-HH-Bio (bearing recombinant peptide inserts) and pACYC184 plasmids. Supplement with 1 % (w/v) glucose and grow overnight at 37 °C with 220 rpm shaking.
2. The next day, transfer 10 mL of starter culture into 1 L of 2× YT+-Amp-Cam and grow at 37 °C with continuous 220 rpm shaking to an OD_{600} of ~0.8.
3. Induce expression of the peptide-VT fusion protein from pVT2-HH-Bio and biotin ligase from pACYC184 by adding IPTG and L-arabinose to final concentrations of 1 mM and 1.5 μM, respectively. Supplement the cultures with 50 μM D-biotin and grow overnight at 28 °C with continuous 220 rpm shaking.
4. The next day, move cultures to 500 mL centrifuge bottles and pellet cells by centrifuging at 6000 rpm (~4000 × *g*) in a JA-10 rotor of a Beckman J2-21 M/E centrifuge for 10 min at 4 °C.
5. Resuspend the harvested cells in 100 mL of ice cold lysis buffer supplemented with one complete EDTA-free protease inhibitor tablet and store at −20 °C overnight.
6. The next day, thaw the harvested cells in lysis buffer at room temperature with occasional shaking or vortexing.
7. Lyse the cells by adding 5 mL of freshly prepared 2 mg/mL lysozyme at room temperature for 30–60 min with occasional shaking or vortexing.
8. Once the lysate becomes viscous, add 300 μL of DNase I (15 U/μL) incubate at room temperature for approximately 20–30 min, until the suspension became more fluid.
9. Move the lysate to oakridge tubes and centrifuge at 12,000 rpm (~15,000 × *g*) in a JA-17 rotor of a Beckman J2-21 M/E centrifuge for 30 min at 4 °C (*see* **Note 4**).
10. Pass the supernatant (soluble fraction containing biotinylated peptide-VT fusion proteins) through a 0.22 μm sterile filter.
11. Dialyze the filtered lysate containing VT-peptide fusion proteins against Buffer A overnight. Use 3 L buffer per 10 mL filtered lysate.
12. Purify the soluble VT-peptide fusion using a 5 mL HiTrap™ chelating HP column on an ÄKTA FPLC system. Briefly, charge the column with 30 mL of 5 mg/mL $NiCl_2$, wash with 15 mL ultrapure H_2O, then calibrate with 15 mL of Buffer

A. Load the filtered lysates at a flow rate of 1 mL/min. Wash the column with Buffer A to remove proteins nonspecifically bound to the column. Once a stable baseline is reached, elute the peptide-VT ligands using a linear 10–500 mM imidazole gradient and collect fractions.

13. Assess the concentration and purity of eluted peptide-VT ligand-containing fractions by SDS-PAGE (*see* Subheading 3.4 below).
14. Pool and dialyze the IMAC fractions containing VT-peptide fusion proteins against PBS supplemented with 3 mM EDTA overnight. Use 3 L buffer per 10 mL IMAC fraction.
15. Measure the absorbance of the pooled protein fractions at 280 nm and use the Expasy Protparam tool [22] to calculate the theoretical extinction coefficient, ε, of each VT-peptide fusion (*see* **Note 5**).
16. Store the purified VT-peptide fusion proteins at 4 °C (*see* **Note 6**).

3.4 Characterization of VT-Peptide Fusion Proteins

Next, verify successful biotinylation as well as the purity and aggregation status of the VT-peptide ligands using SDS-PAGE, Western blotting and size exclusion chromatography (SEC).

3.4.1 SDS-PAGE

1. Clean SDS-PAGE glass plates using 70 % ethanol and Kimwipes and assemble the apparatus.
2. Prepare two 12.5 % resolving gels as follows: 3.1 mL of 30 % acrylamide, 3 mL of 1.5 M Tris–HCl, 1.3 mL of H_2O, 50 μL of 10 % SDS, 36 μL of 10 % APS, and 5 μL TEMED. Invert the tube to mix gently, add between glass plates using a Pasteur pipette to desired level, and then add a layer of isopropanol on top of acrylamide solution.
3. After the resolving gel has polymerized, remove the isopropanol and any excess moisture with Whatman filter paper. Prepare a 7 % stacking gel as follows: 1 mL of 30 % acrylamide, 630 μL of 0.5 M Tris–HCl, 3.6 mL H_2O, 25 μL of 10 % SDS, 25 μL of 10 % APS, and 5 μL TEMED. Invert the tube to mix gently, add between glass plates using a Pasteur pipette and insert a comb into the top of the apparatus.
4. Add running buffer to SDS-PAGE tank and check for leaks.
5. For each peptide-VT protein sample, add protein sample loading dye to a final concentration of 1×, boil at 95–100 °C for 5 min, then centrifuge briefly to collect.
6. Electrophorese at 100–150 V (20–30 mA) for ~1 h, until the dye front has almost migrated to the bottom of the gel. Ensure that the molecular weight markers separate when the current is flowing.

7. Stain the first gel with Coomassie blue for 1 h at room temperature, then destain until the bands can be seen clearly (Fig. 2c, *inset*).

3.4.2 Western Blotting

1. Activate the Immobilon™ PVDF membrane in methanol for 5 min then wash four times with PBS to remove traces of methanol. Immerse the second gel, two pieces of Whatman filter paper cut to size, and Immobilon™ PVDF membrane in transfer buffer for 5–10 min. Assemble the transfer sandwich on the transfer cell (from bottom): filter paper, PVDF membrane, gel, filter paper. Transfer at 15 V for 20 min.
2. Wash the membrane four times with PBS, then block in PBS-M for 30–60 min at room temperature.

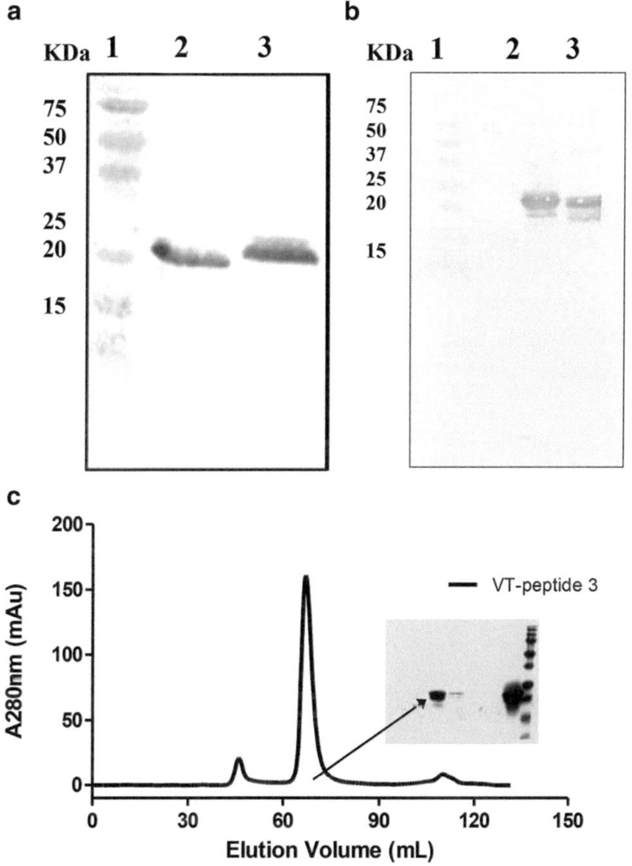

Fig. 2 Characterization of VT-peptide ligands. (**a** and **b**) Western blotting of in vivo biotinylated VT-peptide ligands. His-tagged VT-peptide 3 and VT-peptide 4 ligands were separated by SDS-PAGE, transferred to PVDF membranes then probed with AP-conjugated anti-6×His (**a**) and AP-conjugated streptavidin (**b**). *Lane 1*: molecular weight standards; *Lane 2*: VT-peptide 3 ligand; *Lane 3*: VT-peptide 4 ligand. (**c**) Size exclusion chromatography profiles of VT-peptide 3 ligand. Monomeric VT-peptide ligand 3 was eluted at a volume of ~66 mL

3. Wash the membrane three times with PBS, and then add 10 mL of primary antibody (mouse anti-6×His:AP) diluted 1:1000 in PBS-M for 1 h at room temperature (Fig. 2a). Alternatively, use streptavidin:AP diluted 1:1000 in PBS-M to directly detect the biotinylated VT-peptide fusion (Fig. 2b) and proceed directly to **step 14** (*see* **Note 7**).
4. Wash the membrane three times for 10 min with PBS-T.
5. Wash the membrane five times for 5–10 min with PBS-T.
6. Add 100 μL each of solution A and solution B in 10 mL of AP substrate buffer to the membrane to develop. During this time (approximately 20 min), keep the membrane in the dark but check on it occasionally.
7. When bands appear, stop the reaction by rinsing the membrane with water (Fig. 2a, b).

3.4.3 Size Exclusion Chromatography

1. Wash the Superdex ™ 200 16/60 gel filtration column with two column volumes of filtered and degassed ultrapure H_2O and then equilibrate it with two column volumes of filtered and degassed HBS-EP at a flow rate of 0.5 mL/min.
2. Inject and elute 200 μL of purified VT-peptide ligands, and observe the chromatogram.

3.5 Preparation and Panning of Phage-Displayed sdAb Library

Phage-displayed sdAb libraries are panned separately against VT-peptide ligands (Subheading 3.5.1) and native IGFBP7 protein (Subheading 3.5.2). Although they are presented separately for clarity, both panning experiments may be performed simultaneously.

3.5.1 Panning Against VT-Peptide Ligands

1. Preadsorb 100 μL (~10^{12} cfu/mL) of phage library in solution containing 50 μg VT-irrelevant fusion protein and 20 μg streptavidin in a final volume of 200 μL 0.1 % StartingBlock™. Rotate slowly on a horizontal shaker at 4 °C in a U-bottom Eppendorf tube overnight.
2. Coat four wells of a Maxisorp™ plate with 100 μL of 100 μg/mL streptavidin in PBS. Two wells are for streptavidin subtraction (strep-SUB wells) and two are for the panning experiment (strep-EXP wells). Coat two extra wells with 100 μL of 20 μg/mL VT-irrelevant protein negative control (VT-NEG wells). Seal the wells with Parafilm and incubate overnight at 4 °C.
3. The next morning, centrifuge the preadsorbed library phage at top speed in a microcentrifuge at 4 °C for 15 min. Use the phage supernatant for panning.
4. Discard the protein solutions, wash twice with 200 μL PBS and blot the wells on a paper towel. Add 200 μL StartingBlock™ to the wells, seal and incubate at 37 °C for 2 h.

5. Discard the blocking solution and add 100 μL phage library (~10^{11} cfu) in 100 μL of StartingBlock™, to both the VT-NEG wells. At the same time, add 100 μL of 20 μg/mL VT-peptide 3 and VT-peptide 4 in PBS to the Strep-EXP wells. Seal and incubate at room temperature for 1 h.
6. Transfer the preadsorbed phages from the VT-NEG wells to the strep-SUB wells and incubate for an additional 1 h at room temperature.
7. Empty the Strep-EXP wells and wash three times with PBS. Transfer the collected phage from **step 5** to the Strep-EXP wells. Seal and incubate at room temperature for 2 h (*see* **Note 8**).
8. Discard the unbound phage, rinse the wells five times with 300 μL PBS-T and blot dry. Wash five more times with PBS in a similar manner.
9. Elute the bound phage with 100 μL of 100 mM TEA. Pipette the content of the well up and down several times and incubate at room temperature for up to 10 min (*see* **Note 9**).
10. Transfer the eluted phage to a sterile microcentrifuge tube containing 50 μL of 1 M Tris–HCl, pH 7.4, and vortex to neutralize. Keep the tube on ice until the next step (*see* **Note 10**).
11. Grow a 10 mL LB culture of log-phase *E. coli* TG1 cells (OD ~0.2–0.3) in a sterile 50 mL Falcon tube. Keep 100 μL of cells for a negative control titer (**step 1**). Infect the remaining cells with 150 μL of the eluted phage at 37 °C for 30 min without shaking followed by 30 min slow shaking at 220 rpm.
12. Serially dilute infected cells in 2× YT media (10^{-2}–10^{-6}) and plate 100 μL on 2× YT-Tet plates. Plate 100 μL of uninfected cells as a negative control (*see* **Note 11**). Incubate at 32 °C overnight, then seal the plates with Parafilm and store at 4 °C for DNA sequencing.
13. Centrifuge the undiluted infected TG1 cells at 2000*g* for 12 min at 4 °C. Resuspend the pellet in 500 μL of 2× YT-Tet and plate the cells on large diameter (15 mm) 2× YT-Tet. Incubate at 37 °C overnight.
14. The next morning, scrape colonies from the plate into 50 mL of 2× YT-Tet. Mix 5 mL of cell suspension and 5 mL of fresh 2× YT-Tet broth and incubate at 30 °C for 5 h with 220 rpm shaking.
15. Pellet cells at 3600 ×*g* at 4 °C for 10 min. Pass phage-containing supernatant through a 0.22 μm filter, then precipitate the phage by adding 0.2 volumes of PEG/NaCl solution and incubate on ice for 1 h. Pellet the phage at 3600 ×*g* at 4 °C for 30 min.

16. Resuspend phage pellet in 200 μL of sterile PBS. Use 100 μL of the phage for the second round of panning. Keep the remainder at −20 °C. Serially dilute the phage preparation (10^{-2}–10^{-12}) in PBS and determine the titer as in **step 11** above (*see* **Note 12**).
17. Repeat **steps 1–16** for the second and third rounds of panning and **steps 1–12** for the fourth round of panning, with the following changes: (1) increase the number of washes (**step 8**) to 7, 10 and 12 for the second, third, and fourth rounds, respectively; (2) in the fourth round, sequentially elute the phage (**step 9**) first with ~10 μg/mL of native IGFBP7 protein and then with TEA (*see* **Note 13**).

3.5.2 Panning Against Native IGFBP7 Protein

Perform the panning experiment as in Subheading 3.5.1, with the following modifications.

1. In **step 1** (Subheading 3.5.1), preadsorb 100 μL of the sdAb library with 100 μL of StartingBlock™ (PBS) blocking buffer.
2. In **step 2** (Subheading 3.5.1), coat one well of with 100 μL PBS and a second well with 50 μg of recombinant IGFBP7 in 100 μL PBS.
3. In **step 5** (Subheading 3.5.1), transfer 100 μL of sdAb library to the PBS-coated well and incubate at room temperature for 1 h.
4. In **step 6** (Subheading 3.5.1), skip this step entirely.
5. In **step 6** (Subheading 3.5.1), transfer 100 μL of sdAb library (from **step 5**) from the PBS-coated well to the IGFBP7-coated well and incubate at room temperature for 1.5 h.
6. In **step 17** (Subheading 3.5.1), repeat **steps 1–16** for the second and third rounds of panning and **steps 1–12** for the fourth round of panning, with the following changes: (1) increase the number of washes (**step 8**) to 7, 10 and 12 for the second, third, and fourth rounds, respectively; (2) for the second, third and fourth rounds of panning, the coating concentration of IGFBP7 was decreased to 20 μg in 100 μL PBS.

3.6 Next-Generation DNA Sequencing

Panning of sdAb libraries against peptide antigens inevitably selects for some sdAbs that bind the peptide but not the native protein, and a minority of cross-reactive binders. In some cases such sdAbs can be recovered through traditional panning methods, although this may require extensive screening of peptide-binding clones. An alternative approach is to use NGS to interrogate the library and the eluted phage from panning on VT-peptide ligands and native protein. This strategy allows the identification of sdAbs that are subtly co-enriched during panning on peptide antigens and IGFBP7 and thus presumably represent cross-reactive Abs.

1. Amplify V_HH genes from the library and eluted phage in 25-μL PCR reactions containing 1× ABI Buffer II, 1.5 mM $MgCl_2$, 200 μM each dNTP, 5 pmol each of primers NGS-MJ7 and NGS-MJ8, 1 U of AmpliTaq Gold DNA polymerase and 1 μL of library or eluted phage at ~10^6 phage/μL (*see* **Note 14**).
2. Cycle the reactions as follows: 95 °C for 7 min; 35 cycles of (94 °C for 30 s, 55 °C for 45 s, and 72 °C for 2 min); 72 °C for 10 min.
3. Electrophorese 5-μL aliquots of the PCRs in 1 % (w/v) agarose gels in TAE buffer and confirm amplification of a ~400 bp band.
4. Purify the amplicons using a PureLink® PCR purification kit (300 bp cutoff).
5. Conduct second round "tagging" PCRs in 50-μL reaction volumes containing 1× Phusion HF Buffer, 1.5 mM $MgCl_2$, 200 μM each dNTP, 10 pmol of each primer pair (e.g., P5-seqF and P7-index1-seqR; *see* **Note 15**), 0.25 U Phusion High-Fidelity DNA polymerase and 5 μL first-round PCR as template. Cycle as follows: 98 °C for 30 s; 20 cycles of (98 °C for 10 s, 65 °C for 30 s, and 72 °C for 30 s); 72 °C for 5 min.
6. Electrophorese 5-μL aliquots of the PCRs in 1 % (w/v) agarose gels in TAE buffer and confirm amplification of a ~450–500 bp band.
7. Pool all four amplicons (library; eluted phage from panning on VT-peptide 3, VT-peptide 4 and IGFBP7) and using a PureLink® PCR purification kit (300 bp cutoff). Subsequently, gel purify the pooled library from a 1 % (w/v) agarose gel in TAE buffer using a QIAquick® gel extraction kit and elute in 50 μL EB. Perform final cleanup of the pooled NGS library using 90 μL of AMPure XP beads and elute in 20 μL ultrapure H_2O (*see* **Note 16**).
8. Sequence the library on an Illumina MiSeq Sequencing System using a 500-cycle MiSeq Reagent Kit V2 and a 5 % PhiX genomic DNA spike.
9. Assemble paired forward and reverse reads using FLASH [23], then quality filter the sequence data using the FASTX-Toolkit [24] with a stringency of Q30 over 95 % of bases in the read.
10. Assess enrichment of sdAbs as follows: (1) parse the CDR3 amino acid sequences of all sdAbs in the library and eluted phage; (2) determine the overlapping set of CDR3s, i.e., sdAbs present in the library and eluted phage sequence datasets; (3) for each CDR3, calculate enrichment scores as follows: % representation among eluted phage CDR3s/% representation among library CDR3s.

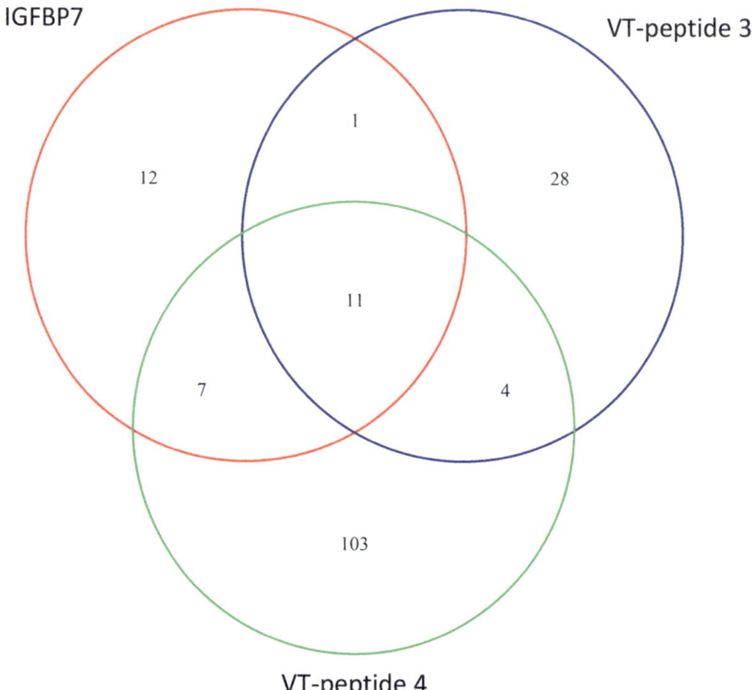

Fig. 3 Next-generation sequencing strategy to identify cross-reactive binders against VT-peptide ligands and native protein. A non-immune phage-displayed sdAb library (LAC-M) was panned in a single round against VT-peptide 3 ligand, VT-peptide 4 ligand, and IGFBP7 protein. The library phage and phage eluted from each panning were used as template to amplify V_HH genes, which were subsequently deep sequenced on an Illumina MiSeq® System (approximately 10^6 sequences each). For each panning target, enrichment scores were calculated for each V_HH observed in both the LAC-M library and the eluted phage. The Venn diagram shows the overlap of sets of V_HHs with enrichment scores of ≥100-fold; the figures in each region indicating the number of V_HH sequences observed. Thus, for example, 28 "peptide-only binding" V_HHs were enriched in panning on VT-peptide 3, while 12 (11+1) were enriched in panning on both VT-peptide 3 and native IGFBP7

11. Determine the extent to which CDR3 sequences with significant panning enrichment scores (e.g., greater than 100-fold) are shared between pannings on VT-peptide ligands and native protein (Fig. 3). These sequences represent putative cross-reactive binders.

3.7 Downstream Characterization of sdAbs

Detailed protocols for expression and purification of sdAbs have been previously published elsewhere [17–19]. DNA encoding each sdAb must be moved into an appropriate bacterial expression vector [17–19, 25], either by subcloning or gene synthesis. Assess binding of sdAbs to VT-peptide ligands and recombinant native protein using ELISA and/or surface plasmon resonance, as described previously [17–19], and identify the best cross-reactive binders (Fig. 4).

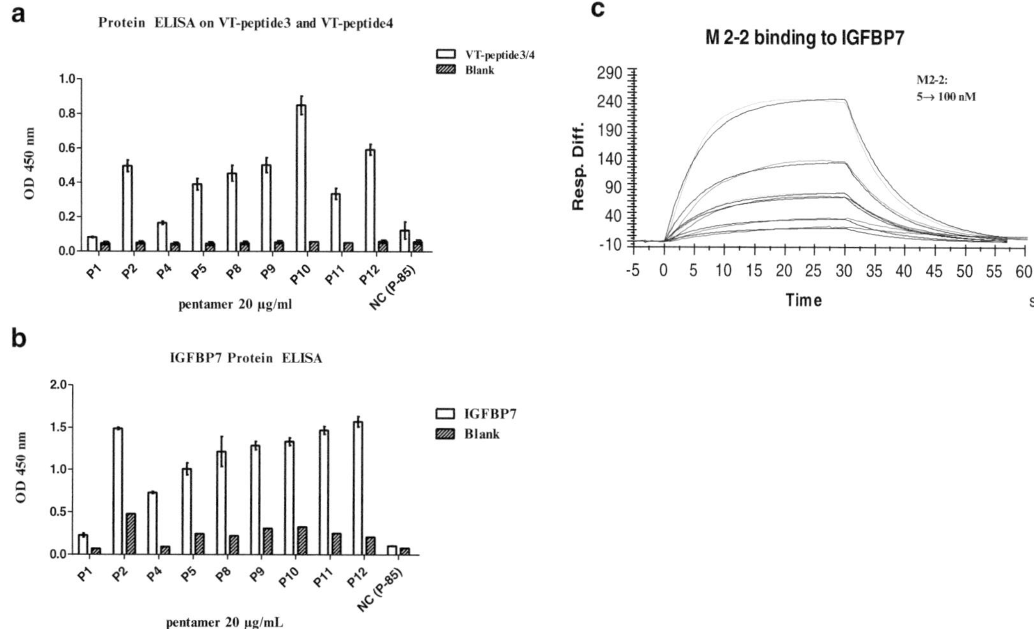

Fig. 4 Characterization of cross-reactive sdAbs obtained by panning against VT-peptide ligands. Binding sdAbs (clones 1, 2, 4, 5, 8, 9, 10, 11, and 12) were produced in pentameric format and tested for binding in ELISA to immobilized VT-peptide 3 or 4 ligands (**a**) or immobilized IGFBP7 protein (**b**). Clone P-85 is a negative control pentamer directed against an irrelevant antigen. (**c**) Surface plasmon resonance of a binding sdAb (clone 2) produced in monomeric format. Native IGFBP7 was immobilized on a CM5 chip and concentrations of monomeric clone 2 were injected at concentrations of 5–100 nM

4 Notes

1. Dilute the chloramphenicol stock solution to 1 mg/mL in ultrapure H_2O before use and store the working stock at 4 °C for up to 30 days. Protect from light to avoid degradation.
2. PBS may be supplemented with 3 mM EDTA to chelate any residual Ni^{2+} and to prevent protein oxidation and degradation, so long as small amounts of EDTA will not adversely affect downstream assays.
3. The PEG solution foams when heated and requires about a day to settle.
4. Repeat the centrifugation of the soluble fraction until it becomes clear.
5. Protein concentrations can be determined using the following formula [protein] = $A_{280}/\varepsilon \times$ molecular weight.
6. The peptide-VT fusion protein is stable at 4 °C for months or years.
7. Incorporation of biotin can also be verified by mass spectrometry.

8. It is suggested to start the *E. coli* TG1 culture to be used for amplification of the eluted phage (**step 10**) at this point.
9. Incubation times can be extended to recover the maximum number of bound phage, but incubation times significantly longer than 10 min are not recommended as this can adversely affect phage infectivity.
10. It is advisable to archive aliquots of eluted phage after each round of panning at −80 °C in case of any mishaps.
11. Repeat the plating of uninfected TG1 cells throughout the panning experiment since phage cross-contamination can occur easily. Generally, pipettors are main source of cross-contamination and must be cleaned frequently. All reagents used for panning should be prepared in small aliquots and discarded after the panning experiment.
12. Phage titers are generally between 10^{12} and 10^{13} cfu/mL. Alternatively, the phage titer can be quickly estimated by measuring the absorbance at 260 nm of phage diluted 1:100 in PBS and using the formula: phage particles/mL = $OD_{260} \times 100 \times 22.14 \times 10^{10}$.
13. The last round of panning can also be performed using wells coated with 10 μg/mL of native IGFBP7 to increase the chance of isolating parental protein-specific sdAbs.
14. Do not exceed 10^6 phage as this can inhibit PCR. Alternatively, phage replicative form DNA can be extracted from infected *E. coli* TG1 cells during phage amplification and used as the template for this PCR.
15. Ensure that the library and eluted phage from each panning (e.g., VT-peptide 3, VT-peptide 4, IGFBP7) use a different reverse primer with a different index sequence.
16. Library purity and concentration can be better determined using a BioAnalyzer and/or quantitative PCR, as per Illumina's recommendations, for determining the optimum amount of library for cluster generation. This is typically two to threefold lower than the concentration measured by spectrophotometry.

Acknowledgements

This is National Research Council of Canada Publication 53286. We gratefully acknowledge the excellent technical assistance of Henk van Faassen, Mary Foss and Shalini Raphael. We thank Dr. Maureen O'Connor for her help and support of the research work. This work was supported by the National Research Council Canada Genomics & Health Initiative. Norah A. Alturki was supported by the King Saud University Scholarship Program.

References

1. Muyldermans S (2001) Single domain camel antibodies: current status. J Biotechnol 74:277–302
2. Harmsen MM, De Haard HJ (2007) Properties, production, and applications of camelid single-domain antibody fragments. Appl Microbiol Biotechnol 77:13–22
3. De Marco A (2011) Biotechnological applications of recombinant single-domain antibody fragments. Microb Cell Factories 10:44
4. Bradbury AR, Sidhu S, Dubel S et al (2011) Beyond natural antibodies: the power of in vitro display technologies. Nat Biotechnol 29:245–254
5. Hutchings CJ, Koglin M, Marshall FH (2010) Therapeutic antibodies directed at G protein-coupled receptors. mAbs 2:594–606
6. Xia L, Willison LN, Porter L et al (2010) Mapping of a conformational epitope on the cashew allergen Ana o 2: a discontinuous large subunit epitope dependent upon homologous or heterologous small subunit association. Mol Immunol 47:1808–1816
7. Cleveland SM, Buratti E, Jones TD et al (2000) Immunogenic and antigenic dominance of a nonneutralizing epitope over a highly conserved neutralizing epitope in the gp41 envelope glycoprotein of human immunodeficiency virus type 1: its deletion leads to a strong neutralizing response. Virology 266:66–78
8. Grant GA (2003) Synthetic peptides for production of antibodies that recognize intact proteins. Curr Protoc Immunol Chapter 9:Unit 9 2
9. Van Regenmortel MH (2001) Antigenicity and immunogenicity of synthetic peptides. Biologicals 29:209–213
10. Larsen JE, Lund O, Nielsen M (2006) Improved method for predicting linear B-cell epitopes. Immunome Res 2:2
11. Slatter DA, Bihan DG, Farndale RW (2011) The effect of purity upon the triple-helical stability of collagenous peptides. Biomaterials 32:6621–6632
12. Minor DL Jr, Kim PS (1996) Context-dependent secondary structure formation of a designed protein sequence. Nature 380: 730–734
13. Bastings MM, Helms BA, Van Baal I et al (2011) From phage display to dendrimer display: insights into multivalent binding. J Am Chem Soc 133:6636–6641
14. Han X, Liu Y, Wu FG et al (2014) Different interfacial behaviors of peptides chemically immobilized on surfaces with different linker lengths and via different termini. J Phys Chem B 118:2904–2912
15. Irving MB, Craig L, Menendez A et al (2010) Exploring peptide mimics for the production of antibodies against discontinuous protein epitopes. Mol Immunol 47:1137–1148
16. Stone E, Hirama T, Tanha J et al (2007) The assembly of single domain antibodies into bispecific decavalent molecules. J Immunol Methods 318:88–94
17. Arbabi-Ghahroudi M, Tanha J, MacKenzie R (2009) Isolation of monoclonal antibody fragments from phage display libraries. Methods Mol Biol 502:341–364
18. Kumaran J, MacKenzie CR, Arbabi-Ghahroudi M (2012) Semiautomated panning of naive camelidae libraries and selection of single-domain antibodies against peptide antigens. Methods Mol Biol 911:105–124
19. Baral TN, MacKenzie R, Arbabi Ghahroudi M (2013) Single-domain antibodies and their utility. Current Protoc Immunol 103: Unit 2 17
20. Parker JM, Guo D, Hodges RS (1986) New hydrophilicity scale derived from high-performance liquid chromatography peptide retention data: correlation of predicted surface residues with antigenicity and X-ray-derived accessible sites. Biochemistry 25:5425–5432
21. Scholle MD, Collart FR, Kay BK (2004) In vivo biotinylated proteins as targets for phage-display selection experiments. Protein Expr Purif 37:243–252
22. Wilkins MR, Gasteiger E, Bairoch A et al (1999) Protein identification and analysis tools in the ExPASy server. Methods Mol Biol 112:531–552
23. Magoc T, Salzberg SL (2011) FLASH: fast length adjustment of short reads to improve genome assemblies. Bioinformatics 27: 2957–2963
24. Schmieder R, Edwards R (2011) Quality control and preprocessing of metagenomic datasets. Bioinformatics 27:863–864
25. Arbabi-Ghahroudi M, To R, Gaudette N et al (2009) Aggregation-resistant V_{H}s selected by in vitro evolution tend to have disulfide-bonded loops and acidic isoelectric points. Protein Eng Des Sel 22:59–66

Chapter 17

Generation of TCR-Like Antibodies Using Phage Display

Brian H. Santich, Hong Liu, Cheng Liu, and Nai-Kong V. Cheung

Abstract

The adaptive immune response against cancer consists of two arms: the humoral response from B cells, and the cell-mediated response from T cells. The humoral response has the advantage of diversity, theoretically recognizing antigens of any type (sugar, protein, lipid, etc.), but is generally limited to surface-expressed targets. T cells on the other hand, can recognize intracellular targets, but only if they are proteins, and presented as small peptide fragments on major histocompatibility complex (MHC) cell surface antigens. However, with advances in protein engineering and phage display, it has become feasible to quickly identify and generate antibodies or single-chain variable fragments against peptide-MHC, thus bridging the two arms, and allowing for recognition, identification, and effector responses against cells expressing intracellular targets.

Key words Phage display, Phage, Human leukocyte antigen, Major histocompatibility complex, Antibody, Protein expression, Fc-fusion protein, Single-chain variable fragment, scFv, T-cell receptor, Peptide MHC

1 Introduction

Thirty years ago, Smith [1] was able to show that bacteriophages could be genetically engineered to express specific polypeptides on their surface. Since his initial findings, the bacteriophage protein expression system, more commonly known as phage display, has been used as a tool to identify, select and optimize genetic sequences for proteins that bind targets with high specificity [2–9]. Phage display has been used extensively in the field of antibody engineering, where this tool can be used to screen large genetic libraries for coding sequences that give rise to specific and high-affinity single-chain variable fragments (scFv) or antigen-binding fragments (Fab) [10–21]. In this way, any antigen can be theoretically targeted, including those that are not very immunogenic [13, 17], or need high stringency to prevent cross-reactivity, as long as one's library is diverse enough.

Here we provide a method for generating peptide-MHC (pMHC) specific scFvs and Fc-fusion proteins using phage display.

Gunnar Houen (ed.), *Peptide Antibodies: Methods and Protocols*, Methods in Molecular Biology, vol. 1348, DOI 10.1007/978-1-4939-2999-3_17,

By screening phages against MHC monomers bound to either a target peptide or an irrelevant control peptide, one can quickly screen and identify scFvs that recognize the target pMHC with high specificity. This method has been used with great success to develop antibodies against a range of intracellular targets, notably WT1 [12] (Wilms tumor 1), a protein expressed in many types of human cancers. By fusing this scFv with other Ig domains, one can create bispecific antibodies, heterodimeric antibodies, and antibody–drug conjugates.

The following protocol describes in detail how to utilize a phage library to generate these TCR-like scFv's, assuming the operator has his/her own phage library. A commercially available Human Single chain Library (HuScL) is used as an example, but the protocol can be applied to any library. Moreover, while this method focuses on scFv libraries and development of scFv-Fc fusions, Fab libraries can be used as well [11, 15, 17, 19], and in fact are more easily remade into the full IgG format.

2 Materials

Prepare all buffer solutions with double deionized water and analytical grade reagents. All reagents can be kept at room temperature unless otherwise specified. Follow all normal waste removal rules and regulations when disposing of reagents. Be sure to use the correct HLA/MHC haplotype for your target peptide, and use the same one for your control peptide. In the following paragraphs, MHC and HLA will also be used interchangeably.

To prevent phage cross contamination, bleach and autoclave all reusable lab wares. Phages are quite robust and can survive normal autoclaving procedures. Similarly, be sure to always use filter tipped pipettes.

2.1 Phage Panning

1. Bovine serum albumin (BSA). Store at 4 °C.
2. Phosphate buffered saline (PBS) (1×): Weigh 8 g of NaCl, 2 g of KCl, 17 g of Na_2HPO_4 and 1.63 g of KH_2PO_4 and dissolve in 1 L of H_2O.
3. Blocking buffer: 2 % BSA in PBS. Weigh 2 g of BSA and dissolve in 100 ml of PBS. Store at 4 °C.
4. Dynabeads® M-280 Streptavidin. Store at 4 °C.
5. HuScL-3(R): Human Single Chain (scFv) Antibody Library. Store at −80 °C.
6. Magnetic Tube Rack.
7. Biotinylated MHC monomer with target peptide (Target pMHC). Store at −80 °C
8. Non-biotinylated MHC monomer with irrelevant peptide (Control pMHC). Store at −80 °C

9. Wash buffer: PBS containing 0.1 % Tween-20 (PBS-T). Add 1 ml Tween-20 to 1 L of PBS.
10. Elution buffer: 0.2 M glycine-HCl. Weigh 1.5 g of glycine and 100 mg of BSA and dissolve in 100 ml of H_2O. Adjust to pH 2.2 with HCl. Store at 4 °C.
11. Neutralization buffer: 2 M Tris–HCl. Weigh out 24.2 g of Tris–HCl and dissolve it in 100 ml of H_2O. Adjust to pH 9.0 with HCl

2.2 Amplification

1. XL1-Blue *E. coli*. Store at −80 °C
2. Tetracycline (Tet) Stock: 5 mg/ml in ethanol. Store at −20 °C
3. Ampicillin (Amp) Stock: 100 mg/ml in H_2O. Sterile filter and store at −20 °C
4. Lysogeny broth (LB): Weigh 10 g tryptone, 5 g yeast extract, and 10 g of NaCl and dissolve in 1 L H_2O. Autoclave and store at 4 °C
5. Shaking Incubator: Temperature set at 37 °C and at 30 °C. Rotating at 200–300 RPM.
6. Spectrophotometer: Capable of reading samples at 600 nm (OD_{600}). Always blank with the correct growth medium.
7. LB with glucose, Amp and Tet (LB-GAT): Prepare LB as described above. Before autoclaving add 20 g of glucose per 1 L of LB. After autoclaving let the LB cool. Once the LB is below 55 °C dilute 1 ml of ampicillin stock and 2 ml of tetracycline stock per 1 L of LB for a final concentration of 100μg/ml Amp and 10μg/ml of Tet. Store at 4 °C.
8. LB-Agar with glucose, Amp and Tet (LB-GAT plates): Prepare LB as described above. Before autoclaving add 15 g of agar and 20 g of glucose per 1 L of LB. After autoclaving let the LB-Agar cool. Once the LB-Agar is below 55 °C dilute 1 ml of ampicillin stock and 2 ml of tetracycline stock per 1 L of LB-Agar for a final concentration of 100 μg/ml Amp and 10 μg/ml of Tet. Pour onto 10 and 15 cm plates at desired thickness. Store at 4 °C.
9. M13KO7 Helper phage. Store at −20 °C
10. Kanamycin Stock: 50 mg/ml in H_2O. Sterile filter and store at −20 °C
11. PEG/NaCl (5×): Weigh 200 g of polyethylene glycol-8000 (20 %) and 150 g of NaCl (2.5 M) in 1 L H_2O and autoclave.

2.3 Clone Selection

1. HB2151 *E. coli*. Store at −80 °C.
2. LB with ampicillin (LB-Amp): prepare LB as described above. After autoclaving let the LB cool. Once the LB is below 55 °C dilute 1 ml of ampicillin stock per 1 L of LB for a final concentration of 100 μg/ml. Store at 4 °C.

3. LB-Agar with ampicillin (LB-Amp plates): Prepare LB as described above. Before autoclaving add 15 g of agar per 1 L of LB. After autoclaving let the LB-Agar cool. Once the LB-Agar is below 55 °C, dilute 1 ml of ampicillin stock for a final concentration of 100 μg/ml. Pour into 10 cm plates at desired thickness. Store at 4 °C.
4. 48-well plate.
5. Isopropyl-1-thio-β-D-galactopyranoside (IPTG): weigh 6.0 g IPTG and dissolve in 50 ml of H_2O. Filter-sterilize and store at −20 °C in 1 ml aliquots.

2.4 Screening of Phage Clones

1. 96-Well EIA microtiter plate.
2. BSA-biotin.
3. Streptavidin.
4. Biotinylated MHC monomer with target peptide (Target pMHC). Store at −80 °C.
5. Biotinylated MHC monomer with irrelevant peptide (Control pMHC). Store at −80 °C.
6. Blocking buffer: 2 % BSA in PBS. Weigh 2 g of BSA and dissolve in 100 ml of PBS. Store at 4 °C.
7. Dilution buffer: 0.5 % BSA in PBS: Weigh 0.5 g of BSA and dissolve in 100 ml of PBS. Store at 4 °C.
8. Mouse anti-V5 antibody. Aliquot 10–12 μl per tube at 1 mg/ml. Store at −80 °C.
9. HRP-conjugated goat anti-human antibody. Store at −80 °C.
10. HRP-conjugated goat anti-mouse antibody. Aliquot 5–10 μl per tube. Store at −80 °C.
11. o-phenylenediamine dihydrochloride (OPD) tablets.
12. Development buffer: 0.05 M phosphate-citrate buffer adjusted to pH 5.0. Weigh 14.2 g of Na_2HPO_4, 4.8 g of $C_6H_8O_7$ and dissolve in 1 L of H_2O.
13. 30 % H_2O_2 (catalog number H325-100).
14. Stopping solution: 5 N H_2SO_4.
15. Optical Plate Reader: Capable of reading samples at 490 nm (OD_{490}).

2.5 Large Scale Expression

1. Sonicator
2. Phenylmethanesulfonylfluoride (PMSF): 1 M PMSF in isopropanol. Weigh out 1.7 g of PMSF and dissolve it in 50 ml of isopropanol. Store at −20 °C.
3. Ni Sepharose High Performance. Store at 4 °C.
4. Elution buffer: 0.5 M imidazole in PBS. Weigh 104.5 g of imidazole hydrochloride and dissolve in 1 L of PBS. Adjust to pH 7.4 with HCl.

5. Spectrophotometer: Capable of reading samples at 280 nm (OD_{280}). Always blank with the correct buffer to match your sample.
6. Slide-A-Lyzer Dialysis Cassettes 10 K MWCO.
7. Sodium Azide: 2 %(w/v) in H_2O. Weight 2 g of NaN_3 and dissolve in 100 ml of H_2O.
8. Gravity column: Poly-Prep Chromatography Columns.

2.6 Fc Fusion Proteins

1. pFUSE-hIgG1-Fc.
2. FreeStyle™ 293-F Cells. Store at <−80 °C.
3. FreeStyle™ MAX Reagent. Store at 4 °C.
4. OptiPRO™ SFM. Store at 4 °C.
5. MabSelect. Store at 4 °C.

3 Methods

3.1 Panning

1. Wash 500 μl of streptavidin paramagnetic dynabeads ten times with PBS-T, using a magnet to isolate the beads each time. Leave the beads in 500 μl of PBS-T after washing. Add 0.5 ml of 2 % BSA in PBS to an Eppendorf tube and mix in 30 μl of the prewashed dynabeads and 150 μg of non-biotinylated Control pMHC. Add about 1E12 pfu phage from the library to the mix and incubate for 1 h at 4 °C with rotation (*see* **Note 1**).
2. After 1 h, apply a magnet to the sample and carefully transfer the supernatant to a new Eppendorf tube. This initial "pre-screening" helps remove phages that bind to the beads or BSA. At this step, the supernatant still contains phages that can bind the Control pMHC as well as the Target pMHC.
3. Add 7.5 μg of the biotinylated Target pMHC to this pre-screened supernatant and incubate for 1 h at 4 °C with rotation. After 1 h add 200 μl of prewashed dynabeads to the supernatant and incubate at 4 °C for another 15 min, with rotation.
4. After the incubation, apply a magnet to the sample, discard the supernatant from the tube, but keep the beads. Be careful not to disturb the dynabeads while aspirating. If necessary, it is acceptable to leave a little residual volume. The beads now contain the phages that bind the Target pMHC, but do not bind the Control pMHC (which remain in the supernatant)
5. Wash the beads ten times with PBS-T and five times with PBS to remove nonspecific or poorly bound phages and transfer the beads to a new Eppendorf tube (*see* **Note 2**). During each

wash, add 1 ml of buffer and incubate for 3–5 min before adding the magnet. After completing the final wash, elute the bound phages by incubating the beads with 500 μl of elution buffer (Glycine-HCl) for 15 min at RT. After elution, transfer the eluent to a new Eppendorf tube and neutralize with 30 μl of neutralization buffer (Tris–HCl). This eluent contains the Target pMHC binding phage, but at a relatively low titer. Therefore, it is necessary to amplify the phage for further downstream applications.

3.2 Amplification

1. Add the phages that have been affinity selected during the panning step to 20 ml of LB + 10μg/ml of Tetracycline with XL1-blue *E. coli* (OD_{600} 0.3–0.6) and incubate in a shaker for 1 h at 37 °C. Next, pellet the cells by spinning the cells at 3000×*g* for 10 min. Discard the supernatant and resuspend cells with 0.5 ml LB, then spread the cells on a 15 cm LB-GAT plate. Incubate overnight at 37 °C.
2. Add 10 ml of LB to the LB-GAT plate and carefully scrape the cells from the plate (*see* **Note 3**). Collect the cells in a 15 ml tube. Calculate the OD_{600} and resuspend the cells in 17 % glycerol/LB to a final density of 50.0 OD_{600}. Freeze down 1 ml of these bacteria as a stock. Using these bacteria, inoculate 20 ml of LB-GAT to a final OD_{600} of 0.1 (for example use 40 μl of the 50.0 OD_{600} stock) and incubate at 37 °C in a shaking incubator (>200 RPM) for 1 h. After this incubation, add 1E11 pfu M13KO7 helper phage, mix well and incubate for another 60 min at 37 °C in a shaking incubator. After this incubation, pellet the cells by spinning them at 3000×*g* for 10 min and resuspend the cells in 20 ml of LB-GAT and add 20 μl of Kanamycin. Incubate the cells overnight at 30 °C in a shaking incubator.
3. The following day, pellet the bacteria by spinning at 3000×*g* for 15 min, and transfer the supernatant (~20 ml) to a new tube and add in 5 ml of PEG-5×. Keep the tube on ice or at 4 °C for 1 h to precipitate the amplified phage. During this incubation, periodically shake the tube. At this time, begin a culture of XL1-Blue (~5 ml per sample). After 1 h, pellet the precipitated phages by spinning at 15,000×*g* for 20 min. Carefully decant the supernatant, taking care not to disturb the phage pellet (*see* **Note 4**).
4. Carefully resuspend the phages in 1 ml of PBS (*see* **Note 5**). Be sure to wash the sides of the tube to remove the phage smear. Spin the 1 ml of phage at 16,000×*g* at 4 °C for 5–10 min in a microcentrifuge to pellet any insoluble particles or remnants of *E. coli*. Titer the phages by inoculating 50 μl of OD_{600} 0.5–1.0 XL1-Blue with 1 μl of 10^{-8}–10^{-12} dilutions of this phage stock. Wait 15 min after inoculation before plating the cells on 10 or 15 cm LB-GAT plates. Incubate the plates at 30 °C O/N

5. The next day, calculate the phage titer by multiplying the number of colonies on each plate by the dilution factor. This is the concentration of phages per microliter (*see* **Note 6**). Freeze down several aliquots of the amplified phages in 15 % glycerol and store at −80 °C, and use ~1E12 pfu phages to continue panning. Repeat the above panning protocol 3–4 times, each time increasing the amount of beads used in the initial screen and decreasing the amount of biotinylated Target pMHC used. This will improve the specificity of the selected phages.

3.3 Clone Selection

1. After the final round of selection, amplify and precipitate and titer the phage as before. After quantifying, freeze down several aliquots of the phage (as explained in Subheading 3.2, **step 5**) and inoculate 1 ml of an OD_{600} 0.3–0.6 culture of HB2151 *E. coli*. Inoculate using a concentration of phages to get between 100 and 200 colonies per 50 μl of culture. Plate the 50 μl per plate on 3–5 LB-Amp plates and incubate O/N at 30 °C.
2. The next day, pick colonies from the plates and inoculate a 48-well plate, with each well containing 400 μl LB-Amp. Be sure to include at least one negative control well per experiment. This well should include the uninfected HB2151 in LB alone (*see* **Note 7**). Incubate the 48-well plate on a shaker at 37 °C for 3–6 h. After the incubation, add 200 μl of 50 % glycerol-LB per well. These now constitute the monoclonal glycerol stocks. Take 5 μl of each stock and inoculate a new 48-well plate with 400 μl of LB-Amp per well, as before. Incubate until the OD_{600} reaches about 0.4 on over half the plate (~3–6 h). Afterwards, freeze down the first 48-well plate −80 °C for long-term storage.
3. Once the cells have reached an OD_{600} ~0.4, add 200 μl of LB-Amp + 0.5 mM IPTG to induce soluble scFv production, and incubate O/N at 28 °C. The next day, centrifuge the plates at 3000 ×*g* for 15 min and transfer supernatant to a new plate for screening. This supernatant now contains soluble scFv from the monoclonal stocks. The next step will be to screen the clones for proper binding activity and specificity.

3.4 Screening Phage Clones

1. For each 48-well plate, coat two ELISA plates with 50 μl/well BSA-biotin at 10 μg/ml in PBS. Cover or seal the plate and incubate O/N at 4 °C (*see* **Note 8**).
2. The next morning, wash the plates 3–5 times with PBS. After washing, add 50 μl/well of streptavidin at 10 μg/ml in PBS and incubate for 1 h at RT.
3. Wash the plates five times with PBS and coat one plate with the Target pMHC and the other with the Control pMHC, 50 μl/well at 5 μg/ml in PBS. Incubate the plates for 1 h at RT (*see* **Note 9**).

4. Wash the plate 3–5 times with PBS and then add 150 μl/well of blocking buffer (2 % BSA) to block the binding of proteins to the plates. Incubate the plates at RT for 30–60 min (*see* **Note 8**).
5. Wash the plate 3–5 times and add 100 μl of each monoclonal stock supernatant, or purified scFv/scFv-Fc diluted in the dilution buffer (0.5 % BSA), to the plate, in duplicate. Incubate the plates at RT for 1 h.
6. Wash the plates five times and add 100 μl/well of mouse anti-V5 antibody at 0.5 μg/ml in dilution buffer if detecting soluble scFv. If detecting scFv-Fc (human Fc) fusion proteins, add 100 μl/well of goat anti-human HRP at 0.5 μg/ml in dilution buffer. Incubate the plates at RT for 1 h.
7. Wash the plates five-times and add 100 μl/well of goat anti-mouse HRP at 0.5 μg/ml, if detecting by V5. Incubate the plates at 4 °C for 1 h. If detecting by Fc skip this step and begin development of the plates.
8. During this incubation make the development buffer by adding two OPD tablets to 40 ml of OPD buffer. Keep at 4 °C until ready to use.
9. Before developing the plates, wash them thoroughly (at least 5×) with PBS. Right before use, add 40 μl of 30 % H_2O_2 to the development buffer and mix well. Immediately add 150 μl/well of this development buffer and incubate at RT in the dark. Check the reaction every 5 min and stop the reaction after 30 min, when positive control wells turn dark yellow, or when negative control wells begin to turn light yellow, whichever comes first. Stop the reaction by adding 30 μl of the stopping solution (5 N H_2SO_4). Add the acid quickly and carefully, being sure to not let too much time pass between the first wells and the final wells being stopped. Once all wells have been stopped, tap the sides of the plate to make sure all the acid mixed well. The acid is denser than the development solution and should mix on its own quite readily.
10. Read the plate on a spectrophotometer set to wavelength 490 nm. If one is using a standard curve to quantify the protein levels, oversaturation can be a problem. In this case, removal of half of the total volume (i.e., 90 μl/well) from the well can lower the OD_{490} by about half, bringing it back within the range for accurate quantitation. Wells with Target pMHC-binding scFvs should have an OD_{490} at least three-times above background (i.e., negative control wells). Compare the binding against the Target pMHC to the Control pMHC to determine specificity (*see* **Note 10**). Select at least 10 clones with sufficient binding and specificity, for further downstream analysis.

3.5 Large Scale Expression

1. Take the monoclonal glycerol stocks from Subheading 3.3, **step 2** and use them to inoculate 3 ml of OD_{600} 0.6 HB2151 *E. coli* in LB and incubate at 37 °C for 1 h in a shaking incubator (>200 RPM). Subculture the 3 ml culture to 1 L of LB-Amp and incubate at 37 °C in a shaking incubator until OD_{600} is 0.4–0.5.
2. Once OD_{600} reaches a range between 0.4 and 0.5, add IPTG to a final concentration of 0.5 mM and incubate O/N in a shaking incubator. The following day, pellet the cells at 3000×*g* for 25 min. Resuspend pellet in 50 ml PBS with 200 μM PMSF and lyse the cells using a sonicator (*see* **Note 11**). Centrifuge the lysates for 30 min at 48,000×*g* at 4 °C. Pass the supernatant through a 0.22 μm filter and collect the filtrate. This filtrate now contains the scFv.
3. Wash approximately 1 ml of Ni-high performance beads three-times in a 15 ml conical tube (*see* **Note 12**). Initially wash in ddH_2O, and then wash twice with PBS, resuspending the beads in a final volume of 1 ml PBS. Add these beads to the filtrate from the previous step and incubate O/N at 4 °C with rotation.
4. The next day, spin down the filtrate at 200×*g* for 5 min to pellet the Ni-NTA beads. Discard the supernatant and transfer the beads to a gravity column. Wash the column with about 10 column volumes (CV) of wash buffer (20 mM imidazole) to remove any irrelevant material binding to the nickel beads. Subsequently, begin eluting the protein in 1.0 CV fractions, in a step-wise fashion, beginning with 50 mM and ending with the 500 mM imidazole elution buffer. Typically the proteins will elute within the 100–300 mM imidazole range. Check the OD_{280} to determine which fractions contain protein. Be sure to blank with the same exact buffer because varying salt concentrations can change the OD_{280} reading. Combine 2–3 consecutive fractions with the highest OD_{280}, if necessary. It is important not to combine all fractions because some will contain aggregates. Similarly, early fractions are essentially wash steps and will likely contain irrelevant proteins.
5. Using a 10 K MWCO Slide-A-Lyzer cassette, dialyze the collected fractions to PBS O/N, at 4 °C. RT is suitable, if necessary (*see* **Note 13**). Measure the OD_{280} after dialysis to determine concentration and run the samples on SDS-PAGE or HPLC to determine purity. Ideally, one can test all fractions by ELISA to validate which ones contain the correct scFv, and compare binding at a specified concentration of scFv.
6. Finally, aliquot the dialyzed protein and freeze at −80 °C. Minimize freeze/thaw cycles to prevent precipitation or degradation of the scFv.

3.6 Generating Fc Fusion proteins

1. Starting from the monoclonal stocks, miniprep the clones of interest and sequence the scFv regions using sequencing primers for the phage library plasmid. After determining the sequence, design primers to PCR amplify out the scFv sequences, adding in relevant 5′ and 3′ restriction sites (i.e., 5′ EcoRI and 3′ BglII). Be sure to check the scFv sequence for these restriction sites, and if necessary use alternative restriction sites found on the pFUSE-hIgG1-Fc vector, or perform Gibson cloning [20]. Digest the PCR fragment and vector for 1 h at 37 °C and gel purify using a 1 % agarose gel. Ligate for 30 min at RT using a 3:1 insert to vector ratio, and transform the plasmid into competent *E. coli*. Plate cells on LB-Amp plates and incubate O/N at 37 °C. Pick 5–10 colonies and miniprep them. Screen by restriction digest to validate the insert and vector bands match the approximate size of the pFUSE vector (4 kb) and scFv (~800 bp). Sequence the screened plasmids and midiprep a selected plasmid with the correct sequence.
2. For transient transfection, begin culturing 293 F cells according to manufacturer's instructions. On the day of transfection, be sure the cells are at a concentration of approximately 1 M (10^6)/ml and viability is above 95 %.
3. For each 30ml of culture, dilute 37.5 μg of midiprepped plasmid into 600 μl of OptiPro™ SFM. Simultaneously mix 37.5 μl of FreeStyle™ MAX with 600 μl of Optipro™ SFM. Incubate for 5 min at RT and before adding the 0.6 ml of diluted DNA from above. Incubate another 20–30 min at RT.
4. After the incubation, add the 1.2 ml of transfection media to the prepared cells, drop wise. Incubate the cells for 5–7 days and harvest when viability drops below ~80 %
5. On the day of harvest, spin down supernatant at >3000 × *g* for 1 h at 4 °C. Store at 4 °C until ready to begin purification. If one will be storing the supernatant above −80 °C for more than 2 days, it is best to add sodium azide (0.05 % final) to prevent bacterial growth. Right before beginning the purification, pass the supernatant through a 0.22 μm filter to remove any precipitate or bacterial/fungal growth.
6. To purify the Fc-Fusion proteins, it is best to use protein-A (e.g., MabSelect). For larger volumes use an FPLC if possible. For smaller volumes, briefly, mix prewashed protein-A resin with the filtered 293 F supernatant and incubate O/N at 4 °C with rotation. The next day load the resin onto a gravity column wash with 10 CV of PBS before eluting, stepwise, with a low pH buffer such as Glycine-HCl (i.e., from pH 7 to pH 2.5). As before, elute into 1.0 CV fractions, 5 per condition, and check OD_{280} on each fraction to identify the optimal

elution conditions. Elute fractions into 0.1 CV 1 M Tris–HCl (pH 8.0) to neutralize the pH and as before, combine and dialyze the fractions of interest in PBS or another suitable buffer (*see* **Note 14**). Aliquot and store samples at −80 °C for long-term storage.

4 Notes

1. Depending on the library used, it may be necessary to generate the phages from *E. coli* stocks. In this case, follow the manufacturer instructions, which should mimic the steps from Subheading 3.2.
2. While PBS and PBS-T are the most common buffers for washing during the panning and screening steps, one can actually implement any buffer. However, the buffer of choice should be similar to any buffers that one hopes to use with scFvs for in vitro or in vivo studies. Each buffer adds a selective pressure to the scFv expressed on the phage, so be sure to choose an optimal buffer or else the selected phage may not bind/fold as well in future experiments. Similarly, the method of washing can add a selective pressure of its own. Longer incubations with wash buffer or using more wash steps, is thought to select for higher affinity binders. Feel free to adjust these steps as necessary.
3. During this step be careful not to brake up the agar when scraping the plate. It is easiest to use a wide scraper to prevent this. The bacteria come off quite easily and it is not necessary to remove all traces of bacteria. If some agar does get into the mix, simply spin down the solution at a low speed (i.e., 100 × *g*) briefly and transfer the supernatant to a fresh tube.
4. Often the phage pellet forms only a faint smear on the side of the tube and is hard to identify. To make it easier to find, mark the bottom of the tube at the location where a pellet is expected to form before spinning. On an angled rotor, this is typically on side of the tube facing away from the rotor. The smear is also apparent if the tube is held to the light and rotated. Typically it gets wider as it approaches the bottom of the tube and is often present even when a phage pellet exists. If no smear or pellet can be seen, simply wash the tube carefully and test for phage by using the tittering scheme explained in Subheading 3.2, **step 4**. Instead starting at 10^1 and going up until 10^{10}.
5. When using a library generated by animal immunization, it is important to know if the target protein was conjugated to another protein or used with an adjuvant. To improve specificity the phage pellet can be dissolved in PBS mixed with the adjuvant

or protein conjugate. For example if the immunization used BSA as an adjuvant, resuspending the phage pellet in 1 % BSA in PBS will negatively select the BSA-binding phages before panning, thus reducing the chance that these phages are selected and amplified. Be sure to not use this same protein in the blocking or dilution buffers during the screening, however.

6. In this protocol, each phage is considered 1 pfu, and therefore, pfu per microliter is equivalent to phages per microliter. Furthermore, if at any stage it is clear that the selected phage are not amplifying or binding with enough specificity, simply go back to a frozen aliquot and continue panning/screening from there. With each subsequent screening, the concentration of phages should increase steadily. To accurately quantify the concentration it is often necessary to dilute the phage even further than 10^{-12}.
7. It is recommended to pick as many colonies as possible to increase the chances of finding a good clone. Be aware, however, picking more than 47 clones per experiment means doing more than two ELISAs per experiment. Always include a negative control well, but feel free to scale up screening as much as necessary. A positive control is only necessary if no binding is detectable after a primary screen.
8. The protein used during the panning steps (in this case BSA) should match the protein used in the blocking and dilution buffers in the screening ELISA; however, it does not need to be BSA. Be sure to use a protein that was not used in generating the phage library. BSA-biotin can be easily replaced with another biotinylated protein, if necessary.
9. For screening it is best to test the monoclonal stocks against both the Target pMHC as well as the Control pMHC to validate the specificity of the selected phage. However, for convenience one can screen a larger selection against the Target pMHC first, and then perform a secondary screen with high binders, against the Control pMHC.
10. Be sure to note that at this stage it is impossible to separate higher affinity binders from more stable sequences. Higher OD_{490} at this stage only means more scFvs were left bound by the end of the ELISA, but it does not distinguish between more efficient expression of the scFvs or more efficient binding to the targets.
11. This step can be replaced by any preferred lysis method (Cell homogenizer, lysis buffer), if necessary. However, try and keep the overall volume small to improve downstream purification efficacy. For a Tekmar sonic disrupter, sonicate on ice for 3 min, pulsing at 50–60 % duty cycle with the output control set at 5.

12. While this protocol uses a nickel purification method, this can easily be replaced with other methods such as protein-A (anti-Fc), protein-L (anti-kappa chain), or anti-V5 affinity chromatography. Each method has its benefits, but nickel purification is only used in this case for its convenience. Protein-L will have improved purity over nickel, but it does not bind all scFv sequences equally. Protein-A can only bind to Fc-fusion or full IgG proteins and anti-V5 requires that the construct has a V5 tag. Similarly, all steps can be performed by FPLC, although small-scale purifications should be limited to gravity columns.
13. To fill the dialysis cassette, it is easiest to use a 22-gauge hypodermic needle fitted onto a small 1 ml syringe. Be sure not to overfill the cassette or let it sink into the dialysis buffer. When adding the sample to the cassette, keep the pointed end of the needle angled slightly downward with the cassette held parallel to the floor. This will help prevent any accidental puncturing of the membrane. The final volume after dialysis can sometimes change quite dramatically from the starting volume, so do not be alarmed if the volume appears to have dropped by up to 50 %.
14. It is difficult to determine the optimal buffer formula for a given protein before enough of it can be successfully purified, but buffer optimization can substantially improve the stability and efficacy of a given protein in vitro and during long-term storage. Similarly, the buffers used during affinity chromatography can have enormous impact on the overall yield and purity of the final product. The buffers listed above should be considered as a starting point but can and should be optimized for each construct produced. For each buffer be sure to test both stability (by HPLC) and activity (ELISA/FACS).

References

1. Smith GP (1985) Filamentous fusion phage - novel expression vectors that display cloned antigens on the virion surface. Science 228(4705):1315–1317. doi:10.1126/Science.4001944
2. Choo Y, Klug A (1994) Toward a code for the interactions of zinc fingers with DNA: selection of randomized fingers displayed on phage. Proc Natl Acad Sci U S A 91(23):11163–11167
3. Cwirla SE, Peters EA, Barrett RW, Dower WJ (1990) Peptides on phage: a vast library of peptides for identifying ligands. Proc Natl Acad Sci U S A 87(16):6378–6382
4. Devlin JJ, Panganiban LC, Devlin PE (1990) Random peptide libraries: a source of specific protein binding molecules. Science 249(4967): 404–406
5. Jamieson AC, Kim SH, Wells JA (1994) In vitro selection of zinc fingers with altered DNA-binding specificity. Biochemistry 33(19): 5689–5695
6. Lowman HB, Bass SH, Simpson N, Wells JA (1991) Selecting high-affinity binding proteins by monovalent phage display. Biochemistry 30(45):10832–10838
7. Rebar EJ, Pabo CO (1994) Zinc finger phage: affinity selection of fingers with new DNA-binding specificities. Science 263(5147):671–673
8. Scott JK, Smith GP (1990) Searching for peptide ligands with an epitope library. Science 249(4967):386–390
9. Wang CI, Yang Q, Craik CS (1995) Isolation of a high affinity inhibitor of urokinase-type

plasminogen activator by phage display of ecotin. J Biol Chem 270(20):12250–12256

10. Barbas CF 3rd (1995) Synthetic human antibodies. Nat Med 1(8):837–839

11. Chames P, Hufton SE, Coulie PG, Uchanska-Ziegler B, Hoogenboom HR (2000) Direct selection of a human antibody fragment directed against the tumor T-cell epitope HLA-A1-MAGE-A1 from a nonimmunized phage-Fab library. Proc Natl Acad Sci U S A 97(14):7969–7974

12. Dao T, Yan S, Veomett N, Pankov D, Zhou L, Korontsvit T, Scott A, Whitten J, Maslak P, Casey E, Tan T, Liu H, Zakhaleva V, Curcio M, Doubrovina E, O'Reilly RJ, Liu C, Scheinberg DA (2013) Targeting the intracellular WT1 oncogene product with a therapeutic human antibody. Sci Transl Med 5(176):176ra133. doi:10.1126/scitranslmed.3005661

13. Mao S, Gao C, Lo CH, Wirsching P, Wong CH, Janda KD (1999) Phage-display library selection of high-affinity human single-chain antibodies to tumor-associated carbohydrate antigens sialyl Lewisx and Lewisx. Proc Natl Acad Sci U S A 96(12):6953–6958

14. Nelson AL, Dhimolea E, Reichert JM (2010) Development trends for human monoclonal antibody therapeutics. Nat Rev Drug Discov 9(10):767–774. doi:10.1038/nrd3229

15. Rauchenberger R, Borges E, Thomassen-Wolf E, Rom E, Adar R, Yaniv Y, Malka M, Chumakov I, Kotzer S, Resnitzky D, Knappik A, Reiffert S, Prassler J, Jury K, Waldherr D, Bauer S, Kretzschmar T, Yayon A, Rothe C (2003) Human combinatorial Fab library yielding specific and functional antibodies against the human fibroblast growth factor receptor 3. J Biol Chem 278(40):38194–38205. doi:10.1074/jbc.M303164200

16. Saggy I, Wine Y, Shefet-Carasso L, Nahary L, Georgiou G, Benhar I (2012) Antibody isolation from immunized animals: comparison of phage display and antibody discovery via V gene repertoire mining. Protein Eng Des Sel 25(10):539–549. doi:10.1093/protein/gzs060

17. Schoonbroodt S, Steukers M, Viswanathan M, Frans N, Timmermans M, Wehnert A, Nguyen M, Ladner RC, Hoet RM (2008) Engineering antibody heavy chain CDR3 to create a phage display Fab library rich in antibodies that bind charged carbohydrates. J Immunol 181(9):6213–6221

18. Vaughan TJ, Williams AJ, Pritchard K, Osbourn JK, Pope AR, Earnshaw JC, McCafferty J, Hodits RA, Wilton J, Johnson KS (1996) Human antibodies with sub-nanomolar affinities isolated from a large non-immunized phage display library. Nat Biotechnol 14(3):309–314. doi:10.1038/nbt0396-309

19. Barbas CF 3rd, Kang AS, Lerner RA, Benkovic SJ (1991) Assembly of combinatorial antibody libraries on phage surfaces: the gene III site. Proc Natl Acad Sci U S A 88(18):7978–7982

20. Gibson DG, Young L, Chuang RY, Venter JC, Hutchison CA 3rd, Smith HO (2009) Enzymatic assembly of DNA molecules up to several hundred kilobases. Nat Methods 6(5):343–345. doi:10.1038/nmeth.1318

Chapter 18

Structural Characterization of Peptide Antibodies

Anna Chailyan and Paolo Marcatili

Abstract

The role of proteins as very effective immunogens for the generation of antibodies is indisputable. Nevertheless, cases in which protein usage for antibody production is not feasible or convenient compelled the creation of a powerful alternative consisting of synthetic peptides. Synthetic peptides can be modified to obtain desired properties or conformation, tagged for purification, isotopically labeled for protein quantitation or conjugated to immunogens for antibody production. The antibodies that bind to these peptides represent an invaluable tool for biological research and discovery. To better understand the underlying mechanisms of antibody–antigen interaction here we present a pipeline developed by us to structurally classify immunoglobulin antigen binding sites and to infer key sequence residues and other variables that have a prominent role in each structural class.

Key words Peptide antibody, Structure, Clustering, Linear epitope

1 Introduction

The usage of peptide antibodies in a plethora of analyses, e.g., identification and purification of proteins, immunodiagnostics purposes, epitope-based vaccine design, has been instrumental in the past and to date for drastic scientific and technological advancements. Nevertheless, a group of critical questions on the principles that govern the way in which an antibody recognizes a peptide remain unanswered. On the antigen side, the B-cell epitope prediction for peptides is currently far from perfection [1], whereas on the antibody side, the repertoire diversity of peptide specific B-cell paratopes still misses its full characterization.

To answer these questions one needs to thoroughly analyze the antibody three-dimensional structure, since it ultimately determines its function. Antibodies are built from four polypeptide chains, two heavy and two light ones, joined by disulphide bonds so that each heavy chain is bound to a light chain and the two heavy chains are linked together. Each chain can be divided on a functional and structural basis in two different regions, called

Gunnar Houen (ed.), *Peptide Antibodies: Methods and Protocols*, Methods in Molecular Biology, vol. 1348,
DOI 10.1007/978-1-4939-2999-3_18, © Springer Science+Business Media New York 2015

Variable region (V_L and V_H) and Constant region (C_L and C_H). Each variable region is in turn composed by four framework regions that surround three hypervariable regions, these latter bearing an extremely variable amino acid composition. Wu and Kabat [2] predicted them to assume a loop conformation arising from the relatively conserved framework. They were subsequently named "complementarity-determining regions" (CDRs) in contrast to the surrounding framework regions (FRs) and have been claimed to be responsible for the selective binding of the antigen.

To better understand the underlying mechanisms of antibody–antigen interaction we developed a method to structurally classify immunoglobulin antigen binding sites and to infer key residues in sequence that have a prominent role in each structural class. It is an integrative approach of bioinformatics (structural superposition, sequence alignments, physiochemical analysis of structures), machine learning (clustering, random forests), and statistical (correlation tests, validation of predictions) techniques.

2 Materials

Prepare the antibody heavy and light chain sequences and structures together with the bound peptide antigens for the analysis. These data can be generated in house or retrieved by querying the publicly available IEDB database [3] (Immune Epitope database) with the key words: "Epitope = Linear peptide" or using any other available/private source of antibodies. In case of IEDB usage you will get a comma-separated values (CSV) file that contains all of the data associated with the records. The CSV format can be easily manipulated using a spreadsheet program, such as Microsoft Excel, or edited with a word processing program, such as Microsoft Word or Notepad.

2.1 Sequence

You will find PDB [4] (Protein Data Bank) codes of the immunoglobulins in the above-mentioned CSV file. For each PDB code retrieve antibody heavy and light chain sequences together with their cognate peptide sequence in FASTA format from Protein Data Bank.

2.2 Structure

Similarly, for each PDB code, retrieve the structure of the antibody–peptide complex in PDB format.

3 Methods

3.1 Structure Renumbering

1. The unambiguous identification of structurally equivalent residues of each immunoglobulin is of crucial importance. Fortunately, the architecture of antibodies have made possible the development of several unified numbering schemes.

Such numberings define portions or specific residues of immunoglobulins that have a similar position in their three-dimensional structure. Currently there are several numbering schemes [5–10]. You can obtain renumbered structures by uploading the PDB files to http://www.bioinf.org.uk/abs/abnum/ or writing a script to perform this step according to the numbering scheme of preference (*see* **Note** in Subheading 4).

2. The resulting file that contains the renumbered antibody structure needs to be saved in PDB format for later analyses.

3.2 ABS Superposition

1. Once all the structures have the same numbering scheme, you need to select the residues that are part of the Antigen Binding Site (ABS). There is a recent article describing the existing ABS definitions [11].
2. After choosing the suitable ABS definition, you need to perform one-against-all ABS comparison. To this aim you can use any algorithm that does comparative analysis of two selected 3D protein structures or fragments of 3D protein structures.
3. A number of different possible subsets of the atoms that make up a protein macromolecule can be used as a reference in a structural alignment (*see* a hint in Subheading 4). When aligning structures bearing very different sequences, the side chain atoms generally are not taken into account because their identities differ between the aligned residues. For this reason the structural alignment methods commonly use only the protein backbone atoms or often, for efficiency, only the C_α atoms, since the peptide bond has a minimally variant planar conformation. When the structures to be aligned are highly similar or identical it's meaningful to align side-chain atoms that will reflect not only the backbone conformation similarity but also the side chain rotameric states. Thus, it is advisable to use C_α atoms as reference in structural alignment, unless the user has a specific case of highly similar antibodies and would like for example to monitor side chains changes correlated to the antigen properties etc.
4. There are numerous measures of model similarity. The most popular one is the RMSD—Root Mean Square Deviation, the measure that is calculated after the best superimposition to show the divergence of one structure from another. Other more sophisticated measures have been developed, such as the GDT—Global Distance Test [12] and the TM-Score—Template-Modeling Score [13]. Each of these measures has its own qualities and drawbacks, and the user should select them according to the specific case (*see* Subheading 4).
5. After choosing an appropriate superimposition algorithm, the user is advised to create a simple script to perform one-against-all superimposition of ABSs of all the structures of the dataset.

This can be done using any of the existing programming/scripting languages (*see* pseudocode in Subheading 4). The resulting distances should be saved in matrix form to facilitate further analysis. The distance between ABSs of antibody i and j will be stored in the cell at position (i,j) of such a matrix.

3.3 ABS Clustering

1. All further analyses will be performed using R, an open-source, free statistical package. R uses a command-line syntax, meaning that you will have to type commands for R. There are several projects that add a graphical user interface (GUI) to R, such as RStudio and R Commander, but currently only limited point-and-click functionality like opening files and viewing charts are available. Nonetheless, the user can decide to use one or more tools to replace R, such as SPSS, Q, Julia or Python.
2. Clustering methods are used to group together samples based on their similarity. Even though in our procedure we use hierarchical clustering, it is important to underline that any clustering method that accepts a distance matrix as input can be adopted.
3. Hierarchical clustering is a method of cluster analysis that builds a hierarchy of clusters showing relations between the individual members based on similarity. There are two strategies for hierarchical clustering: agglomerative (bottom up) or divisive (top down). In the agglomerative approach each observation starts as its own cluster, and pairs of clusters are merged as one moves up the hierarchy whereas in the divisive approach all observations start in one cluster, and splits are performed recursively as one moves down the hierarchy. The results of hierarchical clustering are usually presented in a dendrogram.
4. You are advised to obtain several antibody clusterings by applying different methods. R cluster package (http://cran.r-project.org/web/packages/cluster/) contains diana (divisive) and hclust (agglomerative) methods. The agglomerative clustering can be performed using average, Ward, single and complete joining functions. These functions differ in how the distance between each cluster is measured. To select the method and the distance function that are giving the most compact and well separated cluster definition you might want to use the Silhouette function [14]. The highest value of silhouette is used to identify the optimal cut level for each clustering and the best clustering in general (*see* the example in Subheading 4).

3.4 Correlation Analysis

1. Random Forest [15] is one of the most powerful "black-box" supervised learning methods. It's an ensemble classifier that uses many decision tree models and can be used for classification (categorical variables) or regression (continuous variables) applications.

2. You must use the sequences of immunoglobulins together with the structural clustering as training data for the Random Forest analysis, to see whether a particular sequence position have a prominent role in each structural class. The user is advised to create a unique CSV file containing the name of PDB structure followed by its sequence and by a label that identify the structural cluster it belongs to. For example if you obtained four structural classes while clustering the ABS of immunoglobulins, you might want to name them a, b, c, d.
3. The sequences must be aligned according to a single numbering scheme, so that they all will have the same length (including insertions and deletions) and that conserved residues, insertion and deletion sites will be in the same columns of the CSV file for all the sample sequences.
4. Apart from antibody sequences, additional data can be checked for correlation with the structural clustering. It has to be represented by a small set of labels or by numbers and added to your training data. For example if you have done a mutation analysis, you may want to divide the dataset into a discrete number of groups according to the level of mutation and label each sample in the dataset with a label representing its level of mutation. Similarly, you can perform the labeling of any data that needs to be correlated: peptide size, antigen amino acid residues that are in contact with the antibody, antibody germline, etc. Add these labels to each immunoglobulin in the unique CSV file created earlier.
5. You are advised to perform Random Forest analysis using the R package 'randomForest'. In our protocol, the random forest is trained to predict the structural cluster using the sequence data and the additional variables alone.
6. Random Forests are tuned and trained on the data described above and the Gini Impurity Index [16] is computed to select the most significant sequence positions and variables that display the best correlation with the structural data. The Gini Impurity Index is a measure of the importance of each variable to correctly identify the structural cluster of the corresponding sample and therefore is a strong hint that the given residue or variable is a key element for the ABS to adopt one of the different conformations identified by the structural clusters.
7. The variable importance plot is a critical output of the random forest algorithm. The plot shows each variable on the y-axis, and their importance on the x-axis. They are ordered top-to-bottom as most to least important and an estimate of their importance is given by the position of the dot on the x-axis. You should use the most important variables, as determined from the variable importance plot, to conduct the further analyses or as a support for your hypothesis. In order to decide how many

variables to analyze you are advised to look for a major difference in the values between one variable and the next one ("elbow") in the importance plot to decide how many variables to choose. This is an important step for reducing the number of variables for further data analysis techniques, but you should be careful to have neither too few variables (that won't suffice to explain the structural data) nor too many (having the risk of introducing noise and second-order correlations in the analysis).

4 Notes

1. Structure renumbering needs to be done with caution to prevent errors in the ABS selection and superposition. Since the antibody loops are hypervariable in sequence, special care needs to be taken to correctly align conserved residues, that serve as an additional help in aligning the insertions and deletions.
2. The following pseudocode might help in creating a one-against-all ABS comparison script.

```
FOREACH PDB structure Xi in the dataset A:
  create a directory DXi;
  FOREACH PDB structure Xj≠Xi in the dataset A:
  create a directory DXi/DXj;
  run the superposition software on the
  selected residues and save its output in the
  DXi/DXj folder;
  END OF FOREACH;
END OF FOREACH;
```

3. In order to have a more robust superposition you might want to superimpose the loop regions (following the chosen definition) plus two/three flanking residues at the N- and C- termini of loop.
4. Depending on the similarity measure of choice you might want to convert the calculated pairwise distances into the new distance matrices using the corresponding formulae (*see* GDT, TM-score, RMSD definitions).
5. An example of application

   The protocol described here has been applied to a small set of peptide antibodies to identify whether it is possible to structurally classify their antigen binding sites and subsequently to infer key residues in immunoglobulin sequence that have a prominent role in each structural class. For demonstration purposes we added also the antigen contact area as an additional variable to check its effectiveness in differentiating each structural class.

(a) Querying IEDB database for antibodies binding to linear peptides has retrieved the dataset consisting of 29 peptide antibodies.

(b) The antibodies have been renumbered according to Kabat–Chothia numbering scheme.

(c) We used LGA (Local Global Alignment) [17] for structural superposition of the antigen binding site. The residues used in this analysis were the following:

Chain L: 24: 34, 48: 54, 89: 98

Chain H: 24: 34, 51: 57, 93: 104

The above mentioned residues have been used altogether in the superposition and average C_α RMSD value has been saved in a 29×29 matrix.

(a) Package "cluster" of R has been used to conduct diana and hclust clustering methods. Average, ward, single and complete joining functions were applied to the agglomerative clustering. The best clustering has been defined by silhouette analysis. The highest silhouette value 0.35 has been observed for hclust method with the "Ward" function and total number of seven clusters.

(b) The structural clustering with the highest silhouette value is shown in Fig. 1. The seven identified clusters are outlined by red boxes in the dendrogram (clustering tree).

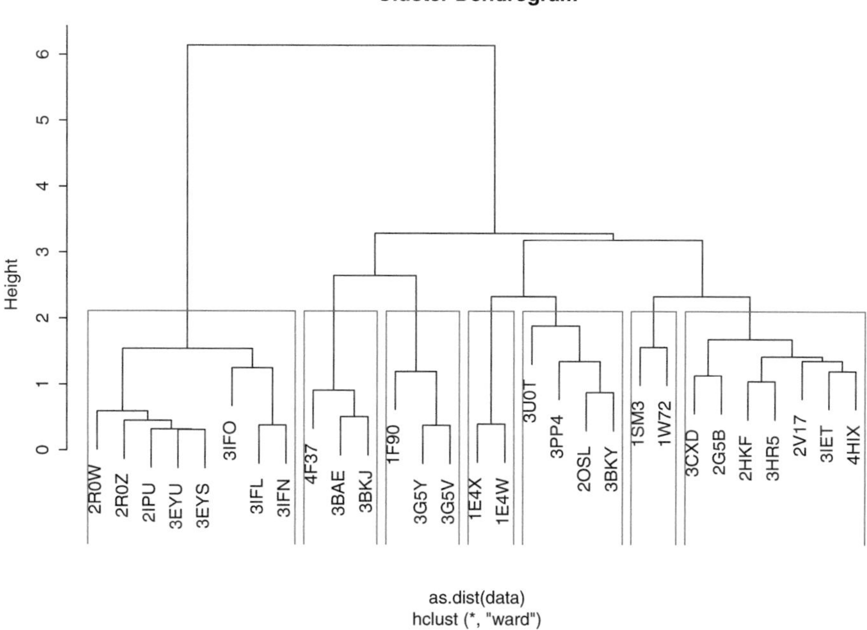

Fig. 1 The structural clustering with the highest silhouette value

Fig. 2 Data RF file containing antibody aligned sequences with structural clustering (*column 2*) and antigen contact area (*last column*) labels

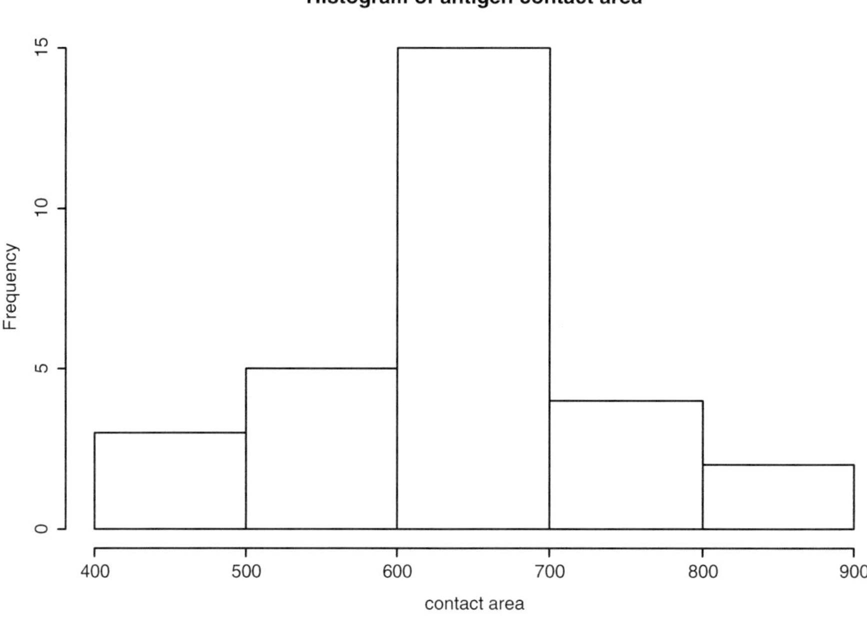

Fig. 3 Antigen contact area distribution

(c) A unique file containing aligned sequences of heavy and light chains of the 29-peptide antibodies has been created (Fig. 2). The contact area for antigen has been retrieved from IEDB query and its distribution is shown in Fig. 3.

(d) In order to create suitable labels for the Random Forest analysis we divided the contact area (ConA) information in four groups as following:

Group A: $\text{ConA} \leq \text{mean} - 1\text{sd}$

Group B: $(\text{mean} - 1\text{sd}) < \text{Con A} \leq \text{mean}$

Group C: $\text{mean} < \text{Con A} \leq (\text{mean} + 1\text{sd})$

Group D: $(\text{mean} + 1\text{sd}) < \text{Con A}$

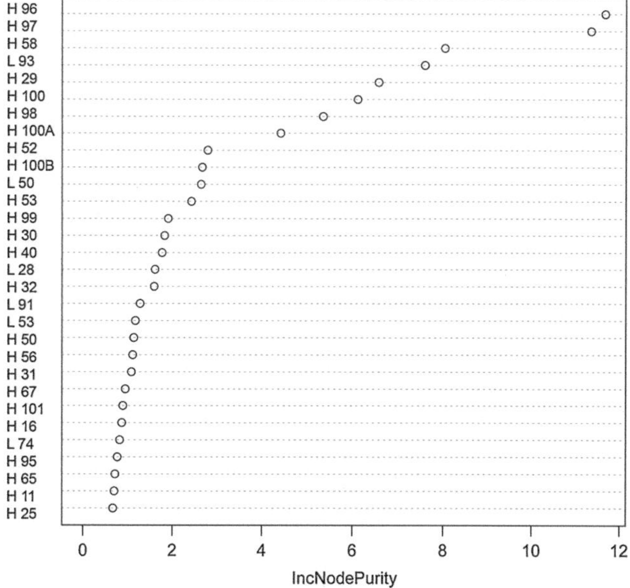

Fig. 4 The Random Forests variable importance plot

Each peptide antibody has been labeled according to the group of contact area it belongs to and this information has been added to the unique file containing aligned sequences (data RF).

(a) Random Forests has been trained on the data RF. The most informative variables have been identified and shown in Fig. 4.

(b) The residues 96, 97, 58 of chain H and the residue 93 of chain L appear to be the most important ones and thus have a prominent role in the structural clustering whereas the antigen contact area is not in the list of important variables and thus doesn't have significant impact.

(c) A thorough analysis of the variable importance plot by means of structural superposition with in deep physio-chemical analysis of structures will permit you to support or reject your hypothesis and thus finalize the antibody characterization.

References

1. Singh H, Ansari HR, Raghava GPS (2013) Improved method for linear B-cell epitope prediction using antigen's primary sequence. PLoS One 8(5). ARTN e62216. doi:10.1371/journal.pone.0062216

2. Wu TT, Kabat EA (2008) An analysis of the sequences of the variable regions of Bence Jones proteins and myeloma light chains and their implications for antibody complementarity (Reprinted from J Exp Med, vol 132,

pg 211–250, 1970). J Immunol 180(11): 7057–7096

3. Vita R, Zarebski L, Greenbaum JA, Emami H, Hoof I, Salimi N, Damle R, Sette A, Peters B (2010) The immune epitope database 2.0. Nucleic Acids Res 38(Database Issue):D854–D862. doi:10.1093/nar/gkp1004
4. Dutta S, Burkhardt K, Young J, Swaminathan GJ, Matsuura T, Henrick K, Nakamura H, Berman HM (2009) Data deposition and annotation at the worldwide protein data bank. Mol Biotechnol 42(1):1–13. doi:10.1007/S12033-008-9127-7
5. Al-Lazikani B, Lesk AM, Chothia C (1997) Standard conformations for the canonical structures of immunoglobulins. J Mol Biol 273(4): 927–948. doi:10.1006/jmbi.1997.1354
6. Kabat EA, Wu TT (1991) Identical V-region amino-acid-sequences and segments of sequences in antibodies of different specificities - relative contributions of Vh and Vl genes, minigenes, and complementarity-determining regions to binding of antibody-combining sites. J Immunol 147(5):1709–1719
7. Lefranc MP, Pommie C, Ruiz M, Giudicelli V, Foulquier E, Truong L, Thouvenin-Contet V, Lefranc G (2003) IMGT unique numbering for immunoglobulin and T cell receptor variable domains and Ig superfamily V-like domains. Dev Comp Immunol 27(1):55–77
8. Honegger A, Pluckthun A (2001) Yet another numbering scheme for immunoglobulin variable domains: an automatic modeling and analysis tool. J Mol Biol 309(3):657–670. doi:10.1006/Jmbi.2001.4662
9. Abhinandan KR, Martin AC (2008) Analysis and improvements to Kabat and structurally correct numbering of antibody variable domains. Mol Immunol 45(14):3832–3839. doi:10.1016/j.molimm.2008.05.022
10. Chothia C, Lesk AM (1987) Canonical structures for the hypervariable regions of immunoglobulins. J Mol Biol 196(4):901–917
11. Kunik V, Peters B, Ofran Y (2012) Structural consensus among antibodies defines the antigen binding site. PLoS Comput Biol 8(2). Artn E1002388. doi:10.1371/Journal.Pcbi.1002388.
12. Zemla A, Venclovas C, Moult J, Fidelis K (1999) Processing and analysis of CASP3 protein structure predictions. Proteins 3:22–29
13. Zhang Y, Skolnick J (2004) Scoring function for automated assessment of protein structure template quality. Proteins 57(4):702–710. doi:10.1002/prot.20264
14. Rousseeuw PJ (1987) Silhouettes - a graphical aid to the interpretation and validation of cluster-analysis. J Comput Appl Math 20:53–65. doi:10.1016/0377-0427(87)90125-7
15. Breiman L (2001) Random forests. Mach Learn 45(1):5–32. doi:10.1023/A:1010933404324
16. Archer KJ, Kirnes RV (2008) Empirical characterization of random forest variable importance measures. Comput Stat Data Anal 52(4):2249–2260. doi:10.1016/J.Csda.2007.08.015
17. Zemla A (2003) LGA: a method for finding 3D similarities in protein structures. Nucleic Acids Res 31(13):3370–3374

Chapter 19

Automated High-Throughput Mapping of Linear B-Cell Epitopes Using a Statistical Analysis of High-Density Peptide Microarray Data

Thomas Østerbye and Søren Buus

Abstract

Detailed information of antibodies' specificity is often missing or inadequate even for continuous (i.e., linear) epitopes. Recent developments in peptide microarray technology has enabled the synthesis of up to two million peptides per array thereby allowing linear peptide epitopes to be examined by a systematic amino acid substitution and positional scanning approach. This kind of analysis generates a very large body of data, which needs to be analyzed and interpreted in a robust and automated manner. Here, we describe a rational systematic approach to define linear antibody epitopes using ANOVA statistics to identify not only significant but also important residues involved in antibody recognition. This statistical approach can be used to perform a comprehensive linear epitope discovery. For polyclonal antibodies, this could be extended to entire proteins pinpointing critical residues for each epitope. We argue that the ANOVA analysis levels out issues of unknown peptide concentration/quality and unknown antibody titers leading to identification of epitopes that otherwise would be neglected if the evaluation was based merely on signal strength.

Key words ANOVA, Peptide chip, Linear epitopes, Eta square, Antibody epitope

1 Introduction

By their very nature, antibodies raised against protein antigens during a natural immune response will be polyclonal and recognize both discontinuous and continuous epitopes; for an in-depth review, see Regenmortel [1]. For many reasons, it may be desirable to identify antibody epitopes, preferably at high resolution, suggesting which specific residues are directly (e.g., contact residues) or indirectly (e.g., framework determinants) involved in the antigen-antibody interaction. Such information could be vital to the identification of antibody targets as well as off-targets. Structural analysis remains the golden standard of epitope analysis in particular of discontinuous epitopes and when detailed information is needed [2]. In contrast, linear

Gunnar Houen (ed.), *Peptide Antibodies: Methods and Protocols*, Methods in Molecular Biology, vol. 1348,
DOI 10.1007/978-1-4939-2999-3_19, © Springer Science+Business Media New York 2015

epitopes can be identified using less cumbersome methods such as synthetic peptide libraries, mix and split libraries, proteolytic fragments, phage display libraries, etc. [3–7]. Recent advances within the field of peptide microarray technologies have generated high-density peptide microarrays expressing up to two million addressable peptides. Here, we use the data from a detailed analysis of linear epitopes recognized by polyclonal rabbit antihuman serum albumin (HSA) antibodies [8] where the capacity of this high-density peptide microarray technology to express and represent HSA by a systematic series of overlapping peptides each including an exhaustive set of single amino acid substitutions was exploited.

In this study [8], the entire HSA protein was synthesized in situ on a peptide microarray as 15-mer peptides with an overlap of 14 amino acids. In addition, every position of each native 15-mer peptide was scanned using single amino acid substitutions including all of the 20 naturally occurring amino acids giving rise to a total of $15 \times (20-1) = 285$ variants and multiple copies of the native peptide. After synthesis and deprotection, the peptide chip was incubated with polyclonal rabbit anti-HSA, washed, subsequently incubated with a Cy3-labeled goat anti-rabbit, washed, and scanned. The resulting image revealed peptides recognized by the polyclonal anti-HSA antibodies. This image was converted to a text file containing peptide sequences and a quantified fluorescence intensity (AU value).

This analysis involved more than 170,000 different peptides. Analysis and interpretation of this amount of data is a challenge. Here, we describe a statistical method based on ANOVA and associated post hoc analysis, which allows an easy, automated, robust, and sensitive method to identify linear epitopes at high resolution (i.e., at the single-residue level).

At a first glance, a scatterplot of the overlapping native peptides and their corresponding signal intensities (AU) revealed 6–7 putative epitopes (Fig. 1a). However, merely evaluating signal intensity is potentially a misleading approach to epitope identification as several crucial parameters, the quantity and quality of the individual synthetic peptide epitopes, the concentration of epitope-specific antibodies, and the K_D of the interaction between peptide epitopes and antibodies, are usually not known in this kind of experiment (*see* **Note 1**). We have proposed an approach that can identify highly specific interactions between linear epitopes and antibodies even if they are detected at low signal intensity. This approach is based on a stringent statistical analysis of the selectivity of the interaction. Using the enormous peptide synthesis capability of current high-density peptide microarray technology, we perform complete single amino acid substitution analyses of every possible linear epitope that can be derived from a given target protein (in the analysis

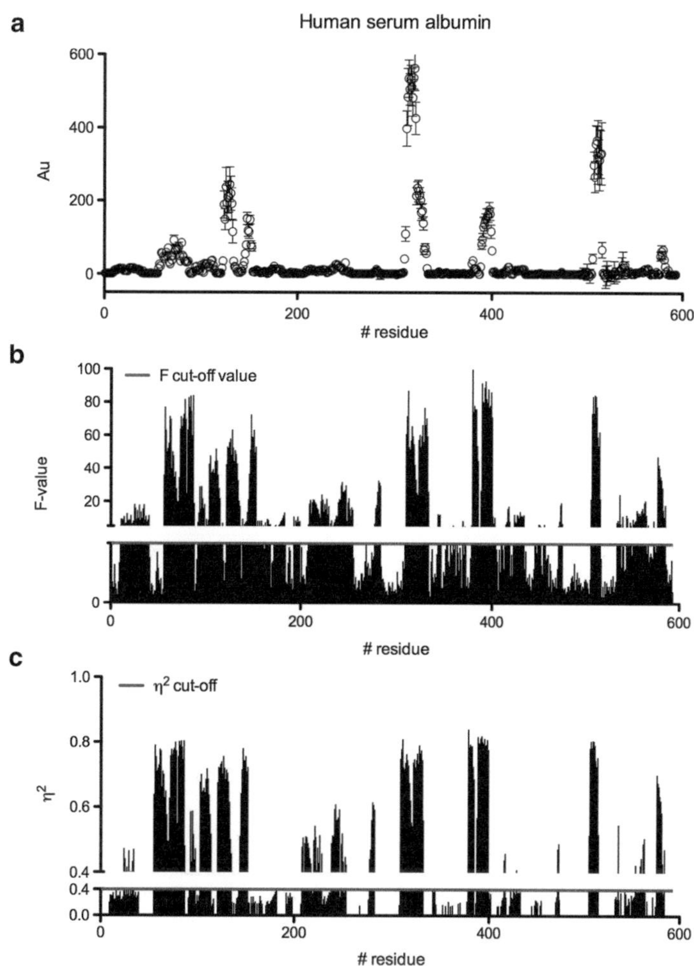

Fig. 1 Quantified data from a peptide chip [8] was downloaded to demonstrate the analysis of variance. (**a**) AU data corresponding to the overlapping native HSA peptides is plotted against each peptide N-terminal position in HSA. (**b**) For each native peptide, the *F*-value of the corresponding PSSMs is plotted against the peptide N-terminal position in HSA. The red line indicates *F*-value cutoff; PSSMs with an *F*-value <3.54 were not considered in the post hoc analysis. (**c**) For PSSMs with $F > 3.54$, η^2 was calculated and plotted against each native peptide's N-terminal position in HSA. The *red line* indicates a η^2 cutoff value 0.4, which indicates that the observed difference in variance is ascribed to variance between groups (amino acid substitutions)

presented below, we have restricted the analysis to linear epitopes up to 15 amino acids in length). This analysis is based on the assumption that the quantities and qualities of synthetic peptides are less variable for a set of single substituted peptide variants of a particular native peptide than across completely different peptide sequences. It is also based on the assumption that the concentration of epitope-specific antibodies is the same for each epitope variant. Finally, it is based on the assumption that normalizing the results of each single amino acid substitution scan to the native 15-mer peptide generates a set of data representing the relative effect of each substitution in each position on the affinity of the antibody in question. Hence, the analysis becomes independent of information about peptide quantity and quality, of the concentration of epitope-specific antibodies, and of the K_D of the parent peptide-antibody interaction. Thus, this approach enables a comprehensive specificity analysis of linear antibody epitopes recognized by polyclonal antibodies. As we shall see below, this allows a simple and automated statistical approach to specificity analyses, which can be reduced to numbers that can be used to identify, compare, and rank epitopes.

2 Materials

To illustrate the statistical methods, we have used the data from Hansen et al. which is publicly available at http://datadryad.org/resource/doi:10.5061/dryad.3003f. This data set represents an analysis of linear epitopes recognized by polyclonal rabbit anti-HSA antibodies and uses a peptide library composed of a systematic set of 15-mer peptides overlapping by 14 amino acids derived from HSA. Each overlapping native 15-mer peptide has been subjected to a complete amino acid substitution analysis at each position thus involving 19 substitutions in 15 positions (Fig. 2), or 285 variant peptides, as well as a number of copies (typically 5–10) of the native peptide.

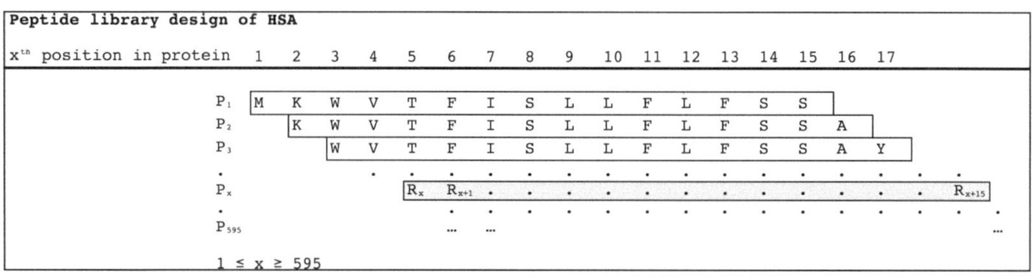

Fig. 2 Design principle of the HSA peptide library arrayed. The entire HSA protein was represented by 595 native overlapping 15-mer peptides (14 amino acid overlap). Overlapping native peptides are designated P_x, where x refers to the starting position of the peptide in HSA. Each peptide is substituted in all positions with all 19 remaining amino acids yielding a total of 285 variants of each native peptide. The native peptide P_x is highlighted in this as subsequent figures to illustrate the calculating principles

3 Methods

3.1 Organization of the Data into a Series of Overlapping Position-Specific Scoring Matrices (PSSMs)

For each native 15-mer peptide, the data is organized into a PSSMs with 15 columns representing the positions of the native amino acids within the peptide and 20 rows representing the 20 naturally occurring amino acids. The general method to organize peptide array data in PSSMs is illustrated in Fig. 3. As an example, raw AU values representing HSA_{510-24} peptide and corresponding substitutions are illustrated in Table 1.

3.2 Data Normalization

For each PSSMs, signals obtained for each substituted peptide are normalized to the mean signal of the corresponding native peptide. For each column (i.e., position), the mean (μ) and standard deviation are calculated as illustrated in Table 2.

If a given position within this peptide does not contribute to the specificity of the interaction (e.g., the amino acid residues present in the native sequence can be freely substituted without

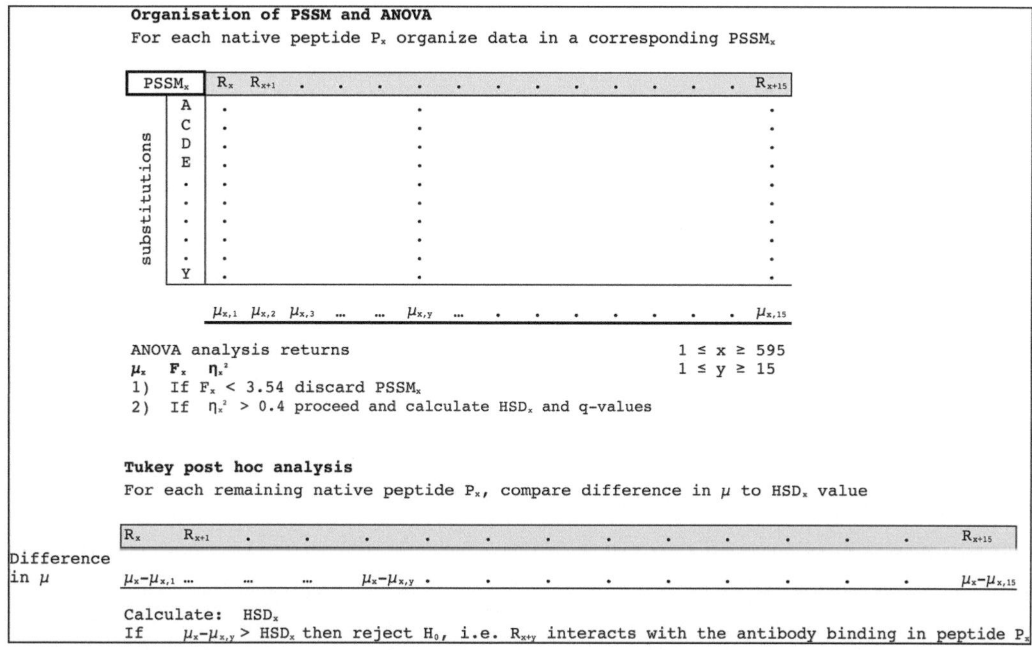

Fig. 3 For each native peptide P_x and corresponding variants, organize data in position-specific scoring matrices ($PSSM_x$) consisting of the scoring values for each peptide and normalize the $PSSM_x$ to the average score (AU value) of the native peptide P_x. Perform an ANOVA of each $PSSM_x$ as described returning the μ_x-, F_x-, and η^2_x-values. Note that the column-specific column means contain two indexes $\mu_{x,y}$, where x always refers to the xth position in the entire protein and y always refers to the yth position within the 15-mer native peptide. This nomenclature is consistent for all subsequent values calculated. Post hoc analysis (Tukey): for PSSMs returning $F > 3.54$ and $\eta^2 > 0.4$, calculate the difference between each position-specific $\mu_{x,y}$ and the mean of the native peptide P_x (normalized to 100) and compare to the HSD value

Table 1
Position-specific scoring matrix (PSSMs). An example of a PSSMs is shown here for peptide PSFSALEPDETYVPK ($HSA_{510–24}$) and all the single-position substituted peptides

	P	**C**	**F**	**S**	**A**	**L**	**E**	**P**	**D**	**E**	**T**	**Y**	**V**	**P**	**K**
A	572	522	633	638	**571**	344	190	495	82	369	101	260	556	528	572
C	492	**571**	522	567	656	321	157	526	82	309	118	245	683	577	562
D	555	487	542	467	453	347	289	432	**571**	410	0	176	586	609	581
E	521	643	548	631	593	236	**571**	442	141	**571**	48	159	697	527	485
F	591	639	**571**	631	615	518	188	468	97	291	77	537	561	539	529
G	612	581	619	611	602	303	141	394	81	322	104	195	480	607	555
H	0	545	595	574	436	357	120	309	105	200	121	200	469	519	524
I	449	628	564	478	506	509	169	541	85	318	87	209	716	604	673
K	555	572	477	505	624	385	142	597	96	310	123	166	547	579	**571**
L	603	603	478	523	561	**571**	131	589	80	381	84	184	507	577	553
M	646	581	616	493	605	432	196	491	93	294	119	315	667	552	570
N	579	600	580	502	473	302	173	404	92	307	159	215	519	607	483
P	**571**	655	496	564	589	346	142	397	0	440	157	184	547	**571**	539
Q	576	509	546	513	518	417	161	532	103	419	104	233	615	559	579
R	591	505	474	627	552	339	122	521	119	218	97	225	608	636	504
S	495	596	527	**571**	528	206	151	407	109	394	525	235	553	497	564
T	594	635	492	574	644	394	205	570	76	350	**571**	188	574	534	541
V	624	634	483	588	573	598	280	571	86	323	112	161	**571**	546	592
W	525	680	501	593	649	428	221	498	112	387	136	330	500	604	550
Y	520	572	515	569	613	291	213	436	96	329	113	**571**	595	540	603
μ	**534**	**588**	**539**	**561**	**568**	**382**	**198**	**481**	**115**	**347**	**148**	**249**	**578**	**566**	**556**
STD	**135**	**54**	**50.9**	**54.1**	**64.5**	**104**	**99.3**	**77.2**	**110**	**80.9**	**141**	**114**	**70.4**	**37.3**	**42.6**

The top row shows the amino acid sequence of the peptide in question. The leftmost column shows all 20 amino acids used to substitute each position in the native peptide. The numbers in the matrix shows the AU value of each peptide corresponding a given substitution (e.g., the AU value of 122 at intersection between the row 15 (R) and column 7 (E) corresponds to a 516E → R substitution in the native peptide)

affecting binding), then the mean normalized binding, μ, of the substitutions at this nonselective position will be 100 %. If, however, this position cannot be freely substituted, then μ will typically be less than 100 % (*see* **Note 2**) and in extreme cases, where the native amino acid is the only acceptable residue of the 20 naturally occurring amino acids, μ will be 1/20 = 5 %. Thus, μ can be used to identify the exact position(s) responsible for selective antibody-peptide interaction(s).

Table 2
Normalized PSSMs. All the AU values from Table 1 are normalized to the AU value of the native peptide, which score is now 100

	P	C	F	S	A	L	E	P	D	E	T	Y	V	P	K
A	100	91	111	112	100	60	33	87	14	65	18	46	97	93	100
C	86	100	91	99	115	56	28	92	14	54	21	43	120	101	98
D	97	85	95	82	79	61	51	76	100	72	0	31	103	107	102
E	91	113	96	111	104	41	100	77	25	100	8	28	122	92	85
F	104	112	100	111	108	91	33	82	17	51	13	94	98	94	93
G	107	102	108	107	106	53	25	69	14	56	18	34	84	106	97
H	0	96	104	101	76	63	21	54	18	35	21	35	82	91	92
I	79	110	99	84	89	89	30	95	15	56	15	37	125	106	118
K	97	100	84	89	109	67	25	105	17	54	22	29	96	101	100
L	106	106	84	92	98	100	23	103	14	67	15	32	89	101	97
M	113	102	108	86	106	76	34	86	16	52	21	55	117	97	100
N	101	105	102	88	83	53	30	71	16	54	28	38	91	106	85
P	100	115	87	99	103	61	25	70	0	77	28	32	96	100	94
Q	101	89	96	90	91	73	28	93	18	73	18	41	108	98	101
R	104	89	83	110	97	59	21	91	21	38	17	39	107	111	88
S	87	104	92	100	93	36	26	71	19	69	92	41	97	87	99
T	104	111	86	101	113	69	36	100	13	61	100	33	101	94	95
V	109	111	85	103	100	105	49	100	15	57	20	28	100	96	104
W	92	119	88	104	114	75	39	87	20	68	24	58	88	106	96
Y	91	100	90	100	107	51	37	76	17	58	20	100	104	95	106
μ	94	103	94	98	100	67	35	84	20	61	26	44	101	99	98
STDEV	23.6	9.46	8.92	9.48	11.3	18.3	17.4	13.5	19.4	14.2	24.8	20	12.3	6.54	7.47

F	76.1
η2	0.79

The calculated mean (μ) of each group (column) and the standard deviations are shown below the PSSMs and subsequently used to calculate the F (76.1)- and η^2 (0.79)-values. The red to green color scale simply highlights which substitutions have the largest effect (no statistical significance can be ascribed to the color grading)

3.3 ANOVA Analysis

To determine whether there is a groupwise (i.e., position-specific) effect of the substitutions, each PSSMs is subjected to an ANOVA analysis. This will determine whether the observations can be explained by random variations.

Calculate the F-values using the formula

$$F = \frac{\text{MS}_{\text{between}}}{\text{MS}_{\text{within}}}$$

where $\text{MS}_{\text{within}}$ is the standard deviation within groups and $\text{MS}_{\text{between}}$ is the standard deviation between groups.

The F-value indicates the likelihood of the observed variance being randomly distributed or not. In the example (Table 2), an F-value of 76.1 is found for the native peptide examined in this manner. Is this a significant finding? The null hypothesis is that the observed distribution of variation between positions can be explained as a result of random variation. Given the degrees of freedom of an ANOVA analysis with 15 columns (peptide positions) and 20 members per column (substitutions), an F-value of more than 2.54 is significant at the 1 % level for a given native 15-mer peptide considered in isolation. However, in this case multiple overlapping peptides are used to represent an entire protein antigen, and consequently this F test is repeated multiple times (to be specific, 595 times). Thus, an F-value of more than 3.54 corresponding to $P = 1.689 \times 10^{-5}$ would be more appropriate as this is still significant at the 1 % level even when considering all 595 experiments as a whole (i.e., at this F-value, there is a $\left(\frac{1-0.00001689}{1}\right)^{595} = 1\%$ risk that one of 595 tests is the result of a random distribution). Thus, an F-value above 3.54 would lead to the rejection of the null whypothesis and to the conclusion that the antibody interacts with the antigen in a position-specific manner. Thus, an F-value of 76.1 in our example (Table 2) is highly significant. It strongly suggests selective antibody recognition, indicating that some positions are more important than others and that the underlying native 15-mer peptide contains a linear epitope.

The calculated F-values for all overlapping native peptides can then be plotted against the N-terminal position of each peptide as shown for the entire HSA in Fig. 1b.

Comparing Fig. 1a, b, it is revealed that already this simple operation identifies several functionally defined (*see* **Note 3**) peptide-antibody interactions with position-specific effects, suggesting the presence of linear epitopes that could not readily be identified by the analysis of Fig. 1a based solely on signal intensity or AU values.

3.4 Post Hoc Analysis: Effect Size

Once significant PSSMs have been identified, the next question is whether these findings are important (*see* **Note 4**). The "effect size," eta² (or η^2), indicates how much of the total variance can be accounted for by the "between-group" (i.e., position-specific) variance. For all native peptides, where the ANOVA analyses yield a significant F-value, the effect size is calculated. The following formula is used:

$$\eta^2 = \frac{SS_{\text{between}}}{SS_{\text{total}}}$$

where SS_{between} is the sum of squares between groups and SS_{total} is the sum of squares in total (*see* **Note 5**).

In our example, an η^2-value of 0.79 was calculated for the native peptide HSA_{510-24} examined in Table 2. This shows that this particular peptide-antibody interaction is strongly affected by substitutions, i.e., that it contains epitope residues critical to antibody recognition. For this kind of analysis, we suggest that values of $\eta^2 > 0.3$–0.4 should be considered important (*see* **Note 6**).

The η^2 is a measure of the effect of the substitutions; an effect, which can be compared across different peptides allowing identification of native peptides where the interaction between antibody and peptide is tangibly affected by substitutions and therefore exhibits appreciable specificity. In Fig. 1c, η^2-values corresponding to each native peptide are plotted against the position of the peptide N-terminal with the η^2 cutoff of 0.4 indicated by the red line.

The η^2 plot reveals that only some of the peptides with high *F*-values are solid epitope candidates. This procedure automates the selection of peptide candidates, which may harbor an epitope, and should be subjected to further post hoc analysis to identify which positions within each peptide candidate are critical to antibody recognition.

3.5 Tukey's Honest Significant Difference

Up to this point, the ANOVA has only indicated which of the 15-mer peptides may contain epitope regions; it has not revealed which positions within the 15-mer peptides may be responsible for the observed nonrandom distribution of variance. Our null hypothesis is that a position is completely nonselective and, as detailed above, such a position would be characterized by a μ of 100 %. In contrast, an absolutely selective position would be characterized by a μ of 5 %. To evaluate whether the observed μ of a given position is significantly different from 100 % (the null hypothesis), perform a Tukey's Honest Significant Difference (Tukey's HSD) post hoc test for all positions of all native 15-mer HSA peptides with significant ANOVA results.

Tukey's Honest Significant Difference (HSD) value can be calculated according to the formula

$$\text{HSD} = q\sqrt{\frac{\text{MS}_{\text{within}}}{n}}$$

where q is the relevant critical value of the studentized range, MSwithin is the mean square error, and n is the number of scores used in calculating the mean of interest.

For each column (i.e., position) within the PSSMs, calculate the difference between the μ of the native peptide (100 after normalization) and the column-specific μ and compare this difference to the LSD value, as illustrated in Fig. 3. At the 1 % significance level, Tukey's HSD will identify one or more positions of selectivity in many of the native 15-mer peptides (*see* **Note 7**).

In our $\text{HSA}_{510\text{–}24}$ example (Table 3), an HSD value of 17.89 at the 1 % level identifies six residues, which functionally (i.e., directly as a contact residue or indirectly as a framework residue) interacts with antibody binding.

3.6 Interactions Assigned to a Single Residue

The Tukey HSD post hoc analysis classifies positions where substitutions have an effect above a certain threshold. A more quantitative estimate can be obtained using the *q*-value for each individual residue within a native 15-mer peptide, according to the formula:

$$q = \frac{(\mu - 100)}{\sqrt{\frac{\text{MS}_{\text{within}}}{n}}}$$

where μ is the mean normalized binding of the substitution of a position, MSwithin is the mean square error within groups, and n is the size of the group.

Table 3
A post hoc (Tukey) analysis of the PSSMs corresponding to the peptide PSFSALEPDETYVPK (HSA_{510-24}) reported a HSD value of 17.89 (least significant difference) at the 1 % level

	P	C	F	S	A	L	E	P	D	E	T	Y	V	P	K
$\mu_{native} - \mu$	6	−3	6	2	0	**33**	**65**	16	**80**	**39**	**74**	**56**	−1	1	2
q	0.86	−0.4	0.74	0.22	0.06	4.38	8.66	2.08	10.6	5.19	9.83	7.46	−0.2	0.12	0.33

The differences (the effect of substitution the native amino acid) are shown below each amino acid with the significant positions highlighted in red and bold numbers. The bottom row shows the calculated *q*-values for each substitution, which indicates the significance of each single amino acid residue as recognized by the rabbit anti-HSA serum

Table 4
Determining the position-specific *q*-values *q*-values representing the overlapping peptides in HSA_{56-93} for PSSMs with $F > 3.54$ are organized according to the principle shown in Fig. 4, with the peptide HSA_{66-80} highlighted by bold borders. Each *q*-value is colored according to value (green to red) showing positions significant to antibody recognition (red). The mean *q* for each position is shown at the bottom, and significant residues in the protein sequence are highlighted in black at the top

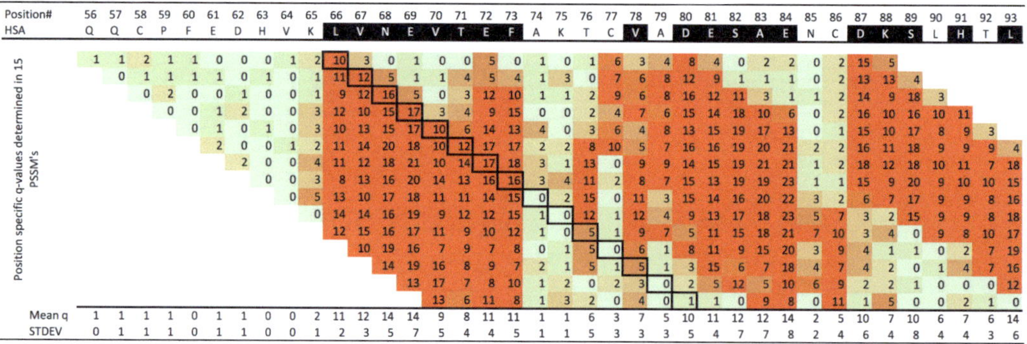

Position#	56	57	58	59	60	61	62	63	64	65	66	67	68	69	70	71	72	73	74
HSA	Q	Q	C	P	F	E	D	H	V	K	L	V	N	E	V	T	E	F	A
Position specific q-values determined in 15 PSSM's	1	1	2	1	1	0	0	0	1	2	10	3	0	1	0	0	5	0	1
		0	1	1	1	1	0	1	0	1	11	12	5	1	1	4	5	4	1
			0	2	0	0	1	0	0	1	9	12	16	5	0	3	12	10	1
				0	0	1	2	0	0	3	12	10	15	17	3	4	9	15	0
					0	1	0	1	0	3	10	13	15	17	10	6	14	13	4
						2	0	0	1	2	11	14	20	18	10	12	17	17	2
							2	0	0	4	11	12	18	21	10	14	17	18	3
								0	0	3	8	13	16	20	14	13	16	16	3
									0	5	13	10	17	18	11	11	14	15	0
										0	14	14	16	19	9	12	12	15	1
											12	15	16	17	11	9	10	12	1
												10	19	16	7	9	7	8	0
													14	19	16	8	9	7	2
														13	17	7	8	10	1
															13	6	11	8	1
Mean q	1	1	1	1	0	1	1	0	0	2	11	12	14	14	9	8	11	11	1
STDEV	0	1	1	1	0	1	1	0	0	1	2	3	5	7	5	4	4	5	1

Position#	75	76	77	78	79	80	81	82	83	84	85	86	87	88	89	90	91	92	93
HSA	K	T	C	V	A	D	E	S	A	E	N	C	D	K	S	L	H	T	L
Position specific q-values determined in 15 PSSM's	0	1	6	3	4	8	4	0	2	2	0	2	15	5					
	3	0	7	6	8	12	9	1	1	1	0	2	13	13	4				
	1	2	9	6	8	16	12	11	3	1	1	2	14	9	18	3			
	0	2	4	7	6	15	14	18	10	6	0	2	16	10	16	10	11		
	0	3	6	4	8	13	15	19	17	13	0	1	15	10	17	8	9	3	
	2	8	10	5	7	16	16	19	20	21	2	2	16	11	18	9	9	9	4
	1	13	0	9	9	14	15	19	21	21	1	2	18	12	18	10	11	7	18
	4	11	2	8	7	15	13	19	19	23	1	1	15	9	20	9	10	10	15
	2	15	0	11	3	15	14	16	20	22	3	2	6	7	17	9	9	8	16
	0	12	1	12	4	9	13	17	18	23	5	7	3	2	15	9	9	8	18
	0	5	1	9	7	5	11	15	18	21	7	10	3	4	0	9	8	10	17
	1	5	0	6	1	8	11	9	15	20	3	9	4	1	1	0	2	7	19
	1	5	1	5	1	3	15	6	7	18	4	7	4	2	0	1	4	7	16
	2	0	2	3	0	2	5	12	5	10	6	9	2	2	1	0	0	0	12
	3	2	0	4	0	1	1	0	9	8	0	11	1	5	0	0	2	1	0
Mean q	1	6	3	7	5	10	11	12	12	14	2	5	10	7	10	6	7	6	14
STDEV	1	5	3	3	3	5	4	7	7	8	2	4	6	4	8	4	4	3	6

In the case of HSA_{510-24}, *q*-values have been included in Table 3. Since each epitope has the opportunity to appear in several consecutive overlapping 15-mer peptides, *q*-values for the same residue, but in different positions of these overlapping 15-mer peptides, should be calculated. As a practical approach to merge the *q*-values of the same residue represented in one or more of these overlapping 15-mer HSA peptides, we propose to use the mean of the *q*-values for a specific residue obtained in overlapping peptides with significant *F*-values and important η^2-fractions. To easily address position-specific interactions, organize the *q*-values from the significant PSSMs as illustrated in Fig. 4, and calculate position-specific *q*-values (mean of *q*s from overlapping native peptides).

By examining the *q*-values found in HSA_{510-24}, it now becomes evident which positions are contributing to the peptide-antibody interactions (-----LE-DETY---) as seen in Table 3. To illustrate the principle of calculating the position-specific (mean *q*) for the entire HSA protein, Table 4 shows the organization of PSSMs specific *q*-values and subsequent assignment of interactions for each residue in HSA_{56-93}.

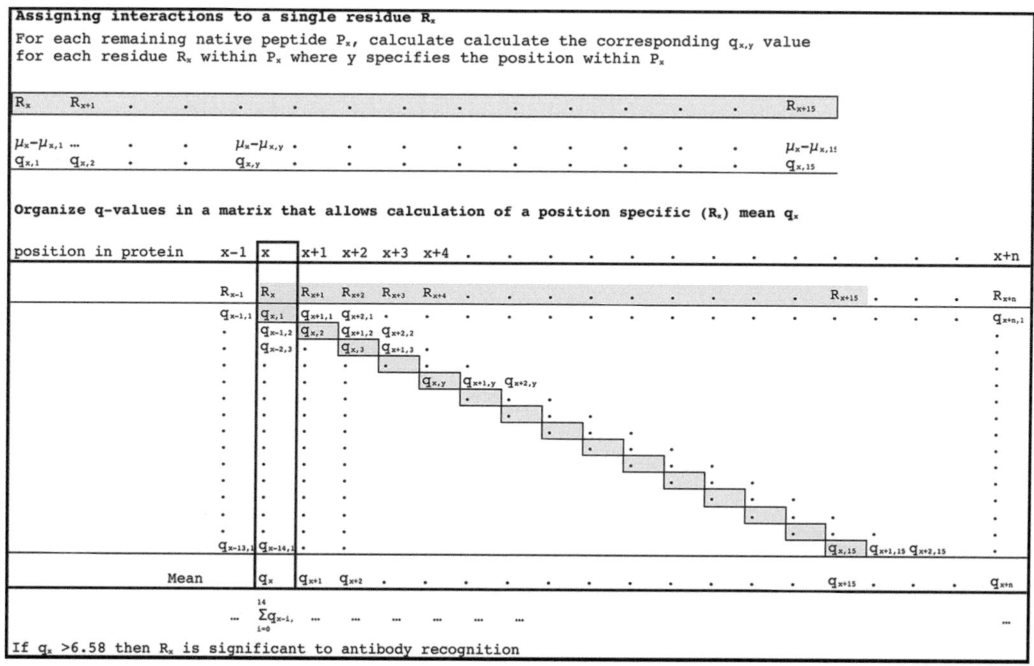

Fig. 4 For each peptide P_x calculate the $q_{x,y}$-values as described and organize the q-values in a new matrix (for HSA a 595 × 15 matrix) so each column contains the 15 q-values calculated for residue R_x. For the highlighted peptide P_x, the corresponding $q_{x,y}$-values are also highlighted. Since each native residue R_x is represented in 15 overlapping native peptides, it is now possible to calculate the mean q_x. Note, that since only significant PSSMs are subjected to post hoc and q-value analysis, the number of q-values may be less than 15

With this operation, we have now resolved the information originating from 15-mer peptides to an assignment to a particular single amino acid residue, relative to the entire protein, Fig. 5.

3.7 Mapping Epitopes onto a Protein Antigen (In Casu HSA) Structure

As a result of the ability to characterize linear epitopes at the single-residue level, it is now possible to map the epitopes onto the 3D structure of the HSA protein with great precision (Fig. 6). Note that this is not possible with the AU, F, or η^2 data since these all emanate from the overlapping 15-mer peptides and therefore will lead to a considerable loss of resolution if this information is mapped onto the 3D structure.

4 Notes

1. With respect to the peptide reactant, P, the quantity and quality of peptides achieved by the photolithographic in situ synthesis process are sequence dependent. In reality, this component of the interaction is unknown. With respect to the antibody reactant, A, the concentration of peptide-specific antibodies within a preparation of polyclonal antibodies is unknown and bound

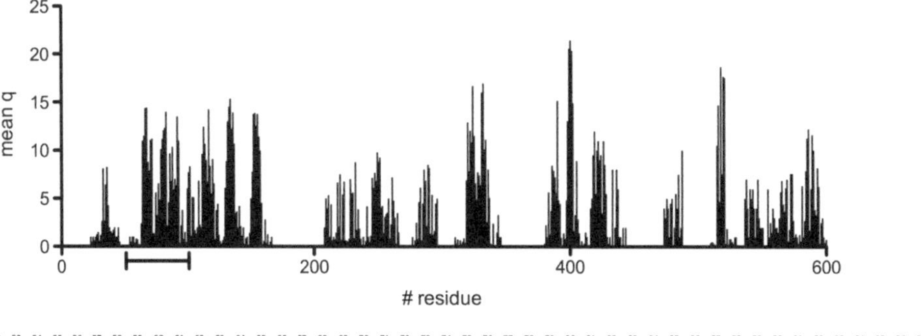

Fig. 5 The position-specific mean q_x-values plotted against x (position in native protein) provide an overview of residues involved in antibody recognition at the single-residue level of HSA. Positions 50–100 in HSA are shown below the plot with residues involved in antibody recognition highlighted

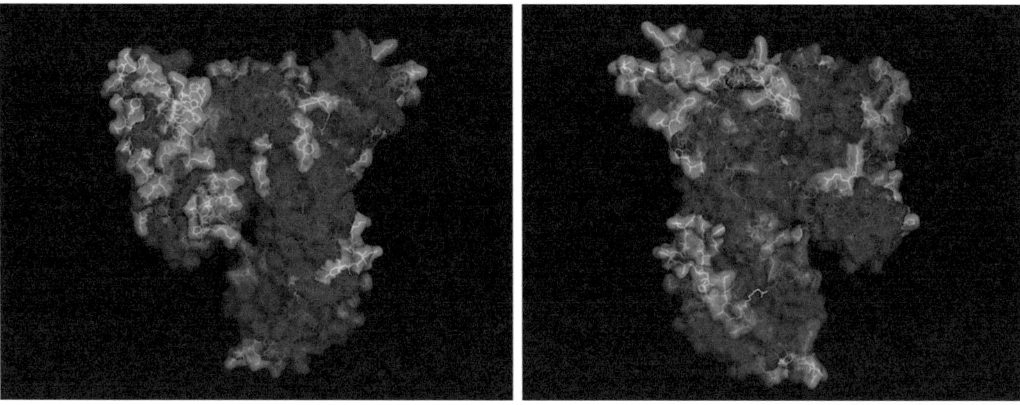

Fig. 6 Residues with a mean $q > 6.5$ are highlighted on the 3D structure of HSA (front and back) providing a detailed target view of the polyclonal antibody response. Note that all residues involved in antibody recognition appear to be located at the surface of the molecule

to be highly variable from preparation to preparation. In reality, this component of the interaction is also unknown. With respect to the affinity of the interaction, K_D, this quality is highly variable and depends on many factors beyond control. Let alone the immunization regime, the affinity can vary by many orders of magnitude. In reality, this component of the interaction is also unknown. Thus, analyzing the results of a peptide microarray solely on the basis of signal intensity is a deceptive evaluation of the affinity of the underlying peptide-antibody interactions. Further compounding this conundrum, even if this was possible to measure affinity and affinity is certainly an important quality of an antibody, the true hallmark of an antibody is its ability to distinguish between antigens, i.e., its specificity. Whereas affinity is a measurable parameter, specificity is conventionally considered as a nonquantifiable quality, which is not easily comparable and defies ranking. This statistical analysis is an attempt to obtain a quantitative measure of specificity.

2. In the event that the native residue is not the preferred amino acid, a situation known as a heteroclitic response, then μ can become more than 100 %. Indeed, a few heteroclitic responses were observed. These will not be further described or discussed here.
3. Strictly speaking, the approach taken here does not identify epitope residues in structural terms; rather, it measures peptide-antibody interactions and identifies epitope residues in functional terms. Using a substitution approach to identify epitope residues does not allow us to discriminate between contact residues between peptide and antibodies or framework determinants that may be needed to present the contact residues in the proper conformation.
4. Whether something is significant or not depends upon the sample size. A very strong (i.e., meaningful and important) effect may not reach a statistical significance level if the sample size is very small, whereas an infinitesimal (i.e., trivial and inconsequential) effect may reach a statistical significance level if the sample size is large enough.
5. Several websites (e.g., http://department.obg.cuhk.edu.hk/researchsupport/OWAV.asp) and software packages (SPSS, GraphPad Prism, SAS, etc.) are able to perform the ANOVA of the PSSMs and automatically report F-, η^2-, and LSD values.
6. What constitutes an important effect is, not surprisingly, somewhat arbitrary. Cohen has suggested that an $\eta^2 > 0.13$ should be considered as a large effect [9]. We have previously reported that linear peptide epitopes tend to be five amino acids long [10]. If we compare epitope length vs. η^2 for our data (Fig. 6), we observe that most epitopes of length ≥ 4 have η^2 of more than 0.3–0.4, which we would like to suggest as thresholds for this kind of analysis (Fig. 7). Whether you select a threshold η^2 of 0.3 or 0.4 depends on how many possible epitopes you will accept in subsequent follow-up experiments. A note of warning: some statistical packages (e.g., SPSS) calculate a partial η2 rather than the η2 used here [11].
7. As expected, peptides with lower F- and η^2-values featured few, if any, significant positions, whereas peptides with higher F- and η^2-values consistently featured one or more significant positions. Applying a 1 % threshold, we identified epitopes containing from 4 to 15 significant positions in 147 (25 %) of the 595 native 15-mer HSA peptides. Note that the same epitopes usually appeared in several consecutive native 15-mer

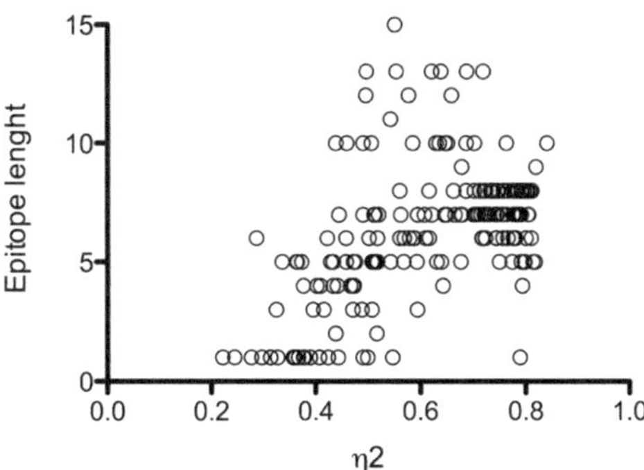

Fig. 7 A scatterplot showing the epitope length (deducted from each significant PSSMs) against the corresponding η^2-value. If an epitope length of four (containing at least three significant residues spaced with one nonsignificant residue) is considered to be the minimum length relevant to antibody recognition, an $\eta^2 > 0.4$ appears to be a relevant criterion to apply in the post hoc analysis

peptides, i.e., the number of different epitopes is considerably smaller than the 147 peptides containing epitopes. In this case, we suggested that this polyclonal rabbit anti-HSA antibody preparation is able to recognize more than 20 different linear HSA epitopes.

References

1. Van Regenmortel MH (2009) What is a B-cell epitope? Methods Mol Biol 524:3–20. doi:10.1007/978-1-59745-450-6_1
2. Gershoni JM, Roitburd-Berman A, Siman-Tov DD, Tarnovitski Freund N, Weiss Y (2007) Epitope mapping: the first step in developing epitope-based vaccines. BioDrugs 21:145–156
3. Benjamin DC, Perdue SS (1996) Site-directed mutagenesis in epitope mapping. Methods 9: 508–515
4. Chandra H, Srivastava S (2010) Cell-free synthesis-based protein microarrays and their applications. Proteomics 10:717–730. doi:10.1002/pmic.200900462
5. Cretich M, Damin F, Pirri G, Chiari M (2006) Protein and peptide arrays: recent trends and new directions. Biomol Eng 23(2-3):77–88. doi:10.1016/j.bioeng.2006.02.001
6. Dhungana S, Williams JG, Fessler MB, Tomer KB (2009) Epitope mapping by proteolysis of antigen-antibody complexes. Methods Mol Biol 524:87–101. doi:10.1007/978-1-59745-450-6_7
7. Pande J, Szewczyk MM, Grover AK (2010) Phage display: concept, innovations, applications and future. Biotechnol Adv 28:849–858. doi:10.1016/j.biotechadv.2010.07.004
8. Hansen LB, Buus S, Schafer-Nielsen C (2013) Identification and mapping of linear antibody epitopes in human serum albumin using high-density peptide arrays. PLoS One 8, e68902. doi:10.1371/journal.pone.0068902
9. Cohen J (1988) Statistical power analysis for the behavioral sciences, 2nd edn. Routledge, Abingdon
10. Buus S, Rockberg J, Forsstrom B, Nilsson P, Uhlen M, Schafer-Nielsen C (2012) High-resolution mapping of linear antibody epitopes using ultrahigh-density peptide microarrays. Mol Cell Proteomics 11:1790–1800. doi:10.1074/mcp.M112.020800
11. Levine TR, Hullett CR (2002) Eta squared, partial eta squared, and misreporting of effect size in communication research. Hum Commun Res 28:612–625. doi:10.1111/j.1468-2958.2002.tb00828.x

Chapter 20

Characterization of Peptide Antibodies by Epitope Mapping Using Resin-Bound and Soluble Peptides

Nicole Hartwig Trier

Abstract

Characterization of peptide antibodies through identification of their target epitopes is of utmost importance. Understanding antibody specificity at the amino acid level provides the key to understand the specific interaction between antibodies and their epitopes and their use as research and diagnostic tools as well as therapeutic agents.

This chapter describes a straightforward strategy for mapping of continuous peptide antibody epitopes using resin-bound and soluble peptides. The approach combines three different types of peptide sets for full characterization of peptide antibodies: (1) overlapping peptides, used to locate antigenic regions; (2) truncated peptides, used to identify the minimal peptide length required for antibody binding; and (3) substituted peptides, used to identify the key residues important for antibody binding and to determine the specific contribution of key residues. For initial screening resin-bound peptides are used for epitope estimation, while soluble peptides subsequently are used for fine mapping. The combination of resin-bound peptides and soluble peptides for epitope mapping provides a time-sparing and straightforward approach for characterization of peptide antibodies.

Key words Peptide antibody, Epitope mapping, Enzyme-linked immunosorbent assay, Synthetic peptides, Resin-bound peptides, Continuous epitope

1 Introduction

Peptide antibodies are powerful tools in experimental biology, and are, in contrast to many other antibodies, usually directed to a specific peptide region of limited size. Characterization of these antibodies is essential in relation to their application. Characterization of peptide antibodies is usually accomplished by epitope mapping, which determines the binding sites of antibodies on their target antigens.

Among several techniques, the use of synthetic peptides has emerged as a powerful tool for identification and characterization of peptide antibody epitopes [1–5]. In fact, the use of synthetic peptides as targets has become an alternative method in identifying

Gunnar Houen (ed.), *Peptide Antibodies: Methods and Protocols*, Methods in Molecular Biology, vol. 1348, DOI 10.1007/978-1-4939-2999-3_20,

individual epitopes, because synthetic peptide fragments can be similar to the native antigen structure, thus allowing the antibody to interact with the peptide antigen. Moreover, peptide antibodies are easily characterized using synthetic peptides, because the origin of these peptide antibodies often is known, as the immunogenic peptide used to generate the antibody is limited in size, usually shorter than 20 amino acids [6], favoring continuous epitopes.

An often applied approach for initial epitope identification is to employ overlapping peptides of varying length, where antibody reactivity is analyzed to peptides covering a protein or an antigenic region of interest [7]. This approach, referred to as peptide scanning or pepscan, is primarily used to identify continuous epitopes, where the key amino acids that mediate contact to the antibody are located within the primary structure, usually not exceeding 15 amino acids in length.

Even though this method primarily is applied to determine continuous epitopes, pepscans have occasionally been applied to indicate components of discontinuous epitopes in the event that two distant peptides each contain sufficient structural elements that allow them to bind the antibody separately, indicating that these peptides contribute to the intact epitope [8].

In recent years, resin-bound peptides in combination with free peptides, which are described in detail in this chapter, have been employed for systematic epitope mapping. This approach relies on the use of different peptide sets. In initial screening steps, resin-bound peptides are employed for rough epitope characterization, while identification of the final epitope and key amino acids are identified using free peptides. The peptide sets used for epitope characterization are described in the following sections. Moreover, details of procedures for detection of continuous epitopes are described. Mapping of continuous epitopes using the protocols provides a rapid, straightforward and cost-effective approach for characterization of peptide antibodies.

1.1 Peptide Sets

1.1.1 Overlapping Peptides

Importantly, the amino acid sequence of the immunogenic peptide used for generation of the peptide antibody must be known, either via protein sequencing or Genbank database, before epitopes can be mapped using synthetic peptides. Given the amino acid sequence, overlapping peptides are generated. Critical is the length of the peptide and the number of residues that is shifted along the protein sequence, also referred to as the offset number or overlap. 15-mer peptides have been recommended for these types of studies, since most continuous peptides do not exceed this length [9]. Nevertheless, studies using peptides of four to eight amino acids have been reported for identification of short epitopes [7]. However, smaller peptides may experience problems in obtaining stable conformations. As the majority of epitopes are longer than eight amino acids, the application of these peptide sets may be limited [9, 10]. The advantages in using longer peptides, for example peptides of 15–20 amino acids, are that longer epitopes can be identified and that longer peptides often adopt a structure similar to the native structure [5, 11, 12]. The most critical disadvantage in employing longer peptides is that they are more difficult to synthesize, which may result in lower purity.

Dependent on the location of the epitope within the protein, it may be an advantage to employ a relatively short offset number. One strategy is to apply peptides containing an offset of a single or two amino acids [1, 7]. This requires almost as many peptides as there are amino acids in the protein sequence or the immunogen. The advantage in applying a small offset is that terminal amino acid boundaries essential for antibody reactivity are readily determined. Employing a greater amino acid offset provides lower resolution and may need to be followed up with a more defined study to determine the minimal epitope. Examples include peptides of 15–17 amino acids with an offset of five residues [5, 13]. In general, efficient epitope mapping has been reported when applying peptides of 8–20 amino acids in length, with an overlap of 1–10 amino acids [5, 14–16].

1.1.2 Truncated Peptides

After identification of the antigenic region, truncated peptides are used to identify the epitope and key interacting amino acid residues. Using the immunogen or the antigenic region as template the peptides are truncated systematically from the N- or C-terminal end. This type of peptide set contains n peptides, where n represents the number of amino acid residues within the original peptides. The advantage of using N-terminally truncated resin-bound peptides for epitope mapping is that the peptides can be synthesized in the same batch as the peptides used for immunization for generation of peptide antibodies. Alternatively, the N-terminal screening results can be used as template for generation of C-terminally truncated peptides [10, 14, 16]. Determination of the minimal epitope requires testing of peptides of various lengths from 4-mers to the maximum size [16–18]. Alternatively, if the essential amino acids are known, the direction of truncation can be selected around them, as opposed to systematic truncation from both ends of the peptide sequence.

1.1.3 Substituted Peptides

The final step in epitope mapping determines the amino acids essential for antibody reactivity using single modified peptides [12, 14, 19, 20]. Those amino acids that mediate direct contact with the antibody and cannot be substituted without loss of binding are essential for antibody binding. Initially the peptides are substituted systematically with Ala in each position [11, 12, 14]. Alascan determines the roles of individual amino acid side chains for antibody reactivity, as the substitution with alanine removes all side-chain atoms past the β-carbon. The amino acid Ala is used because of its non-bulky, chemically inert, methyl functional group which nevertheless mimics the secondary structure preferences that many other amino acids possess. This type of peptide set consists of Ala single-site substitution analogs. Ala scan may be supported by functionality scans, where amino acids containing similar side-chain functionality are substituted in the identified epitope [10, 20, 21]. Here each amino acid is substituted with an amino acid of similar functionality, generating a peptide set of single substitution analogs. Alternatively, each amino acid is substituted with all 20 amino acids, generating a library consisting of all possible single-site substitution analogs. However, this complete substitutional peptide analysis is rather time consuming and requires many peptides [7]. Although rarely applied, this strategy determines whether the side-chain functionality rather than the specific amino acid side chain is essential for activity at key positions within the epitope.

1.2 The Principle of Synthetic Resin-Bound Peptides for Rough Epitope Identification

When using synthetic peptides for epitope characterization, peptide presentation is of utmost importance. Several approaches for peptide presentation have been described in the literature, e.g., membranes, biotin-streptavidin, and pins. This chapter, however, describes the use of resins for peptide presentation. Resin or solid support is a commonly used term applied to denote the matrix upon which peptide synthesis is conducted, which was introduced in peptide synthesis by Merrifield [22].

An obstacle when conducting peptide enzyme-linked immunosorbent assays (ELISA) is that the conformation of the peptide may be altered when coated onto the surface of microtiter plates, hiding essential amino acid side chains for reactivity. Moreover, the amino acids essential for reactivity may interact with the microtiter plate and become unable to interact with the antibody. Similarly, it may be difficult to ensure binding of small peptides, due to a limited number of possible interactions between the microtiter plate and the peptide. One way to circumvent these issues is to conduct peptide screening directly on the solid support. When using resin-bound peptides, only a single amino acid is bound to the solid support, making the remaining amino acids capable of achieving a flexible structure, whereas in ELISA several amino acid residues interact with the solid surface, presumably making some of the peptides sterically hindered in adopting the conformations necessary for epitope presentation.

Several resins may be used for peptide presentation. The choice of resin only determines how the peptide is cleaved from the resin; hence, the actual screening approach is not influenced by the resin applied. The advantage of using resins as solid support is that time consuming steps, such as peptide cleavage, peptide purification and coating steps are avoided. A potential drawback of using

01 amino acids 648-667: TANLAAFLYLDRPEERITGI

02 amino acids 658-677: DRPEERITGINDPRLRNPSD

03 amino acids 668-687: NDPRLRNPSDKFIYATVKQS

04 amino acids 678-697: KFIYATVKQSSVDIYFRRQV

Peptide	1	2	3	4
% Reactivity	0	64	100	0

Fig. 1 Scan of overlapping peptides analyzed by modified ELISA using resin-bound peptides. A section of the human NMDAR sequence containing the immunogentic peptide amino acids 657–678 (LDRPEERITGINDPRLRNPSDK), used for production of the monoclonal antibody, is dissected into 20-mer overlapping peptides, containing 10-amino acid residue overlap. Two overlapping peptides were recognized by the NMDAR mAb. The sequence common for these peptides is *underlined*

resin-bound peptides is that the purity of the peptides is reduced with increasing amino acid number. However, screening of antibody reactivity using 20-mer resin-bound peptides has proved to be successful [10, 14, 21].

In the present approach, the peptides are screened on a resin with an acid-stabile linker, which allows cleavage of the side-chain protecting groups but leaves the peptide linked to the resin upon completion of peptide synthesis [10]. Alternatively, the peptides may be cleaved from the solid support and tested in an antibody capture ELISA using 0.1 M NaOH. No obvious difference between traditional peptide ELISA and resin-bound peptide ELISA is found, except from the microtiter plate used and final detection.

2 Materials

2.1 Peptides

A total of four overlapping peptides were designed based on the human *N*-methyl-D-aspartate receptor (NMDAR) protein sequence to cover the immunogenic peptide (amino acids 657–678) used for generation of the anti-human NMDAR peptide antibody. Every peptide was 20 amino acids long, containing a 10 amino acid overlap to the next peptide and remained on the solid support during antibody screening.

Following identification of the antigenic peptide, N-terminally truncated peptides were generated as seen in Fig. 1. Based on screening results C-terminally truncated peptides generated. Following, screening of N- and C-terminally truncated peptides, peptides truncated around essential amino acid residues were screened for identification of the final epitope (Fig. 2). Finally, systematically alanine-substituted peptides were screened for reactivity, followed by functionality substituted reactivity (Fig. 3).

1. The resin-bound peptides are supplied by in 10 % ethanol, 100 mg resin per 1 ml solvent.
2. The free peptides are supplied as a lyophilised product. Dissolve the free peptides in a suitable solvent (following the manufacturer's instructions) at a concentration of 1 mg/mL (*see* **Notes 1** and **2**).

2.2 Modified ELISA Using Resin-Bound Peptides and Competitive Inhibition Assay

1. 96-Well multiscreen filter plate (*see* **Note 3**).
2. Nunc Immuno® Maxisorp 96-well microtiter plate.
3. Nunc Immuno® round bottom 96-well microwell plate.
4. Tris-Tween-NaCl buffer (TTN): 0.05 M Tris, 0.3 M NaCl, 1 % Tween 20, pH 7.5 (*see* **Note 4**) Store at 4 °C.
5. Carbonate buffer: 15 mM Na_2CO_3, 35 mM $NaHCO_3$, 0.001 % phenol red, pH 9.6. Store at 4 °C. Discard if changes in pH occur.
6. Alkaline phosphatase (AP)-substrate buffer: 1 M diethanolamine, 0.5 mM $MgCl_2$, pH 9.8. Store at 4 °C.
7. Primary antibody: Rabbit anti-human NMDAR1 monoclonal antibody, clone D65B7, generated against a synthetic peptide (amino acids, 657–678) [23] surrounding Pro660 of the human NMDA receptor. Keep the reagent in the refrigerator or freezer according to the manufacturer's recommendations (*see* **Note 5**).
8. Secondary antibody: AP-conjugated anti-rabbit IgG antibody.
9. AP buffer: Dissolve one phosphatase substrate tablet (4-nitrophenyl phosphate) (5 mg) in 5 mL AP-substrate buffer. Any remains should be discarded. The substrate is light sensitive. Should be prepared immediately before use and kept in the dark or wrapped.

3 Methods

3.1 Modified ELISA Using Resin-Bound Peptides

It is strongly recommended to optimize concentrations of reagents such as peptides and antibodies for each ELISA system. The concentration mentioned does not apply to all antibodies examined.

1. Prepare dilutions of resin-bound peptides of the stock solutions. Resin-bound peptides (100 mg/mL resin) are diluted 1:30 in TTN buffer.
2. 100 μL Resin-bound peptides are added to a 96-well multiscreen filter plate (3.3 μg/μL).

a

N-terminal truncation

```
01.GINDPRLRNPSDKFI
02.-INDPRLRNPSDKFI
03.--NDPRLRNPSDKFI
04.---DPRLRNPSDKFI
05.----PRLRNPSDKFI
06.-----RLRNPSDKFI
07.------LRNPSDKFI
08.-------RNPSDKFI*
09.--------NPSDKFI
10.---------PSDKFI
11.----------SDKFI
12.-----------DKFI
13.------------KFI
14.-------------FI
15.--------------I
```

C-terminal truncation

```
16.NPSDKFI
17.NPSDKF-
18.NPSDK--
19.NPSD---
20.NPS----
21.NP-----
22.N------
```

Center truncation

```
23.--------NPSDK--
24.-------RNPSDK–
25.-------RNPSDKF-
26.-------RNPSDKFI
27.------LRNPSDKFI
```

b

Peptide	23	24	25	26	27	3
Reactivity%	42	100	100	100	90	100

Fig. 2 Scheme for truncation analysis of peptides. (**a**) Peptide 3 (*see* Fig. 1) of the overlapping peptides was subjected to N, C-, and bidirectional truncations. For generation of C-terminally truncated peptides, the shortest N-terminally truncated peptide (marked *asterisk*) yielding highest reactivity was used as template. (**b**) Reactivity to bidirectional truncations analysed by modified ELISA using resin-bound peptides, identifying the final epitope as peptide 24 (*underlined*)

Wt epitope	**Ala (a)**	**Functional**
RNPSDK	**A**NPSDK	**K**NPSDK
RNPSDK	R**A**PSDK	R**Q**PSDK
RNPSDK	RN**A**SDK	RNPSDK
RNPSDK	RNP**A**DK	RNP**T**DK
RNPSDK	RNPS**A**K	RNPS**E**K
RNPSDK	RNPSD**A**	RNPSD**R**

Peptide	RNPSDK	ANPSDK	RAPSDK	RNASDK	RNPADK	RNPSAK	RNPSDK
Inhibition (%)	90	37.5	71	75	100	60	100

Fig. 3 Substitutional analysis of the NMDAR mAb epitope. Alanine-substituted peptides and functionality substituted peptides were synthesized and screened for NMDAR binding in a competitive inhibition ELISA using free peptides. As an example, alanine-substituted peptides are depicted at a peptide concentration of 500 μg/mL. Inhibition in % is measured relative to inhibition of the control peptide, corresponding to peptide 3

3. Remove excess buffer by placing the 96-well filter plate on a vacuum suction manifold or on an equivalent device removing buffer from each well without disrupting the membrane. Wash the plates three times with TTN buffer (250 μL/well).
4. Add 200 μL TTN as blocking buffer to each well to block non-binding sites and incubate for 30 min at room temperature.
5. Dilute the primary antibody (1:1000) in TTN just before use (*see* **Note 6**). Add 100 μL of the diluted antibody to each well. The plate is incubated on a platform shaker at low speed for 1 h at room temperature.
6. After incubation, discard the solution and repeat the washing procedure described in **step 3**.
7. Immediately before use, dilute the secondary antibody (AP-conjugated anti-rabbit IgG antibody) 1:1000 in TTN.
8. Dispense 100 μL of the diluted secondary antibody to each well and incubate as described in **step 5**.
9. Following incubation with secondary antibody repeat washing steps described in **step 3**.
10. Detect the presence of bound antibodies by adding 100 μL of freshly prepared AP buffer solution to each well. Place the plates on a platform shaker and read the plate when the solution within the wells turns yellow.

11. Tip the plate and transfer 90 μL AP-buffer from each well to a Maxisorp microtiter plate and measure the absorbance at 405 nm, with background subtraction at 650 nm, on a Thermomax microtiter plate reader, or on an equivalent instrument using a wavelength of 405 nm and a reference wavelength of 650 nm (*see* **Note 7**).

3.2 Competitive Inhibition Enzyme-Linked Immunosorbent Assay

1. Prepare dilutions of free peptides of the stock solutions. Free peptides (1 mg/ml) are diluted 1:1000 in carbonate buffer of TTN buffer.
2. Coat Nunc-Immuno Maxisorp F96-well microtiter plates with 100 μL of the free peptide to each well. Incubate overnight at 4 °C (*see* **Note 8**).
3. Incubate primary antibody (1:1000 dilution) with varying peptide concentrations (2, 4, 20, 125, 500 μg/mL) in a total volume of 100 μL in a round bottom 96-microwell plate for 1 h at room temperature on a platform shaker.
4. Remove the solution from the coated plate by slapping the plate (well side down) on a clean towel or absorbent paper. Wash the plates three times with TTN buffer (250 μL/well).
5. Add 200 μL TTN as blocking buffer to each well to block free non-binding sites and incubate for 30 min at room temperature.
6. Remove the solution from the plate. Remove the solution and repeat the washing procedure described in **step 4**.
7. Add 100 μL of peptide-antibody complex from each well of the round-bottom microwell plate to each well of the coated microtiter plate. The microtiter plate is incubated on a platform shaker at low speed for 1 h at room temperature. The round-bottom microwell plate is discarded.
8. After incubation, remove the solution and repeat the washing procedure described in **step 4**.
9. Immediately before use, dilute the secondary antibody (anti-rabbit IgG alkaline phosphatase-conjugated antibody) 1:1000 in TTN.
10. Dispense 100 μL of the diluted secondary antibody to each well and incubate as described in **step** 7.
11. Following incubation with secondary antibody repeat washing steps described in **step 4**.
12. Detect the presence of bound antibodies by adding 100 μL of freshly prepared AP buffer solution to each well. Place the plates on a platform shaker and read the plate when the solution within the wells turns yellow.

13. The absorbance is measured at 405 nm, with background subtraction at 650 nm, on a Thermomax microtiter plate reader, or on an equivalent instrument using a wavelength of 405 nm and a reference wavelength of 650 nm.
14. Absorbance values are depicted in graphs or histograms relative to control peptides.

4 Notes

1. A common issue with synthetic peptides, especially those containing hydrophobic amino acid residues, is insolubility in aqueous solutions. Other solvents recommended for peptide solvation include 10 % dimethylsulfoxide, dimethylformamide or acetonitrile in water.
2. After lyophilization, peptides retain significant amounts of water. Peptides are oxidized over time at −20 °C and slowly degrade. Thus, the peptide stock solution should be stored in small aliquots upon arrival to prevent degradation caused by repeated freezing and thawing.
3. An alternative to membrane plates is to conduct the experiment in 1 mL tubes. After each washing and incubation step, the sample supernatant is removed and the resin is retained in the tube. However, this is rather time consuming and is not recommended.
4. Alternatively, TTN buffer can be replaced with phosphate-buffered saline.
5. It is strongly recommended to optimize concentration of reagents such as peptides and antibodies for each ELISA system.
6. This dilution is not necessarily ideal if other types of antibodies or sera are being tested, because optimum dilution of serum depends on the source and the amount of antibodies present in the sample.
7. Alternatively, transfer the buffer back for further development and read the plate as described.
8. Coat with a peptide containing the complete epitope, in this case the 20-mer peptide. Alternatively, coat with the complete immunogen if this is known.

References

1. Geysen HM, Rodda SJ, Mason TJ, Tribbick G, Schoofs PG (1987) Strategies for epitope analysis using peptide synthesis. J Immunol Methods 102:259–274
2. Houghten RA, Pinilla C, Blondelle SE, Appel JR, Dooley CT, Cuervo JH (1991) Generation and use of synthetic peptide combinatorial libraries for basic research and drug discovery. Nature 354:84–86
3. Lam KS, Lake D, Salmon SE, Smith J, Chen ML, Wade S, Abdul-Latif F, Knapp RJ, Leblova Z, Ferguson RD, Krchnak V, Sepetov VNF,

Lebl M (1996) A one-bead one-peptide combinatorial library method for B-cell epitope mapping. Methods 9:482–493

4. Mahler M, Mierau R, Bluthner M (2000) Fine-specificity of the anti-CENP-A B-cell autoimmune response. J Mol Med (Berl) 78: 460–467
5. Van Der Geld YM, Simpelaar A, Van Der Zee R, Tervaert JW, Stegeman CA, Limburg PC, Kallenberg CG (2001) Antineutrophil cytoplasmic antibodies to proteinase 3 in Wegener's granulomatosis: epitope analysis using synthetic peptides. Kidney Int 59:147–159
6. Trier NH, Hansen PR, Houen G (2012) Production and characterization of peptide antibodies. Methods 56:136–144
7. Geysen HM, Meloen RH, Barteling SJ (1984) Use of peptide synthesis to probe viral antigens for epitopes to a resolution of a single amino acid. Proc Natl Acad Sci U S A 81:3998–4002
8. Gershoni JM, Roitburd-Berman A, Siman-Tov DD, Tarnovitski FN, Weiss Y (2007) Epitope mapping: the first step in developing epitope-based vaccines. BioDrugs 21:145–156
9. Rubinstein ND, Mayrose I, Halperin D, Yekutieli D, Gershoni JM, Pupko T (2008) Computational characterization of B-cell epitopes. Mol Immunol 45:3477–3489
10. Amrutkar SD, Trier NH, Hansen PR, Houen G (2012) Fine mapping of a monoclonal antibody to the N-Methyl D-aspartate receptor reveals a short linear epitope. Biopolymers 98:567–575
11. Timmerman P, Beld J, Puijk WC, Meloen RH (2005) Rapid and quantitative cyclization of multiple peptide loops onto synthetic scaffolds for structural mimicry of protein surfaces. Chembiochem 6:821–824
12. Timmerman P, Puijk WC, Meloen RH (2007) Functional reconstruction and synthetic mimicry of a conformational epitope using CLIPS technology. J Mol Recognit 20:283–299
13. Behan KA, Johnston PG, Allegra CJ (1998) Epitope mapping of a series of human thymidylate synthase monoclonal antibodies. Cancer Res 58:2606–2611
14. Trier NH, Hansen PR, Vedeler CA, Somnier FE, Houen G (2012) Identification of continuous epitopes of HuD antibodies related to paraneoplastic diseases/small cell lung cancer. J Neuroimmunol 243:25–33
15. Welner S, Trier NH, Frisch M, Locht H, Hansen PR, Houen G (2013) Correlation between centromere protein-F autoantibodies and cancer analyzed by enzyme-linked immunosorbent assay. Mol Cancer 12:95
16. Welner S, Trier NH, Houen G, Hansen PR (2013) Identification and mapping of a linear epitope of centromere protein F using monoclonal antibodies. J Pept Sci 19:95–101
17. Paterson Y (1985) Delineation and conformational analysis of two synthetic peptide models of antigenic sites on rodent cytochrome c. Biochemistry 24:1048–1055
18. Petersen NH, Hansen PR, Houen G (2011) Fast and efficient characterization of an anti-gliadin monoclonal antibody epitope related to celiac disease using resin-bound peptides. J Immunol Methods 365:174–182
19. Lok SM, Kostyuchenko V, Nybakken GE, Holdaway HA, Battisti AJ, Sukupolvi-Petty S, Sedlak D, Fremont DH, Chipman PR, Roehrig JT, Diamond MS, Kuhn RJ, Rossmann MG (2008) Binding of a neutralizing antibody to dengue virus alters the arrangement of surface glycoproteins. Nat Struct Mol Biol 15: 312–317
20. Tian Y, Ramesh CV, Ma X, Naqvi S, Patel T, Cenizal T, Tiscione M, Diaz K, Crea T, Arnold E, Arnold GF, Taylor JW (2002) Structure-affinity relationships in the gp41 ELDKWA epitope for the HIV-1 neutralizing monoclonal antibody 2F5: effects of side-chain and backbone modifications and conformational constraints. J Pept Res 59:264–276
21. Tronstrom J, Draborg AH, Hansen PR, Houen G, Trier NH (2013) Identification of a linear epitope recognized by a monoclonal antibody directed to the heterogeneous nucleoriboprotein A2. Protein Pept, Lett
22. Merrifield RB (1963) Solid phase peptide synthesis. I. The synthesis of a tetrapeptide. J Am Chem Soc 85:2149–2154
23. Yamamori S, Itakura M, Sugaya D, Katsumata O, Sakagami H, Takahashi M (2011) Differential expression of SNAP-25 family proteins in the mouse brain. J Comp Neurol 519:916–932

Chapter 21

Screening and Characterization of Linear B-Cell Epitopes by Biotinylated Peptide Libraries

Ida Rosenkrands and Anja Olsen

Abstract

Identification of B-cell epitopes is important for the use of antibodies as therapeutic agents, the design of epitope-based vaccines against infectious diseases, and immunological assays based on peptide antibodies. A large number of methods are available for epitope mapping, but many of them require specialized laboratories and are expensive. In this chapter, we describe a high-throughput approach for epitope mapping of peptide antibodies by use of a library of soluble, overlapping, biotinylated peptides. As example, we present characterization of monoclonal and polyclonal antibodies specific for peptides of *Mycobacterium tuberculosis* acyl carrier protein AcpM and the *Chlamydia trachomatis* chaperone Ct043 by ELISA.

Key words ELISA, B-cell epitope, Antibody, Mapping, Overlapping, Peptide library, Biotinylated

1 Introduction

B-cell epitope mapping is an important step in the characterization of antibodies—monoclonal as well as polyclonal. The epitope is the part of an antigen that is recognized by the immune system, here the antibodies. Epitopes may be linear (continuous) or conformational (discontinuous), and mapping of the epitope can be performed by various techniques, such as X-ray crystallography, hydrogen-deuterium exchange mass spectrometry (HDX-MS), nuclear magnetic resonance (NMR), electron microscopy (EM), surface plasmon resonance (SPR), as well as libraries of antigen-derived peptides displayed on the surface of phage particles (phage display) or synthetic peptides for screening in ELISA. Algorithms for B-cell epitope prediction are also available, but there is still a way to go for reliable B-cell epitope prediction [1].

Information about the epitope has a wide range of applications and the literature about this subject is quite comprehensive as reviewed [2]. Examples are use of epitope-mapped monoclonal antibodies for functional analysis and structure elucidation as shown for *Brome mosaic virus* RNA polymerase [3], inhibition of the ricin

Gunnar Houen (ed.), *Peptide Antibodies: Methods and Protocols*, Methods in Molecular Biology, vol. 1348, DOI 10.1007/978-1-4939-2999-3_21, © Springer Science+Business Media New York 2015

toxin by a monoclonal antibody recognizing a linear epitope [4] and numerous applications in the cancer field for therapeutic or diagnostic use, e.g., Trastuzumab which recognizes human epidermal growth factor receptor 2 (HER2) that is frequently overexpressed in breast cancer [5], and immunoassays of plasma tumor markers for early detection of prostate, ovarian, and liver cancers [6].

The focus of the research in our laboratory is interventions against infectious diseases. Antibodies are a first line of defense and can inactivate a pathogen before it attaches to its target cell and thereby block invasion. Knowledge of the B-cell epitopes in antibodies protective against an infectious disease may lead to development of therapeutic monoclonal antibodies as shown, e.g., for rabies [7, 8] and respiratory syncytial virus (RSV) [9, 10]. In addition, epitope discovery by characterization of the antibody response raised after infection with different serovars, after vaccination and in individuals who have resolved the infection, is an important tool in the design and development of prophylactic vaccines against infectious diseases, e.g., as shown by B-cell epitope mapping of both viral and bacterial pathogens by overlapping biotinylated peptides [11–13]. Information of the B-cell epitopes may also lead to new peptide-based diagnostics against infectious diseases, as shown for *Herpes simplex* and tick-borne encephalitis patients [14, 15]. Finally, polyclonal and monoclonal antibodies are useful analytical tools in the detailed characterization of antigens of interest for interventions against infectious diseases and may be used for western blot analysis, immunofluorescence microscopy, ELISA, etc. Also in this context, B-cell epitope mapping is valuable, and the information may aid the assay development, e.g., in the selection of antibodies for a sandwich ELISA.

Epitope mapping by screening of synthetic peptides does not require specialized instruments and can be performed in any laboratory setup for ELISA, and in this chapter, we describe in detail a protocol for linear B-cell epitope discovery by a library of biotinylated overlapping peptides based on the amino acid sequence of the antigen.

2 Materials

All solutions are prepared from autoclaved Milli-Q water and analytical grade reagents. The buffer solutions are stored at 4 °C unless other instruction is indicated. Be aware of local regulations for disposal of waste materials after the experiment.

2.1 Biotinylated Overlapping Peptides

The biotinylated overlapping peptides are ordered as typically 1–3 mg of overlapping crude (i.e., non-purified) peptides of desired length. The format described here is biotin-SGSG-PEPTIDE-amide, except for the N-terminal peptide which is synthesized as amine-PEPTIDE-GSG-biocytin amide (biocytin is a lysine residue with a biotin group attached to its side chain), and the C-terminal peptide which is synthesized as biotin-SGSG-PEPTIDE-acid. SGSG and GSG serve as spacer arms to ensure that all peptide residues are available for antibody binding in ELISA (*see* **Note 1**).

2.2 Antibodies and Conjugates

1. Primary antibody: Antiserum obtained after immunization of mice with the antigen of interest; here we used recombinant *Mycobacterium tuberculosis* AcpM [16] and recombinant *Chlamydia trachomatis* Ct043 [17] adjuvanted in aluminum hydroxide or CAF01 [18], respectively. Furthermore, a purified, mouse monoclonal antibody raised against AcpM was used (*see* **Notes 2** and **3**).
2. Secondary antibody: Peroxidase-conjugated rabbit-anti-mouse immunoglobulin (*see* **Note 4**).

2.3 ELISA Reagents

1. Clear flat-bottom immuno nonsterile 96-well plates.
2. Multichannel pipette (100–200 μl volumes)
3. Microplate washer, optional.
4. Streptavidin: Prepare stock in Milli-Q water, 2 mg/ml. Store at −20 °C.
5. Coating buffer: 50 mM carbonate/bicarbonate buffer, pH 9.6. 1.59 g Na_2CO_3, and 2.93 g $NaHCO_3$ are dissolved in Milli-Q water. 2 ml 0.5 % Phenol red is added, and Milli-Q water is added to a final volume of 1 l.
6. Phosphate-buffered saline (PBS), pH 7.2: 1.44 g $Na_2HPO_4{\cdot}2H_2O$, 0.20 g KH_2PO_4, 8.0 g NaCl, and 0.20 g KCl are dissolved in Milli-Q water, which is added to a final volume of 1 l.
7. Blocking solution: 2 % skimmed milk powder in PBS (*see* **Notes 5** and **6**).
8. Dilution buffer: 1 % bovine serum albumin in PBS. Sterile filtration (1000 ml flask, 0.2 μm).
9. Washing buffer: PBS with 0.2 % Tween 20 (PBS-T).
10. Substrate: TMB ready-to-use substrate. Store at 4 °C, but adjust to room temperature before use.
11. Stop solution: 0.2 M H_2SO_4.
12. ELISA reader: A microplate reader equipped with a 405 or 450 nm filter.

3 Methods

This method outlines a standard ELISA where biotinylated peptides are immobilized onto a streptavidin-coated plate followed by incubation with peptide-capturing primary antibody and then the secondary conjugated antibody (*see* **Note 7**).

3.1 Streptavidin Coating

The streptavidin stock is diluted 400 times in the coating buffer; the final concentration is 5 μg/ml. 100 μl is applied in each well of the ELISA plates (*see* **Note 8**), and the plate is incubated overnight at 4 °C. The coating solution is removed and the wells are washed once with PBS-T. Blocking is performed with 200 μl blocking solution in each well for 1½ h, followed by washing three times with PBS-T.

3.2 Binding of Biotinylated Peptides

Peptides are dissolved in an appropriate solvent, e.g., 200 μl dimethyl sulfoxide (DMSO) corresponding to a peptide concentration of approx. 10 mg/ml, and the dissolved peptides are stored at −20 °C or lower (*see* **Note 9**). The peptides are diluted 1000 times in dilution buffer, and 100 μl diluted peptide is applied to each well followed by incubation for 1 h on a rocking table at room temperature (*see* **Notes 10** and **11**).

3.3 Primary Antibody Reaction

The peptide solution is removed and the wells are washed three times with PBS-T. The primary antibody is diluted to 1 μg/ml or antiserum diluted normally 100–2000 times in dilution buffer (*see* **Note 12**). 100 μl is applied in each well and the plate is incubated for 2 h at room temperature on a rocking table.

3.4 Secondary Antibody Reaction

The primary antibody solution is removed and the wells are washed three times with PBS-T. The secondary antibody is diluted 4000 times in dilution buffer (*see* **Note 12**). 100 μl is applied in each well and the plate is incubated for 1 h at room temperature on a rocking table.

3.5 Development

The secondary antibody solution is removed and the wells are washed five times with PBS-T. TMB substrate is applied, 100 μl to each well. Color development takes place for 1–30 min, and the reaction is stopped by adding 100 μl stop solution to each well. Absorbance is read in a microplate reader.

3.6 Interpretation of the Results

ELISA readings are imported in appropriate software for preparing a bar graph as shown in Fig. 1 (here GraphPad Prism version 6.00 for Windows, GraphPad Software, La Jolla, CA, USA). The following interpretation depends on the purpose of the screening. In Fig. 1a, serum from three different mice immunized with recombinant *M. tuberculosis* AcpM protein was mapped with a peptide library (13mers, overlap of 11). The purpose was selection of which mouse to proceed with for obtaining monoclonal antibodies; the mouse designated Pab Mtb AcpM 1 appeared to have the broadest recognition of the peptide library, and was selected for the hybridoma generation. From another project, the resulting anti-AcpM monoclonal antibody was characterized with the same peptide library (Fig. 1b), and it appeared that two apparently independent peptide regions were recognized. The antibody was

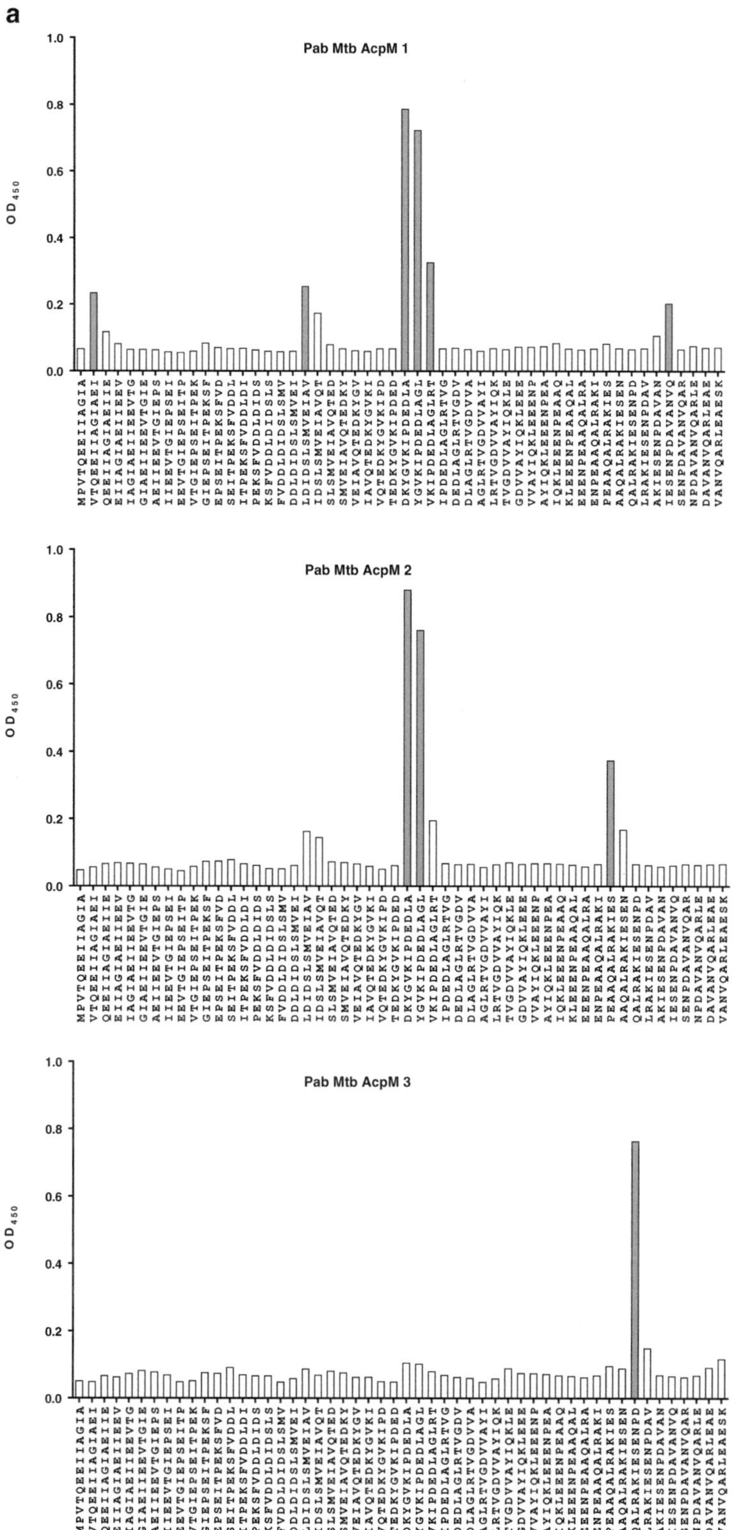

Fig. 1 Screening of peptides with mouse antibodies: Each bar represents the absorbance detected for the peptides at 450 nm. The peptides were biotinylated and synthesized with an offspring of two and immobilized on streptavidin-coated plates. The peptide response profile of three different polyclonal mouse antisera to *Mycobacterium tuberculosis* AcpM represented as 13mer peptides (**a**). The response to AcpM pepset by the mouse monoclonal antibody TB-3 (**b**). The response of mouse polyclonal antiserum to *Chlamydia trachomatis* Ct043 represented as 15mer peptides (**c**)

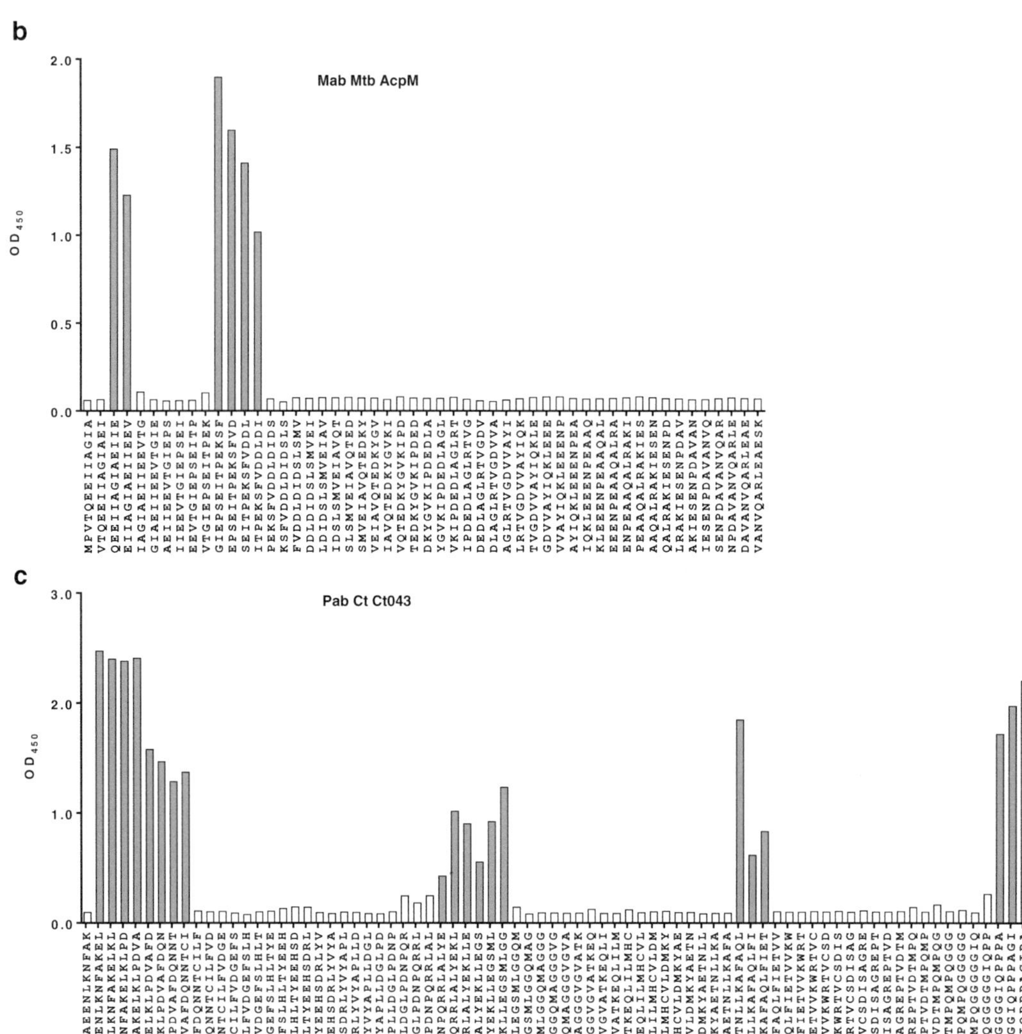

Fig. 1 (continued)

confirmed to be monoclonal, and we therefore suspected that the epitope recognized by this monoclonal antibody is discontinuous, and the three-dimensional structure of AcpM supports that these two peptide regions are in proximity in the AcpM protein as they form an α-helix [16].

Finally, we characterized serum from a pool of eight mice immunized with recombinant Ct043, an experimental subunit vaccine candidate against *Chlamydia trachomatis* infection [17, 19]. Responses to a peptide library (15mers, overlap of 13) are shown in Fig. 1c, and it revealed that four overall peptide regions are recognized after immunization.

It should be emphasized that the peptide library approach only identifies linear epitopes. Lack of response to a biotinylated peptide library can be explained by the presence of conformational epitopes; this is in particular a relevant consideration when characterizing monoclonal antibodies. If possible, include always a relevant (polyclonal) positive control sample to exclude any technical issues for the lack of response.

3.7 Further Analyses

Based on the results from the first peptide library screening further analyses can be undertaken. First of all, if the ELISA was based on crude biotinylated peptides, it is highly recommended to confirm the findings by testing selected purified and characterized peptides.

If the overall purpose of evaluation of the peptide library was to do a full epitope mapping, the next step would be a more narrow mapping with shorter peptides using a truncated peptide library to define the minimal epitope, i.e., the shortest amino acid sequence needed for antibody binding. To further identify the contribution from individual amino acids to antibody binding, each residue can be substituted with alanine one at a time with an alanine-scanning peptide library. Finally, further epitope characterization could be a positional scanning library which replaces the selected residues with all other natural amino acids, one at a time.

4 Notes

1. A library of overlapping biotinylated peptides is usually provided with a length of 6–15 amino acids with an offset of 1–4 amino acids; as a rule of thumb, peptides for mapping of monoclonal antibodies are designed with an overlap of at least 11 amino acids, e.g., 13mers with an offset of two as shown in Fig. 1b, whereas peptides for characterization of polyclonal sera are designed with an overlap of at least seven amino acids. The number of peptides required depends on the size of the protein and the desired offset. Other considerations are the economy of the project and how many peptides can be handled in the analyses, e.g., if all the peptides can be analyzed in one 96-well plate (or in one half plate) it is advantageous for the following analyses. On most of the websites of the providers of synthetic peptides one can design a peptide library of desired length and overlap, e.g., http://www.sigmaaldrich.com/life-science/custom-oligos/custom-peptides/learning-center/pepscreen-calculator/pepscreen-calculator.html or http://www.mimotopes.com/peptideLibraryScreening.asp?id=90. Other considerations when designing a peptide library are to include known posttranslational modifications such as phosphorylation or acetylation, and exclusion of sequence repeats if present to minimize the library. Finally, it can be considered to

exclude membrane-spanning regions (predicted by http://www.cbs.dtu.dk/services/TMHMM/) which are not expected to include epitopes and further the peptides may give rise to solubility problems.

2. Various sources of primary antibodies can be used for screening a peptide library, ranging from highly purified monoclonal antibodies to crude serum or plasma from immunized or infected animals or humans. Be aware of appropriate safety considerations when handling human sera and sera from subjects exposed to infectious diseases.
3. Include also pre-immune or other negative serum to confirm that the observed response to any of the peptides is due to specific antibodies. For purified monoclonal antibodies, another "irrelevant" purified monoclonal antibody can serve as negative control.
4. Use appropriate secondary antibody corresponding to the primary antibody.
5. Prepare this solution fresh each time.
6. Alternative blocking solutions: In our laboratory we use either skimmed milk powder as described in the protocol or BSA in the blocking solution to block non-occupied sites on the surface, and both work well in our hands with no background problems. Other protein-blocking alternatives exist such as casein or whole normal serum. It should be considered if the blocking agent is compatible with the specific analysis; for example if BSA has been used as carrier for the peptide immunization it should not be used as blocking agent. Also commercially available blocking agents are available.
7. For consistent results, use the same incubation times in all steps for related experiments.
8. Precoated streptavidin-coated ELISA plates can be acquired from different local suppliers.
9. Peptides provided in powder form are dissolved in an appropriate solvent. As a standard procedure we have dissolved peptides in 200 μl DMSO followed by storage at −20 °C or lower. Other options are dimethyl formamide, acetonitrile, or isopropanol. If aqueous solutions are preferred, try dissolving in 0.1 % acetic acid and sonicate if the peptides are not dissolved. However, for hydrophobic peptides (>50 % content of A, F, I, L, M, P, V, W, or Y) organic solvents are usually required. For large peptide libraries it is not feasible to dissolve peptides on an individual basis and it is preferred to find a solvent or solution which will work for all peptides in the library. It is recommended to analyze the amino acid composition for each peptide with respect to acidic, basic, and hydrophobic residues by a peptide property calculator beforehand (there are several Web-based tools,

e.g., https://www.genscript.com/ssl-bin/peptide_mw). Be aware of appropriate safety considerations when dissolving peptides in organic solvents.

10. A biotinylated overlapping peptide library (e.g., from Mimotopes, Clayton, Victoria, Australia) can be provided already immobilized on streptavidin-coated 96-well plates to avoid extensive pipetting of numerous individual peptides in the coating step.
11. The optimal dilution of the peptide should be determined using a titration assay.
12. The optimal dilution of the primary and secondary antibody should be determined by titration.

Acknowledgements

The authors appreciate the funding received from Foundation for Innovative New Diagnostics (Geneva, Switzerland) and we thank Patricia Grenés and Annette Hansen for excellent technical assistance.

References

1. Sun P, Ju H, Liu Z, Ning Q, Zhang J, Zhao X, Huang Y, Ma Z, Li Y (2013) Bioinformatics resources and tools for conformational B-cell epitope prediction. Comput Math Methods Med 2013:943636
2. Ladner RC (2007) Mapping the epitopes of antibodies. Biotechnol Genet Eng Rev 24: 1–30
3. Dohi K, Mise K, Furusawa I, Okuno T (2002) RNA-dependent RNA polymerase complex of Brome mosaic virus: analysis of the molecular structure with monoclonal antibodies. J Gen Virol 83:2879–2890
4. McGuinness CR, Mantis NJ (2006) Characterization of a novel high-affinity monoclonal immunoglobulin G antibody against the ricin B subunit. Infect Immun 74:3463–3470
5. Molina MA, Codony-Servat J, Albanell J, Rojo F, Arribas J, Baselga J (2001) Trastuzumab (herceptin), a humanized anti-Her2 receptor monoclonal antibody, inhibits basal and activated Her2 ectodomain cleavage in breast cancer cells. Cancer Res 61:4744–4749
6. Meany DL, Sokoll LJ, Chan DW (2009) Early detection of cancer: immunoassays for plasma tumor markers. Expert Opin Med Diagn 3: 597–605
7. Both L, Banyard AC, van Dolleweerd C, Horton DL, Ma JK, Fooks AR (2012) Passive immunity in the prevention of rabies. Lancet Infect Dis 12:397–407
8. Kaur M, Garg R, Singh S, Bhatnagar R (2014) Rabies vaccines: where do we stand, where are we heading? Expert Rev Vaccines 14:369–381
9. Arbiza J, Taylor G, Lopez JA, Furze J, Wyld S, Whyte P, Stott EJ, Wertz G, Sullender W, Trudel M et al (1992) Characterization of two antigenic sites recognized by neutralizing monoclonal antibodies directed against the fusion glycoprotein of human respiratory syncytial virus. J Gen Virol 73:2225–2234
10. Gomez RS, Guisle-Marsollier I, Bohmwald K, Bueno SM, Kalergis AM (2014) Respiratory syncytial virus: pathology, therapeutic drugs and prophylaxis. Immunol Lett 162:237–247
11. Kam YW, Lee WW, Simarmata D, Harjanto S, Teng TS, Tolou H, Chow A, Lin RT, Leo YS, Renia L, Ng LF (2012) Longitudinal analysis of the human antibody response to Chikungunya virus infection: implications for serodiagnosis and vaccine development. J Virol 86:13005–13015
12. Kollipara A, Polkinghorne A, Beagley KW, Timms P (2013) Vaccination of koalas with

a recombinant *Chlamydia pecorum* major outer membrane protein induces antibodies of different specificity compared to those following a natural live infection. PLoS One 8: e74808

13. Radford AD, Willoughby K, Dawson S, McCracken C, Gaskell RM (1999) The capsid gene of feline calicivirus contains linear B-cell epitopes in both variable and conserved regions. J Virol 73:8496–8502
14. Bhullar SS, Chandak NH, Baheti NN, Purohit HJ, Taori GM, Daginawala HF, Kashyap RS (2014) Identification of an immunodominant epitope in glycoproteins B and G of herpes simplex viruses (HSVs) using synthetic peptides as antigens in assay of antibodies to HSV in herpes simplex encephalitis patients. Acta Virol 58:267–273
15. Kuivanen S, Hepojoki J, Vene S, Vaheri A, Vapalahti O (2014) Identification of linear human B-cell epitopes of tick-borne encephalitis virus. Virol J 11:115
16. Wong HC, Liu G, Zhang YM, Rock CO, Zheng J (2002) The solution structure of acyl carrier protein from *Mycobacterium tuberculosis*. J Biol Chem 277:15874–15880
17. Olsen AW, Andersen P, Follmann F (2014) Characterization of protective immune responses promoted by human antigen targets in a urogenital *Chlamydia trachomatis* mouse model. Vaccine 32:685–692
18. Agger EM, Rosenkrands I, Hansen J, Brahimi K, Vandahl BS, Aagaard C, Werninghaus K, Kirschning C, Lang R, Christensen D, Theisen M, Follmann F, Andersen P (2008) Cationic liposomes formulated with synthetic mycobacterial cord factor (CAF01): a versatile adjuvant for vaccines with different immunological requirements. PLoS One 3:e3116
19. Meoni E, Faenzi E, Frigimelica E, Zedda L, Skibinski D, Giovinazzi S, Bonci A, Petracca R, Bartolini E, Galli G, Agnusdei M, Nardelli F, Buricchi F, Norais N, Ferlenghi I, Donati M, Cevenini R, Finco O, Grandi G, Grifantini R (2009) CT043, a protective antigen that induces a CD4+ Th1 response during *Chlamydia trachomatis* infection in mice and humans. Infect Immun 77:4168–4176

Chapter 22

Bead-Based Peptide Arrays for Profiling the Specificity of Modification State-Specific Antibodies

Angela Filomena, Yvonne Beiter, Markus F. Templin, Thomas O. Joos, Nicole Schneiderhan-Marra, and Oliver Poetz

Abstract

Modification state-specific antibodies are powerful tools for investigating posttranslational modifications in proteins. The majority of these antibodies have been generated against peptide-antigen conjugates. They are useful in a plethora of methods, such as Western blotting, immunohistochemistry, sandwich immunoassay, immunoprecipitation, and immunoprecipitation coupled with mass spectrometry. Phosphorylation, acetylation, methylation, sulfation, nitrosylation, ubiquitination, and sumoylation are some of the modifications that can be studied using such antibodies. However, investigating the on- and off-target binding of antibodies is crucial to the interpretation of experimental data. Peptide arrays are excellent tools for such in-depth studies of off-target and on-target binding of antibodies. Dozens or even hundreds of modified peptides can be integrated into a single experimental setup to analyze the antibody's binding behavior. Here, we propose three different protocols for peptide bead array generation and describe their suitability for such types of assay.

Key words Antibody characterization, Acetylation, Methylation, Peptide bead array, Phosphorylation, Posttranslational modification

1 Introduction

In general, two types of antibodies are used in studying the modification of proteins: modification state-specific antibodies, which are capable of detecting posttranslational modifications (ptm) generically; and antibodies, which recognize the posttranslationally modified amino acid in combination with neighboring amino acid side chains. Either type of antibody is an integral part of a vast number of applications such as Western blotting [1], immunohistochemistry, sandwich immunoassay [2, 3], immunoprecipitation, and immunoprecipitation coupled with mass spectrometry [4–6]. The antibodies are basically employed as capture or detection tools. Because of a lack of generic methods, it is often the assay itself which serves as quality control for the antibodies.

Gunnar Houen (ed.), *Peptide Antibodies: Methods and Protocols*, Methods in Molecular Biology, vol. 1348, DOI 10.1007/978-1-4939-2999-3_22, © Springer Science+Business Media New York 2015

For instance, a protein extract from a cell culture treated with a stimulus triggering the respective modification, can be analyzed by using such an antibody in a Western blot experiment. The antibody is assumed to be specific if a signal is raised at the estimated molecular weight area in the treated sample. However, since the signal in a Western blot can be influenced by various factors, an orthogonal method is recommended for evaluating a reagent.

Since the generation of ptm-specific antibodies is solely done against synthetic peptides, they comprise linear epitopes. These epitopes can be excellently characterized in detail by applying protein and peptide arrays (Figs. 1 and 2) [7, 8]. Thus, hundreds of different features can be investigated in a single experiment.

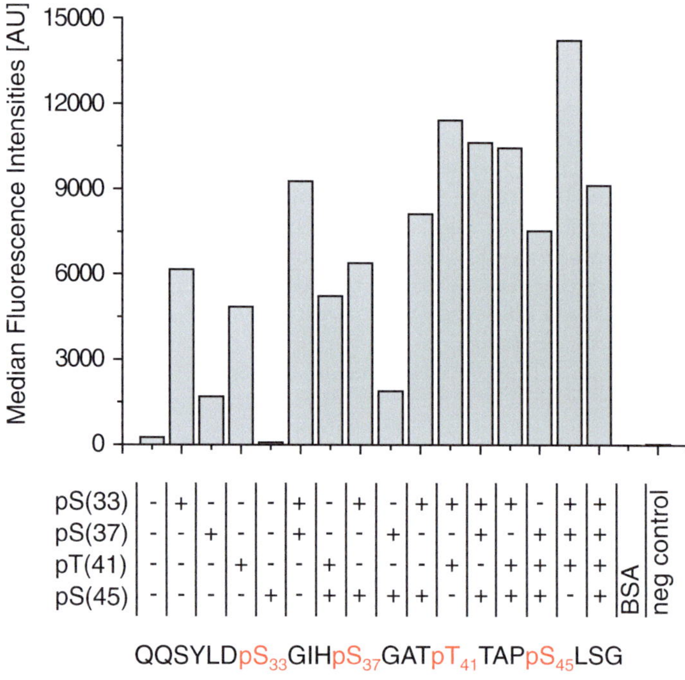

Fig. 1 Evaluation of the specificity of a posttranslational modification-specific anti-beta catenin antibody. A commercially available polyclonal antibody specific for beta-catenin, phosphorylated at serine 33, serine 37, and threonine 41, was probed with a low-plex bead-based peptide array containing peptides which harbor the SSTS phosphorylation motif. The fully phosphorylated SSTS sequence is a ubiquitin-ligase binding motif [16] and tags beta-catenin [17, 18] for proteasomal degradation. A set of 16 biotinyl peptides containing either one, two, three, four, or no phosphorylations were used for array generation. The antibody binds only to peptides containing the phosphorylated amino acids serine 33, serine 37 and threonine 41. However, the antibody binds to single- as well as double- and triple-phosphorylated sequences. Therefore, the results of a Western blot using this antibody can basically confirm phosphorylation of the motif but cannot indicate the number of modified sites. Binding intensity is shown as median fluorescence intensities

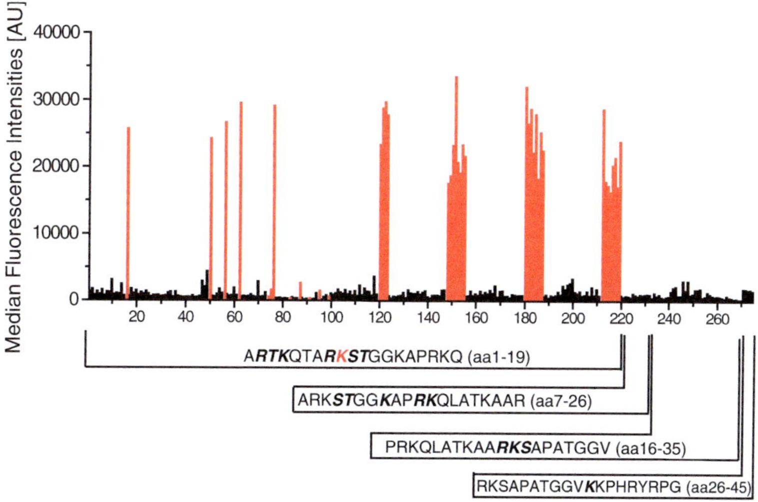

Fig. 2 Evaluation of the specificity of an anti-acetyl-histone H3 (lysine9) antibody (CST 9649). The monoclonal rabbit (clone C5B11) antibody was probed with a medium-plex bead-based peptide array containing 384 peptides (only results for histone H3-related peptides are shown). The peptide array included the N-terminal regions of histones H2A, H2B, H3, and H4, with overlapping sequences (detailed peptide listing can be found here [12]). For array generation, up to fourfold modified (amino acids displayed in *italics* and **bold**) and unmodified peptides were synthesized by SPOT synthesis and covalently immobilized on distinct color-coded beads. Peptides contained serine and threonine phosphorylation, lysine acetylation, and arginine methylation. The anti-acetyl-histone H3 (lysine9) antibody was probed with the array and peptides containing an acetylation at lysine 9 (*red*, *italics*, and **bold** sequence) showed signals. Phosphorylation of serine 10 or threonine 11 next to lysine 9 suppressed the binding of the antibody. Results for peptides acetylated at lysine 9 are highlighted in *red*. Binding intensity is shown as median fluorescence intensities

The peptides immobilized in an array can range from 2 to 10,000 different molecules. In situ array synthesis even allows the production of ultrahigh-density arrays comprising a whole proteome [9, 10]. However, in situ peptide array synthesis chemistry or recombinant protein technologies still do not allow the generation of peptide arrays or protein arrays involving modifications such as phosphorylation, acetylation, methylation, sulfation, or nitrosylation. Therefore, synthesized peptides using Fmoc chemistry combined with array printing technology [11] or bead array technology [12] offer feasible alternatives for generating arrays comprising hundreds of posttranslationally modified peptides. Here, we propose three different protocols for the generation of bead-based peptide arrays.

Depending on the array complexity, we recommend three different immobilization strategies using (1) the streptavidin biotin system in combination with biotinylated peptides, (2) directed chemical immobilization with a bifunctional linker and cysteinyl peptides [13, 14], or (3) cellulose peptide conjugates [15] (synthesized by spot synthesis) in combination with EDC/NHS chemistry [12].

2 Materials

2.1 Preparation of Low-Plex Bead Arrays Using Biotinylated Peptides

2.1.1 Preparation of Neutravidin Beads (Manual Protocol)

1. Activation buffer: 100 mM disodium hydrogen phosphate, pH 6.2.
2. Coupling buffer: 50 mM 2-(*N*-morpholino) ethanesulfonic acid (MES), pH 5.0.
3. Washing buffer: phosphate-buffered saline, pH 7.4, 0.05 % (v/v) Tween 20.
4. EDC: 1-ethyl-3-[3-dimethylaminopropyl] carbodiimide hydrochloride dissolved in activation buffer at a concentration of 50 mg/ml.
5. Sulfo-NHS: N-hydroxysulfosuccinimide dissolved in dimethyl sulfoxide (DMSO) at a concentration of 50 mg/ml.
6. Neutravidin dissolved in coupling buffer at a concentration of 100 μg/ml.
7. Storage buffer: Phosphate-buffered saline, pH 7.4, 1 % (w/v) BSA, 0.1 % (w/v) sodium azide (NaN_3).
8. Beads: Carboxy-modified paramagnetic color-coded polystyrene beads (Magplex microspheres), one population per peptide.
9. 1.5 ml Eppendorf protein LoBind tubes.
10. Magnetic separator (e.g. DynaMag Spin).
11. Rotator for (micro)tubes.

2.1.2 Non-covalent Immobilization of Biotin Peptides on to Neutravidin Beads (Manual Protocol)

1. Washing buffer: Phosphate-buffered saline, pH 7.4, 0.1 % (v/v) Tween 20.
2. Storage buffer: Phosphate-buffered saline, pH 7.4, 1 % (w/v) BSA, 0.1 % (w/v) sodium azide.
3. Neutravidin-coated beads from subheading 3.1.1.
4. Synthetic biotinyl peptides dissolved in DMSO at a concentration of 500 μM.
5. Peptide dilution buffer: phosphate-buffered saline, pH 7.4, 1 % (w/v) BSA.
6. 1.5 ml Eppendorf protein LoBind tubes.
7. Magnetic separator (e.g., DynaMag-Spin).
8. Rotator for (micro)tubes.

2.2 Preparation of Medium- and High-Plex Bead Arrays Using BSA Carrier and Cysteinyl Peptides

2.2.1 Preparation of BSA Beads (Semiautomated Protocol)

1. 0.055 % (v/v) Triton X-100.
2. Activation buffer: 100 mM Na_2HPO_4, pH 6.2, 0.005 % (v/v) Triton X-100.
3. Coupling buffer: 50 mM MES, pH 5.0, 0.005 % (v/v) Triton X-100.
4. Washing buffer: Phosphate-buffered saline, pH 7.4, 0.005 % (v/v) Triton X-100.
5. EDC dissolved in activation buffer at a concentration of 50 mg/ml.
6. Sulfo-NHS dissolved in DMSO at a concentration of 50 mg/ml.
7. BSA: Bovine serum albumin dissolved in coupling buffer at a concentration of 100 μg/ml.
9. Storage buffer: Phosphate-buffered saline, pH 7.4, 1 % (w/v) BSA, 0.1 % (w/v) sodium azide.
8. Beads: Carboxy-modified paramagnetic color-coded polystyrene beads (MagPlex microspheres).
9. Magnetic particle processor (e.g., KingFisher 96).
10. KingFisher 96 KF plate (200 μl).
11. KingFisher 96 tip comb for deep-well magnets.
12. Adhesive foil (e.g., Axygen® AxySeal).

2.2.2 Covalent Immobilization of Cysteinyl Peptides on BSA Beads (Semiautomated Protocol)

1. 0.1 % (v/v) Triton X-100.
2. Sulfo-SMPB: Sulfosuccinimidyl 4-(*p*-maleimidophenyl) butyrate dissolved in DMSO at a concentration of 15 mg/ml. Dilute solution 1:10 in PBS.
3. Dilution buffer 1: Phosphate-buffered saline, pH 7.4, 0.01 % (v/v) Triton X-100.
4. Washing buffer: Phosphate-buffered saline, pH 7.4, 0.005 % (v/v) Triton X-100.
5. TCEP: Tris(2-carboxyethyl)phosphine dissolved in phosphate-buffered saline at a concentration of 1 mM (0.286 mg/ml).
6. Dilution buffer 2: Phosphate-buffered saline, pH 7.4, 0.0125 % (v/v) Triton X-100.
7. Storage buffer: Phosphate-buffered saline, pH 7.4, 1 % (w/v) BSA, 0.1 % (w/v) NaN_3.
8. BSA beads from Subheading 3.2.1.
9. Synthetic cysteinyl peptides dissolved in phosphate-buffered saline, pH 7.4, 10 % (v/v) DMSO at a concentration of 1 mM.
10. Magnetic particle processor (e.g., KingFisher 96).
11. KingFisher 96 KF plate (200 μl).

12. KingFisher 96 tip comb for deep-well magnets.
13. Plate shaker (e.g., Thermomixer comfort with exchangeable MTP thermoblock).
14. 1.5 ml Eppendorf protein LoBind tube.
15. Magnetic separator (e.g., DynaMag-Spin).

2.3 Preparation of Medium- and High-Plex Bead Arrays Using Cellulose Peptides (Semiautomated Protocol)

1. Activation buffer: 100 mM Na_2HPO_4, pH 6.2, 0.005 % (v/v) Triton X-100.
2. Coupling buffer: 50 mM 4-(2-hydroxyethyl)-1-piperazine-ethanesulfonic acid (HEPES), pH 5.0, 0.005 % (v/v) Triton X-100, 10 % (v/v) DMSO.
3. Washing buffer: Phosphate-buffered saline, pH 7.4, 0.005 % (v/v) Triton X-100.
4. EDC dissolved in activation buffer at a concentration of 50 mg/ml.
5. Sulfo-NHS dissolved in DMSO at a concentration of 50 mg/ml.
6. Storage buffer: Phosphate-buffered saline, pH 7.4, 1 % (w/v) BSA, 0.1 % (w/v) NaN_3.
7. Beads: Carboxy-modified paramagnetic color-coded polystyrene beads (Magplex microspheres).
8. Cellulose peptide conjugates dissolved in DMSO at a concentration of 100 μM.
9. Magnetic particle processor (e.g., KingFisher 96).
10. KingFisher 96 KF plate (200 μl).
11. KingFisher 96 tip comb for deep-well magnets.
12. 1.5 ml Eppendorf protein LoBind tube.
13. Magnetic separator, e.g., DynaMag-Spin.

2.4 Antibody Characterization

1. Assay buffer: Phosphate-buffered saline, pH 7.4, 1 % (w/v) BSA, 0.05 % (v/v) Tween 20.
2. Washing buffer: Phosphate-buffered saline, pH 7.4, 0.05 % (v/v) Tween 20.
3. Readout buffer: Phosphate-buffered saline, pH 7.4.
4. Assay plate: 96-wellSigma #71507-250G half-area microplate, non-binding surface, flat bottom.
5. Adhesive foil (e.g., Axygen® AxySeal).
6. Magnetic bead separator plate.
7. Plate shaker (e.g. Thermomixer comfort with MTP thermoblock).
8. Luminex FlexMap 3D instrument, alternatively Luminex 100/200 or MagPix. Local distributors are listed on the Luminex web page www.luminexcorp.com).

9. Bead-based peptide array, e.g., biotin-peptide array, cysteinyl peptide array, or cellulose peptide array.
10. Ptm-specific antibodies for characterization (commercial or in-house).
11. Detection antibody: Phycoerythrin (PE)-conjugated anti-species antibody, e.g., goat anti-mouse IgG (H+L)-RPE, donkey anti-rabbit IgG (H+L)-RPE, donkey anti-goat IgG (H+L)-RPE.

3 Methods

Depending on the number of peptides to be used for generating an array, different protocols can be used. For instance, if only one phosphorylation-state-specific antibody must be characterized, it is not necessary to generate a mid- or high-plex peptide array. It is generally adequate to immobilize the phosphorylated peptide as well as the non-phosphorylated peptide to determine the selectivity of the antibody. A similarity search for homologous phosphorylation sequences should be done first. Based on this search, similar sequences should be included in the investigation. In this case, the protocol for low-plex arrays (2–20 peptides) using biotinylated peptides is suitable. No automation is required and each step can be easily and rapidly performed manually. Should the experiments require higher complexity and a larger number of arrays, a covalent immobilization strategy as described in the protocol for cysteinyl peptides or cellulose peptides is recommended. Furthermore, robotic equipment is suggested for the production of medium- and high-plex arrays (<20–100>, <100–500>). Moreover, the number of arrays/tests to be generated should be considered first, as the amount of beads in the array generation protocol can be adapted accordingly.

Due to limitations of the peptide synthesis chemistry, the length of peptides should not exceed 20 amino acids. Moreover, the peptides should contain spacer molecules to increase distance between bead surface and actual peptide sequence. Sequences such as gly-ser-gly or amino-caproic acid are suitable for minimizing steric hindrance during antibody-peptide interaction.

3.1 Preparation of Low-Plex Bead Arrays Using Biotinylated Peptides

3.1.1 Preparation of Neutravidin Beads (Manual Protocol)

1. Vortex (for at least 10 s) and sonicate bead stocks, transfer 200 μl of each bead stock solution (1.25×10^7 beads/ml) to 1.5 ml protein LoBind tubes.
2. Place the tubes in a magnetic separator to collect the beads (for at least 30 s).
3. Discard the supernatant with a pipette and wash beads twice with 500 μl activation buffer. Vortex beads between the washing steps (for at least 30 s) and collect the beads again using the magnetic separator.

4. Prepare an activation mix during the last washing step. For one tube, mix 120 μl activation buffer, 15 μl sulfo-NHS solution (50 mg/ml), and 15 μl EDC solution (50 mg/ml).
5. Add 150 μl activation mix to the beads and incubate for 20 min at room temperature (RT) on a rotator in the dark.
6. During activation, prepare a dilution of Neutravidin (100 μg/ml).
7. Place the tubes in a magnetic separator (for at least 30 s) to collect the beads.
8. Discard the activation mix solution with a pipette and wash the beads three times with 500 μl coupling buffer. Vortex beads between the washing steps (for at least 30 s) and collect the beads again using the magnetic separator.
9. Add 250 μl of the diluted Neutravidin solution to the beads and incubate for 2 h at RT on a rotator in the dark.
10. Place the tubes in a magnetic separator (for at least 30 s) to collect the beads.
11. Discard the protein solution with a pipette and wash the beads three times with 500 μl washing buffer. Vortex beads between the washing steps (for at least 30 s) and collect the beads again using the magnetic separator.
12. Add 100 μl storage buffer and store the beads at 4 °C in the dark.

    The bead recovery should be determined as follows:

13. Dilute the beads 1:500 in assay buffer.
14. Transfer 100 μl/well of diluted beads to a microtiter plate and incubate the diluted beads for 30 min in a 96-well plate shaker, at 750 rpm at room temperature.
15. Count the number of beads with a Luminex reader using the following settings:

Sample size:	50 μl
Time-out:	80 s
Total beads:	10,000

16. Calculate the bead concentration as follows:

    Beads per μl = (number of beads/30 μl) × 500

3.1.2 Non-covalent Immobilization of Biotinyl Peptides onto Neutravidin Beads (Manual Protocol)

1. Dissolve biotinyl peptides in DMSO to achieve a 1 mM concentration.
2. Vortex Neutravidin beads for at least 10 s (1 population required per peptide).
3. Transfer 100,000 beads to a new tube (sufficient for approximately 100 assays/tests).

4. Place the tubes in a magnetic separator (for at least 30 s) to collect the beads.
5. Discard the supernatant with a pipette and wash beads twice with 500 μl washing buffer. Vortex beads between the washing steps (for at least 30 s) and collect the beads again using the magnetic separator.
6. Dilute biotin peptides to 1 μM in dilution buffer.
7. Add 250 μl of one diluted biotinyl peptide solution to one Neutravidin bead population and incubate for 60 min at RT on a rotator in the dark.
8. Place the tubes in a magnetic separator (for at least 30 s) to collect the beads.
9. Remove the biotin peptide solution with a pipette and wash the beads three times with 500 μl washing buffer. Vortex beads between the washing steps (for at least 30 s) and collect the beads again using the magnetic separator.
10. Add 100 μl storage buffer.
11. Combine the single-bead populations to generate a biotinyl peptide array:
    (a) Resuspend bead solutions and transfer solutions into one 1.5 ml LoBind tube.
    (b) Place the tube in a magnetic separator (for at least 30 s) to collect the beads.
    (c) Discard supernatant.
    (d) Start over again with the next biotinyl peptide Neutravidin bead population.
    (e) Repeat (a)–(c) as often as required.
    (f) Add 1 ml storage buffer.
12. Store biotinyl peptide bead array at 4 °C in the dark until required.

3.2 Preparation of Medium- and High-Plex Bead Arrays Using BSA Carrier and Cysteinyl Peptides

For the immobilization of proteins on more than ten different bead populations, a magnetic particle processor (King Fisher 96) can be used. This system uses a magnetic head and plastic comb, which enable the paramagnetic beads to be transferred between microtiter plates.

3.2.1 Preparation of BSA Beads (Semiautomated Protocol)

1. Prepare eight KingFisher 96 KF plates (200 μl) as follows:

Plate 1:	Vortex (for at least 10 s) and sonicate bead stocks, transfer 100 μl of each bead stock solution (1.25×10^7 beads/ml) to the respective wells. Add 10 μl volume of 0.055 % Triton X-100 to reach a final concentration of 0.005 % (v/v).
Plate 2:	Add 250 μl activation buffer to the respective wells.

(continued)

(continued)

Plate 3:	Add 150 μl activation mix: 120 μl activation buffer, 15 μl sulfo-NHS solution (50 mg/ml) and 15 μl EDC solution (50 mg/ml) (*see* **Note 1**).
Plate 4:	Add 250 μl coupling buffer to the respective wells.
Plate 5:	Add 250 μl coupling buffer to the respective wells.
Plate 6:	Add 125 μl of diluted BSA solution (100 μg/ml).
Plate 7:	Add 250 μl washing buffer to the respective wells.
Plate 8:	Add 250 μl storage buffer to the respective wells.

2. Place the eight plates into the magnetic particle processor (KingFisher 96).
3. Place a magnetic particle processor tip comb for deep-well magnets in plate 1.
4. Start the KingFisher 96 routine as shown in Table 1
5. Table 1 (protocol must be created in advance with the BindIt Software 3.1 for KingFisher Instruments). At the end of every step, the beads are collected 5 × 30 s and transferred to the next plate.
6. Cover plate 8 with adhesive foil and store at 4 °C in the dark until required.

Table 1
KingFisher 96 routine for the immobilization of BSA on paramagnetic beads

Step	Plate	Function	Duration
1	Plate 1 (beads)	Collect beads	5 × 30 s
2	Plate 2 (washing plate 1)	Washing (mixing)	1 min
3	Plate 3 (activation plate)	Activation (4 cycles of 0:30-min mixing/4:30-min pause)	20 min
4	Plate 4 (washing plate 2)	Washing (mixing)	1 min
5	Plate 5 (washing plate 3)	Washing (mixing)	1 min
6	Plate 6 (BSA solution)	Immobilization (24 cycles of 0:30-min mixing/4:30-min pause)	2 h
7	Plate 7 (washing plate 4)	Washing (mixing)	1 min
8	Plate 8 (release)	Release of beads	30 s

3.2.2 Covalent Immobilization of Cysteinyl Peptides on BSA Beads (Semiautomated Protocol)

1. Prepare eight KingFisher 96 KF plates (200 µl) as follows:

Plate 1:	Take plate 8 of the semiautomated BSA immobilization (250 µl BSA beads per well) and add 10 µl of 0.1 % (v/v) Triton X-100 to the respective wells.
Plate 2:	Add 100 µl freshly prepared sulfo-SMPB solution and 100 µl dilution buffer 1 to the respective wells.
Plate 3:	Add 250 µl washing buffer to the respective wells.
Plate 4:	Add 250 µl washing buffer to the respective wells.
Plate 5:	Add 50 µl cysteinyl peptide (1 mM) and 50 µl TCEP (1 mM), incubate for 20 min at RT with 350 rpm in a plate shaker. Add 150 µl dilution buffer 2 to the respective wells (*see* **Note 2**).
Plate 6:	Add 250 µl washing buffer to the respective wells.
Plate 7:	Add 250 µl washing buffer to the respective wells.
Plate 8:	Add 100 µl storage buffer to the respective wells.

2. Place all plates—except plate 5—into the KingFisher 96.
3. Place a KingFisher 96 tip comb for deep-well magnets in plate 1.
4. Start the KingFisher 96 routine shown in Table 2 (must be created in advance with the BindIt Software 3.1 for KingFisher Instruments). At the end of every step, the beads are collected 5 × 30 s and transferred to the next plate.
5. During **step 1**, prepare plate 5 (*see* **Note 2**) and place it in due course into KingFisher 96.

Table 2
KingFisher 96 routine for the immobilization of cysteinyl peptides on BSA beads

Step	Plate	Function	Duration
1	Plate 1 (BSA beads)	Collect beads	5 × 30 s
2	Plate 2 (activation plate)	Activation (12 cycles of 0:30-min mixing/4:30-min pause)	1 h
3	Plate 3 (washing plate 1)	Washing (mixing)	1 min
4	Plate 4 (washing plate 2)	Washing (mixing)	1 min
5	Plate 5 (peptide solution)	Immobilization (12 cycles of 0:30-min mixing/4:30-min pause)	1 h
6	Plate 6 (washing plate 3)	Washing (mixing)	1 min
7	Plate 7 (washing plate 4)	Washing (mixing)	1 min
8	Plate 8 (release)	Release of beads	30 s

6. The bead recovery should be determined as described under subheading 3.1.1 starting with **step 13**.
7. Generate a cysteinyl peptide array:
   (a) Resuspend the bead solutions from plate 8 (one million beads per population, sufficient for 1000 assays) and transfer into one 1.5 ml LoBind tube.
   (b) Place the tube in a magnetic separator (for at least 30 s) to collect the beads.
   (c) Discard the supernatant.
   (d) Repeat (a)–(c) as often as necessary.
   (e) Add 1 ml storage buffer and store beads at 4 °C in the dark.
8. The bead population concentration in the array should be determined as described under subheading 3.1.1 starting with **step 13**.

3.3 Preparation of Medium- and High-Plex Bead Arrays Using Cellulose Peptide Conjugates (Semiautomated Protocol)

1. Prepare eight KingFisher 96 KF plates (200 μl) as follows:

Plate 1:	Vortex (for at least 10 s) and sonicate bead stocks, transfer 100 μl of each bead stock solution (1.25×10^7 beads/ml) to the respective wells. Add 10 μl of 0.055 % of Triton X-100 to reach a final concentration of 0.005 % (v/v).
Plate 2:	Add 250 μl activation buffer to the respective wells.
Plate 3:	Add 150 μl activation mix: 120 μl activation buffer, 15 μl sulfo-NHS solution (50 mg/ml), and 15 μl EDC solution (50 mg/ml) (*see* **Note 1**).
Plate 4:	Add 250 μl coupling buffer to the respective wells.
Plate 5:	Add 250 μl coupling buffer to the respective wells.
Plate 6:	Add 125 μl of dissolved cellulose peptides (10 μM).
Plate 7:	Add 250 μl washing buffer to the respective wells.
Plate 8:	Add 100 μl storage buffer to the respective wells.

2. Place the eight plates into the KingFisher 96.
3. Place a KingFisher 96 tip comb for deep-well magnets in plate 1.
4. Start the KingFisher 96 routine shown in Table 3 (protocol must be created in advance with the BindIt Software 3.1 for KingFisher Instruments). At the end of every step, the beads are collected 5 × 30 s and transferred to the next plate.
5. Combine the single-bead populations to generate a cellulose peptide array:
   (a) Resuspend the bead solutions (one million beads per population, sufficient for 1000 assays) from plate 8 and transfer into one 1.5 ml LoBind tube.

Table 3
KingFisher 96 routine for the immobilization of cellulose peptides on paramagnetic beads

Step	Plate	Function	Duration
1	Plate 1 (beads)	Collect beads	5 × 30 s
2	Plate 2 (washing plate 1)	Washing (mixing)	1 min
3	Plate 3 (activation plate)	Activation (4 cycles of 0:30-min mixing/ 4:30-min pause)	20 min
4	Plate 4 (washing plate 2)	Washing (mixing)	1 min
5	Plate 5 (washing plate 3)	washing (mixing)	1 min
6	Plate 6 (peptide solution)	Immobilization (12 cycles of 0:30-min mixing/4:30-min pause)	1 h
7	Plate 7 (washing plate 4)	Washing (mixing)	1 min
8	Plate 8 (release)	Release of beads	30 s

(b) Place the tube in a magnetic separator (for at least 30 s) to collect the beads.

(c) Discard supernatant.

(d) Repeat (a)–(c) as often as necessary.

(e) Add 1 ml storage buffer and store beads at 4 °C in the dark.

6. The bead population concentration in the array should be determined as described under subheading 3.1.1 starting with **step 13**.

3.4 Antibody Characterization (Figs. 1 and 2)

1. Dilute the antibodies in assay buffer. Test various concentrations of antibody to avoid saturation affects. Typically, 50, 100, and 500 ng/ml are suitable concentrations.
2. Thoroughly vortex the bead-based peptide array and prepare a working solution containing 20,000 beads per population per ml assay buffer.
3. Add 50 μl bead solution (corresponding to 1000 beads per population) to the respective wells of an assay plate.
4. Place the assay plate on a magnetic bead separation plate (for at least 30 s) to collect the beads.
5. Discard supernatant by inverting the assay plate fixed to the magnetic bead separation plate.
6. Remove all liquid from the wells by tapping the assay plate fixed to the magnetic bead separation plate on a paper towel.
7. Add 50 μl antibody solution to the respective wells of the assay plate and incubate for 2 h at RT with 650 rpm in a plate shaker.

8. Place the assay plate on a magnetic bead separation plate (for at least 30 s) to collect the beads. Discard the supernatant by inverting the assay plate fixed to the magnetic bead separation plate and wash the beads twice with 100 μl washing buffer.
9. Dilute phycoerythrin-conjugated secondary antibody in assay buffer to reach a concentration of 2.5 μg/ml during the last washing step.
10. Remove all liquid from the wells by tapping the assay plate fixed on the magnetic bead separation plate on a paper towel.
11. Add 50 μl detection antibody solution to the respective wells of the assay plate and incubate for 1 h at RT with 650 rpm in a plate shaker.
12. Place the assay plate on a magnetic bead separation plate (for at least 30 s) to collect the beads. Discard supernatant by inverting the assay plate fixed to the magnetic bead separation plate and wash the beads twice with 100 μl washing buffer.
13. Remove all liquid from the wells by tapping the assay plate fixed on the magnetic bead separation plate on a paper towel.
14. Add 100 μl readout buffer and place the assay plate in a Luminex instrument (FlexMap 3D, Luminex 100/200 or MagPix). Standard settings for readout are doublet discriminator 7500–15,000, sample volume 80 μl, time-out 60 s, and minimal bead count per sort (region) 100.

4 Notes

1. Since sulfo-NHS and EDC hydrolyze in aqueous solution very quickly, this plate should be prepared last.
2. Thiol groups of cysteinyl peptides tend to oxidize very quickly. Therefore, this plate has to be prepared during the 60-min incubation time of plate 2. Start 35 min after the 60-min incubation. Place plate 5 into KingFisher 96 before the end of the 60-min incubation time.

References

1. Ferrell JE Jr, Bhatt RR (1997) Mechanistic studies of the dual phosphorylation of mitogen-activated protein kinase. J Biol Chem 272: 19008–19016
2. Du J, Bernasconi P, Clauser KR, Mani DR, Finn SP, Beroukhim R, Burns M, Julian B, Peng XP, Hieronymus H, Maglathlin RL, Lewis TA, Liau LM, Nghiemphu P, Mellinghoff IK, Louis DN, Loda M, Carr SA, Kung AL, Golub TR (2009) Bead-based profiling of tyrosine kinase phosphorylation identifies SRC as a potential target for glioblastoma therapy. Nat Biotechnol 27:77–83
3. Poetz O, Henzler T, Hartmann M, Kazmaier C, Templin MF, Herget T, Joos TO (2010) Sequential multiplex analyte capturing for phosphoprotein profiling. Mol Cell Proteomics 9:2474–2481

4. Steen H, Kuster B, Fernandez M, Pandey A, Mann M (2002) Tyrosine phosphorylation mapping of the epidermal growth factor receptor signaling pathway. J Biol Chem 277: 1031–1039
5. Choudhary C, Kumar C, Gnad F, Nielsen ML, Rehman M, Walther TC, Olsen JV, Mann M (2009) Lysine acetylation targets protein complexes and co-regulates major cellular functions. Science 325:834–840
6. Peach SE, Rudomin EL, Udeshi ND, Carr SA, Jaffe JD (2012) Quantitative assessment of chromatin immunoprecipitation grade antibodies directed against histone modifications reveals patterns of co-occurring marks on histone protein molecules. Mol Cell Proteomics 11:128–137
7. Nilsson P, Paavilainen L, Larsson K, Odling J, Sundberg M, Andersson AC, Kampf C, Persson A, Al-Khalili Szigyarto C, Ottosson J, Bjorling E, Hober S, Wernerus H, Wester K, Ponten F, Uhlen M (2005) Towards a human proteome atlas: high-throughput generation of monospecific antibodies for tissue profiling. Proteomics 5:4327–4337
8. Poetz O, Schwenk JM, Kramer S, Stoll D, Templin MF, Joos TO (2005) Protein microarrays: catching the proteome. Mech Ageing Dev 126:161–170
9. Buus S, Rockberg J, Forsstr Oumlm BO, Nilsson P, Uhlen M, Schafer-Nielsen C (2012) High-resolution mapping of linear antibody epitopes using ultrahigh-density peptide microarrays. Mol Cell Proteomics 11:1790–1800
10. Breitling F, Felgenhauer T, Nesterov A, Lindenstruth V, Stadler V, Bischoff FR (2009) Particle-based synthesis of peptide arrays. Chem BioChem 10:803–808
11. Bock I, Kudithipudi S, Tamas R, Kungulovski G, Dhayalan A, Jeltsch A (2011) Application of Celluspots peptide arrays for the analysis of the binding specificity of epigenetic reading domains to modified histone tails. BMC Biochem 12:48
12. Heubach Y, Planatscher H, Sommersdorf C, Maisch D, Maier J, Joos TO, Templin MF, Poetz O (2013) From spots to beads-PTM-peptide bead arrays for the characterization of anti-histone antibodies. Proteomics 13: 1010–1015
13. Kurzeder C, Koppold B, Sauer G, Pabst S, Kreienberg R, Deissler H (2007) CD9 promotes adeno-associated virus type 2 infection of mammary carcinoma cells with low cell surface expression of heparan sulphate proteoglycans. Int J Mol Med 19:325–333
14. Bauer M, Chicca A, Tamborrini M, Eisen D, Lerner R, Lutz B, Poetz O, Pluschke G, Gertsch J (2012) Identification and quantification of a new family of peptide endocannabinoids (Pepcans) showing negative allosteric modulation at CB1 receptors. J Biol Chem 287:36944–36967
15. Frank R (1992) Spot-synthesis: an easy technique for the positionally addressable, parallel chemical synthesis on a membrane support. Tetrahedron 48:9217–9232
16. Liu C, Li Y, Semenov M, Han C, Baeg GH, Tan Y, Zhang Z, Lin X, He X (2002) Control of beta-catenin phosphorylation/degradation by a dual-kinase mechanism. Cell 108: 837–847
17. Aberle H, Bauer A, Stappert J, Kispert A, Kemler R (1997) beta-catenin is a target for the ubiquitin-proteasome pathway. EMBO J 16:3797–3804
18. Orford K, Crockett C, Jensen JP, Weissman AM, Byers SW (1997) Serine phosphorylation-regulated ubiquitination and degradation of beta-catenin. J Biol Chem 272:24735–24738

Chapter 23

Surface Plasmon Resonance Method to Evaluate Anti-citrullinated Protein/Peptide Antibody Affinity to Citrullinated Peptides

Feliciana Real-Fernández, Giada Rossi, Filomena Panza, Federico Pratesi, Paola Migliorini, and Paolo Rovero

Abstract

Surface plasmon resonance (SPR) technique is extremely interesting in immunology because it has the potential to directly visualize biomolecular interactions in real-time monitoring antibody affinity, one of the parameters affecting pathogenicity in autoimmune diseases. Herein we describe the affinity evaluation of anti-citrullinated peptide antibodies (ACPA) to a peptide-based biosensor by SPR. The method describes the purification of ACPA isolated from rheumatoid arthritis (RA) patients using affinity columns, the strategy employed for the immobilization of citrullinated peptides onto a sensor chip, and the evaluation of the specific binding of purified ACPA to immobilized peptides.

Key words Antibody affinity, Kinetic, Citrullinated peptides, Biacore, Surface plasmon resonance (SPR)

1 Introduction

Biosensor technology based on surface plasmon resonance (SPR) has become increasingly popular for monitoring binding interactions. SPR technique has been largely applied to characterize analyte-ligand affinity in different fields of biology and medicine including immunology [1–4]. The technique allows both the screening for candidates and a detailed kinetic and mechanistic analysis of molecular interactions [5]. Comparison of the association and dissociation rates for anti-citrullinated protein/peptide antibodies (ACPA) recognizing diagnostic peptides was performed in the particular case or rheumatoid arthritis (RA), suggesting possible cross-reactivity with potential clinical relevance [6].

Herein, we present a detailed description of an SPR method to characterize and evaluate affinity of purified RA antibodies to diagnostic citrullinated peptides. A similar procedure has been recently used to study ACPA profiles directly on sera of early arthritis

Gunnar Houen (ed.), *Peptide Antibodies: Methods and Protocols*, Methods in Molecular Biology, vol. 1348, DOI 10.1007/978-1-4939-2999-3_23,

patients without providing, however, an evaluation of affinity [7]. The affinity of autoantibodies for their target has a relevant role in determining their pathogenicity. In fact, previous studies on ACPA affinity and avidity, employing the classical immunoenzymatic ELISA, indicate that high-avidity ACPA are produced in RA prior to disease onset and are detectable in a symptomatic patients [8, 9].

The protocol herein described includes three steps: (a) purification of ACPA from RA patients' sera by affinity chromatography, (b) immobilization of diagnostic citrullinated peptides onto gold sensor chip, and (c) peptide-antibody binding experiments and affinity evaluation. The first step provides purified ACPA to be further used in SPR avoiding nonspecific interactions due to the presence of other serum components including immunoglobulins of other specificity. Thus, we obtain clearer experiments/signals decreasing bulk effects. Thanks to the second step we can assess the orientation of peptides not only in the case of linear sequences but also when immobilizing citrullinated multiple antigenic peptides (MAPs) in the sensor chip. MAPs are synthetic peptides built upon a core resulting in three-dimensional molecules with highly localized peptide density, and classic amine coupling protocol did not guarantee a well-known orientation. These steps supply the optimal conditions to further evaluate the interaction of ACPA with citrullinated MAPs and the calculation described in the third step.

2 Materials

Conduce experiments using a SPR instrument (e.g., Biacore T100) and prepare all solutions and buffers with Milli-Q water or similar quality.

Commercially available sensor chips CM5, amine coupling kit, thiol-coupling reagent, and running buffer (e.g., HBS–EP+10×) were used.

Eject sensor chip from the instrument, place it into the appropriate plastic case, and store at +4 °C; gently wash chip surface with Milli-Q water and dry under a nitrogen stream before inserting it into the instrument. Filter buffers daily with a 0.22 μm system. Store enriched IgG fractions from RA patients' sera and purified antibody samples at −20 °C avoiding thawing and freezing cycles. Follow the waste disposal regulation for biological and chemical waste.

2.1 Affinity Chromatography

1. Enriched IgG fraction: Purify IgG antibodies using the protein G-sepharose column as reported [10].
2. Phosphate buffer: 20 mM Na_2HPO_4, 150 mM NaCl, pH 7.2. Weigh 567.84 mg of Na_2HPO_4 and 1.77 g of NaCl, add 200 mL of water, and adjust pH at 7.2. Store at room temperature.

3. Elution buffer: 100 mM glycine, pH 2.8. Weigh 750.7 mg glycine, add 100 mL water, and adjust pH at 2.8. Store at +4 °C.
4. Tris buffer: 1 M Tris, pH 8.0. Weigh 12.11 g of Tris, dissolve in 100 mL water, and adjust pH at 8.0. Store at +4 °C.
5. D-PBS buffer 10×: Place 50 mL of buffer in a graduate cylinder, add 450 mL water, and adjust pH at 7.2. Store at room temperature.

2.2 PH Scouting Buffers (*See* Note 1)

1. Running buffer HBS–EP+10× (0.1 M Hepes, 1.5 M NaCl, 30 mM EDTA, and 0.5 % [v/v] P20): Place 50 mL of buffer in a graduate cylinder, dilute ten times with water, and adjust pH at 7.4. Store at +4 °C.
2. D-PBS buffer 10×: *See* Subheading 2.1.
3. Sodium acetate buffers: 5 mM $CH_3COO^-Na^+$, pH 4.5, 5.0, and 6.0. Weigh 68.04 mg of $CH_3COO^-Na^+$, prepare 100 mL solution, and adjust pH at 4.5; repeat the same procedure for pH 5.0 and 6.0. 0.5 mM $CH_3COO^-Na^+$ pH 4.5 and 5.0, measure 10 mL of the previously prepared buffers at the desired pH, and add water up to 100 mL. 0.1 mM $CH_3COO^-Na^+$ pH 4.5 and 6.0. Measure 2 mL of the 5 mM buffers at the desired pH and add water up to 100 mL.
4. Regeneration solution: 0.1 M NaOH. Weigh 2 g of NaOH and dissolve in 500 mL water.
5. Peptide stock solutions: Dissolve each peptide in water to obtain a final concentration of 1 mg/ml (weigh 500 μg of peptide and dilute it in 500 μL of water).

2.3 Citrullinated Peptide Immobilization Reagents

1. Amine coupling kit: 0.1 M *N*-hydroxysuccinimide (NHS) stock solution, dissolve 115 mg in 10 mL water. 0.4 M 1-ethyl-3-(3-dimethylaminopropyl)-carbodiimide (EDC) stock solution, dissolve 750 mg in 10 mL water. Dispense stock solutions separately in aliquots and store frozen at −18 °C. 1 M Ethanolamine-HCl pH 8.5, store at +4 °C.
2. Thiol-coupling reagent: 120 mM 2-(2-pyridinyldithio)ethaneamine (PDEA) stock solution, dissolve 100 mg of PDEA in 3.74 mL water, store stock solution in aliquots at −18 °C.
3. Sodium borate buffer: 0.1 M $Na_2B_4O_7 \cdot 10\ H_2O$ pH 8.5. Weigh 381.37 mg of $Na_2B_4O_7 \cdot 10\ H_2O$, dissolve in 10 mL water, and adjust pH at 8.5.
4. Disulfide blocking buffer: 50 mM cysteine—HCl, 1 M NaCl, 0.1 M $CH_3COO^-Na^+$ pH 4.3. Weigh 394.05 mg of cysteine-HCl, 2.92 g of NaCl, and 680.45 mg of $CH_3COO^-Na^+$, add 50 mL of water, adjust pH at 4.3, and store at room temperature.

2.4 Solutions for Kinetic and Affinity Studies

1. Running buffer (RB) (*see* Subheading 2.2, **item 1**).
2. Anti-citrullinated peptide antibody fractions: Use different final concentrations of each fraction (1000, 500, 250, 125, 62.5, 31.25, and 0 nM). Prepare 120 μL of every concentrations, by serial dilution in running buffer (*see* **Note 2**): take the appropriate initial volume from each fraction according to the calculated initial concentration (*see* Subheading 3.1, **step 3**) (*see* **Note 3**).
3. Regeneration solutions: 0.1 M NaOH (*see* Subheading 2.2, **item 4**). 10 mM glycine pH 2.5. weigh 75.07 mg, dissolve in 100 mL water, adjust pH at 2.5.

3 Methods

Conduce all the experimental work at room temperature.

3.1 Anti-peptide Antibody Purification

1. Conjugate 1 mg of the citrullinated peptide to 100 mg of CNBr-activated sepharose according to the standard manufacturer instructions. Insert the functionalized resin into a chromatography column of 5 mL volume. Repeat this procedure for each single peptide separately.
2. Mix 1 mL of the enriched IgG fraction with 4 mL of phosphate buffer pH 7.2, apply the solution to the column three times, wash with 10 mL of phosphate buffer pH 7.2, elute anti-citrullinated peptide antibodies with 5 mL of elution buffer, and immediately neutralize with 50 μL of Tris buffer (*see* **Note 4**). Dialyze overnight against D-PBS buffer (*see* **Note 5**).
3. Calculate the concentration of each fraction using a quartz cuvette to measure the sample UV absorbance: C[mg/ml] = UV absorbance at $280_{\text{nm}}/1.4$.

3.2 pH Scouting: Selection of Immobilization Buffer

1. Mix 1 μL of peptide stock solution and 99 μL of each immobilization buffer.
2. Inject the first sample onto the chip surface for 120 s at a flow rate of 10 μL/min.
3. Regenerate chip surface with a pulse of 0.1 M NaOH for 30 s at a flow rate of 10 μL/min.
4. Repeat **steps 2** and **3** for each ligand in each immobilization buffer (*see* **Note 6**).
5. Select the buffer with the highest sensorgram slope as immobilization buffer.

3.3 Peptide Immobilization: Ligand Thiol Coupling Strategy

1. Chip surface activation: Mix 80 μL of NHS stock solution and 80 μL of EDC stock solution, and inject for 420 and 60 s at a flow rate of 10 μL/min (*see* **Note** 7).
2. Active disulfide group introduction: Mix 67 μL of PDEA stock solution and 33 μL of sodium borate buffer, and inject for 420 s at a flow rate of 10 μL/min (*see* **Note 8**).
3. Ligand injection: Mix 4 μL of each peptide stock solution with 396 μL of the previously selected immobilization buffer, and inject for 420 and 60 s (*see* **Note 9**) at a flow rate of 10 μL/min (*see* **Note 10**).
4. Unreacted disulfide group block: Inject 200 μL of disulfide blocking buffer for 420 s at a flow rate of 10 μL/min.
5. Unreacted succinimide group block: Inject 200 μL of 1 M ethanolamine–HCl pH 8.5 for 420 s at a flow rate of 10 μL/min.
6. Reference channel: Repeat steps from Subheading 3.3, **steps 1–5**, except step in Subheading 3.3, **step 3** (*see* **Note 11**).

3.4 SPR Kinetic and Affinity Studies

1. Sensor chip surface start-up: Take 3 mL of running buffer and inject for 120 s, take 1 mL of 10 mM glycine pH 2.5 and inject for 30 s, take 1 mL of 0.1 M NaOH and inject for 60 s, and use a flow rate of 30 μL/min. Repeat this step at least four times.
2. Cycle of analysis: Take 120 μL of each concentration of the first anti-citrullinated peptide antibody fraction and contemporary inject over up to the three different immobilized peptides for 120 s; flush running buffer for 200 s; regenerate chip surface by injecting 10 mM glycine pH 2.5 for 30 s and 0.1 M NaOH for 60 s; use a flow rate of 30 μL/min. Repeat this cycle for each purified antibody fraction (*see* **Note 12**).
3. Result elaboration: Use the evaluation software (e.g., Biacore evaluation software version 2.0) to fit experimental data to mathematic models (Fig. 1) and to calculate the kinetic parameters of the association k_a (M^{-1} s^{-1}) and the dissociation rate k_d (s^{-1}), and the affinity constants K_D (M).
4. Fitting validation: Verify that the residual values between the experimental points and the theoretical ones are closely distributed along the zero, the chi-square value is as lower as possible, and the standard errors are less than 10 % of the referred parameter value (*see* **Note 13**).
5. Results comparison: Use the Biacore T100 kinetic summary application to obtain the k_a/k_d plot (Fig. 2) showing all kinetic and affinity parameters (*see* **Note 14**).

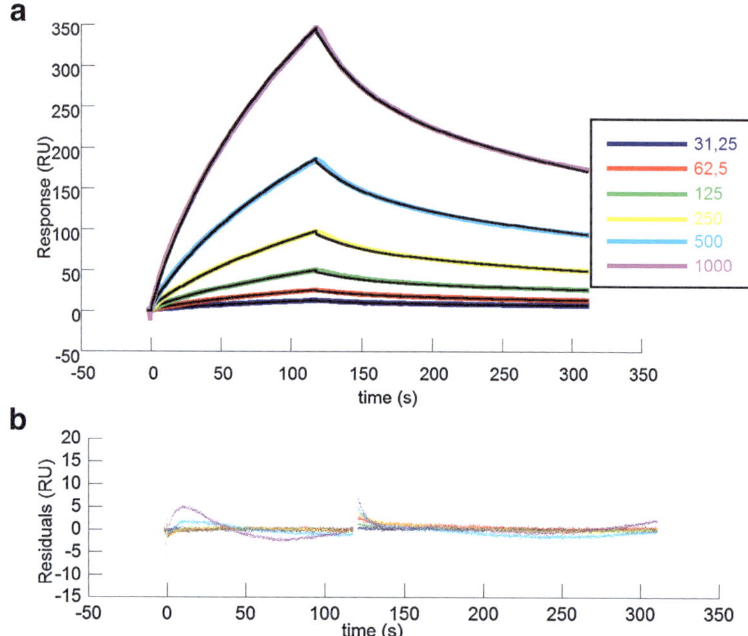

Fig. 1 (**a**) Sensorgrams resulting from the kinetic and affinity study of purified antibodies from one RA patient and one immobilized citrullinated peptide. Each colored curve corresponds to a different sample concentration (nM); the fitted binding 1:1 model is reported in *black lines*. (**b**) Residual values indicating the difference between experimental and mathematical points

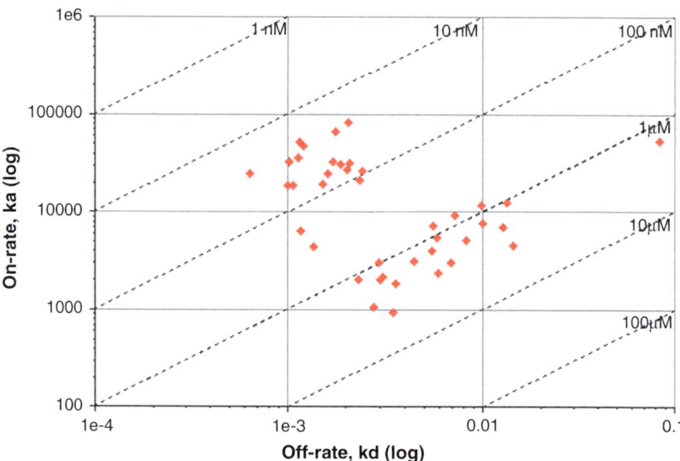

Fig. 2 k_a/k_d plot for the 1:1 binding model summarizing interactions among purified ACPA and the citrullinated peptides. Each point summarizes k_a, k_d, and K_D calculated values for each ACPA-citrullinated peptide kinetic study performed. Interactions characterized by same affinity and different kinetics are indicated by points lying on the same *diagonal line*

4 Notes

1. To achieve the electrostatic pre-concentration of ligands in the dextran matrix of CM5 chip, use immobilization buffers at pH between 3.5 and the isoelectric point of the ligands; pre-concentration is favored by low ionic strength in the buffer.
2. Dilution series play an important role in kinetics and affinity studies, preparing the samples by serial dilution where each dilution is used as the stock solution for the next step is useful to reduce the experimental error. Avoid bubble formation when diluting samples, or spin them to remove bubbles.
3. To prepare 120 μL of each sample concentration follow this scheme: prepare 240 μL of the 1000 nM concentration, take 117 μL and mix with 117 μL of RB for the 500 nM concentration, take 113 μL and mix with 113 μL of RB for the 250 nM concentration, take 105 μL and mix with 105 μL of RB for the 125 nM, take 90 μL and mix with 90 μL of RB for the 62.5 nM concentration, take 60 μL and mix with 60 μL of RB for the 31.25 nM concentration, and use 120 μL of RB for the zero concentration.
4. Collect the flow through (not retained fraction) and test it and anti-peptide antibody content in SP-ELISA, to confirm the success of the purification procedure.
5. This dialysis step is needed to match the refractive index of sample solution with RB in order to perform SPR studies avoiding the bulk effect which could affect the first part of the sensorgram of the higher injected concentration and that could be responsible for a wrong fitting to kinetic models during the association phase.
6. Discard buffers that give irregular sensorgrams or signals with irregular slopes, probably due to ligand aggregation/precipitation or chip saturation.
7. Mix equal volumes of EDC and NHS immediately before use.
8. The most appropriate covalent immobilization strategy has to be selected according to the reactive groups present in the ligand. Each citrullinated peptide herein used contains only one cysteine residue; thus the ligand thiol coupling strategy was chosen to obtain a well-defined and reproducible orientation of the ligands on the chip.
9. The best concentration will vary according to the nature of the ligand; higher concentrations and/or additional injections may be needed for ligands that do not pre-concentrate efficiently, probably due to their low isoelectric point.
10. A lower flow rate could be chosen for the immobilization of higher molecular weight ligands.

11. One channel without any immobilized ligand is used as reference, to remove the nonspecific signal depending on interactions between molecules present in the biological samples and gold on sensor chip surface.
12. A replicate concentration can be inserted to assess the surface regeneration performance and the reproducibility of the assay; start the analysis cycle by injecting the lower sample concentration at first.
13. A high-quality fitting could be obtained with more than one kinetic model. Consider that SPR signal is directly dependent on change in reflective index and thus in mass over the chip surface and consequently interactions between multivalent molecules do not produce an instrumental signal; if possible choose the more reliable binding 1:1 model.
14. Plot the logarithm of calculated k_d and k_a on *X*- and *Y*-axis, respectively, for each purified peptide-antibody interaction to summarize results if kinetic summary software Biacore 2.0 is not available.

References

1. Kausaite A, Ramanaviciene A, Mostovojus V et al (2007) Surface plasmon resonance and its application to biomedical research. Medicina (Kaunas) 43:355–365
2. Kaymakcalan Z, Sakorafas P, Bose S et al (2009) Comparisons of affinities, avidities, and complement activation of adalimumab, infliximab, and etanercept in binding to soluble and membrane tumor necrosis factor. Clin Immunol 131:308–316
3. Wittenberg NJ, Wootla B, Jordan LR et al (2014) Applications of SPR for the characterization of molecules important in the pathogenesis and treatment of neurodegenerative diseases. Expert Rev Neurother 14:449–463
4. Hearty S, Conroy PJ, Ayyar BV et al (2010) Surface plasmon resonance for vaccine design and efficacy studies: recent applications and future trends. Expert Rev Vaccines 9: 645–664
5. Cimitan S, Lindgren MT, Bertucci C et al (2005) Early absorption and distribution analysis of antitumor and anti-AIDS drugs: lipid membrane and plasma protein interactions. J Med Chem 48:3536–3546
6. Rossi G, Real-Fernández F, Panza F et al (2014) Biosensor analysis of anti-citrullinated protein/peptide antibody affinity. Anal Biochem 465C:96–101
7. van Beers JJ, Willemze A, Jansen JJ et al (2013) ACPA fine-specificity profiles in early rheumatoid arthritis patients do not correlate with clinical features at baseline or with disease progression. Arthritis Res Ther 15:R140
8. Suwannalai P, van de Stadt LA, Radner H et al (2012) Avidity maturation of anti-citrullinated protein antibodies in rheumatoid arthritis. Arthritis Rheum 64:1323–1328
9. Ossipova E, Cerqueira CF, Reed E et al (2014) Affinity purified anti-citrullinated protein/peptide antibodies target antigens expressed in the rheumatoid joint. Arthritis Res Ther 16:R167
10. Grodzki AC, Berenstein E (2010) Antibody purification: affinity chromatography—protein A and protein G Sepharose. Methods Mol Biol 588:33–41

Chapter 24

Specificity Analysis of Histone Modification-Specific Antibodies or Reading Domains on Histone Peptide Arrays

Goran Kungulovski, Ina Kycia, Rebekka Mauser, and Albert Jeltsch

Abstract

Histone posttranslational modifications (PTMs) have a crucial role in chromatin regulation and dynamics. They are specifically bound by so-called reading domains, which mediate the biological effects of histone PTMs. On a similar note, antibodies are invaluable reagents in chromatin biology for the detection, characterization, and mapping of histone PTMs. Despite these central roles in chromatin research and biology, the specificity of many antibodies and reading domains has been insufficiently characterized and documented. Here we describe in detail the application of the MODified™ Histone Peptide Array for the investigation of the binding specificity of histone binding antibodies or domains. The array contains 384 histone tail peptides carrying 59 posttranslational modifications in different combinations which can be used to study the primary binding specificity, but at the same time also allow to determine the combinatorial effect of secondary marks on antibody or reading domain binding.

Key words Histone posttranslational modification, Histone antibody, ChIP-seq, Antibody quality control, Reading domain

1 Introduction

Chromatin is a very dynamic biological entity, which plays an essential role in gene regulation [1]. The basic unit of chromatin is the nucleosome, consisting of ~147 bp of DNA wrapped around the highly basic histone protein octamer [2]. Histones harbor unstructured, flexible N-terminal tails, which project outwards of the core nucleosome and these tails along with the lateral surfaces of histones can be massively posttranslationally modified and these modifications (PTMs) play crucial roles in gene regulation, transcription, DNA replication and repair, development, and disease [3–7]. Proper characterization of histone marks can provide valuable insights into the inner workings of chromatin. Numerous biochemical methods have been developed for molecular dissection of other epigenetic modifications such as DNA methylation and noncoding RNAs [8, 9]. In contrast to this, histone marks can be globally

Gunnar Houen (ed.), *Peptide Antibodies: Methods and Protocols*, Methods in Molecular Biology, vol. 1348, DOI 10.1007/978-1-4939-2999-3_24,

characterized by both mass spectrometry and antibody-based approaches, but the locus specific characterization of histone marks is reliant on antibodies as the only single molecular tool for analysis. Although, antibodies are quite powerful molecular reagents, their target specificity is not always sufficiently and rigorously scrutinized. This is pivotal in chromatin biology where subtle and chemically related epitopes (often with different downstream biological effects) must be recognized differentially. Many labs have raised concerns regarding the quality of histone modification-specific antibodies making thorough and rigorous quality control an absolute necessity [10–16]. Histone PTMs exert their biological effects mainly through the regulated binding of reading domains, which are critical constituents of many chromatin-modifying complexes [17, 18]. Hence, the characterization of the PTM-specific binding of such reading domains is another important field of current research.

In order to obtain detailed specificity information, which in turn is needed for correct data interpretation, it is an imperative to test the binding of these proteins (antibodies or reading domains) to a large library of potential substrate peptides which carry different modification patterns. Such experiments can be ideally conducted with modified peptide arrays, and to this end, herein we describe the application of the MODified™ Histone Peptide Array as a first step of characterization and screening of antibodies or reading domains. This histone peptide array features 384 peptides and 59 identified or hypothetical histone PTMs in duplicates, to ensure reproducibility. All the peptides on the array reflect the sequence of eight different regions of the N-terminal tails of histones, viz. H3 1–19, 7–26, 16–35, and 26–45; H4 1–19 and 11–30; and H2A 1–19 and H2B 1–19. A representative annotation of all histone tail peptides over the binding profile of anti-H3K4me1 antibody (Active Motif, AM 39297) is shown in Fig. 1.

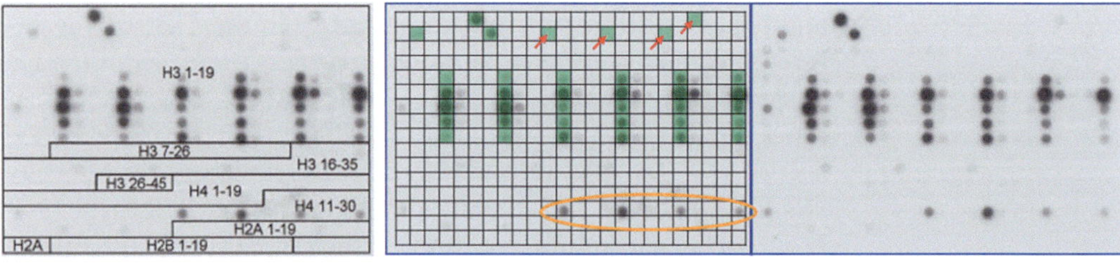

Fig. 1 Design of the peptide array and binding pattern of an H3K4me1 antibody. On the *left*, one side of the peptide array is shown and the peptides are annotated. On the *right*, the entire array is shown and the peptide modifications are annotated on one part of it. *Green regions* indicate peptides containing the H3K4me1 modification. *Red arrows* point to spots which also contain H3R2me2s, H3R2me2a, or H3T3Ph and are not bound by the antibody although they contain H3K4me1 (false negatives). The *orange circle* indicates false positives which all contain H4K20me1. The figure has been created using data published in ref. 10. A detailed list of the peptides and their corresponding modifications is given in http://www.activemotif.com/catalog/667/histone-peptide-array

The MODified™ Histone Peptide Array is produced using the CelluSpots technology [19]. This particular array provides a cost-efficient way to obtain specificity information for histone tail peptide-binding proteins. It has been validated for the specificity analysis of antibodies [10] and histone-reading domains [20] and it has been successfully used in many publications to characterize similar interactions afterwards (*see* for example refs. 21–32). However, when interpreting the results one needs to keep in mind that peptide-based specificity analyses of histone PTM binders may not always fully reflect their behavior on modified full-length histone proteins or on native nucleosomes. Differences arising between peptide and histone binding of antibodies or reading domains may be due to the fact that peptides are too short, and as such that they present the modification close to one of the termini. Peptide presenting internal histone tail modifications also contain artificial ends when compared to histone proteins, which may influence the proper recognition. Furthermore, many reading domains may form additional (secondary) interactions to the H3 outside of the peptide region or even to other tails as well, which could influence the binding specificity to peptides versus histones and nucleosomes, and finally nucleosomal epitopes may not be always accessible and hinder binding. However, modified peptides are one of the few ways to define the specificity of an antibody or a reading domain by first principle, which makes this approach an important reference point when analyzing the specificity of histone modification-binding reagents.

2 Materials

Prepare all solutions with ultrapure Milli-Q water and molecular biology-grade reagents and chemicals.

2.1 Array Binding and Analysis Components

1. MODified™ Histone Peptide Arrays. CelluSpots™ arrays are manufactured under license by INTAVIS Bioanalytical Instruments AG and sold through Active Motif Inc.
2. List of peptides and their corresponding modifications and the Array Analyze Software (http://www.activemotif.com/catalog/667/histone-peptide-array).
3. Tween-20-containing Tris-buffered saline (TTBS 10×): 100 mM Tris–Cl pH 7.5, 0.5 % Tween-20, and 1.5 M NaCl (*see* **Note 1**).
4. Tween-20-containing Tris-buffered saline (TTBS 1×): 10 mM Tris–Cl pH 7.5, 0.05 % Tween-20, and 150 mM NaCl.
5. Blocking solution: 5 % nonfat milk in TTBS (*see* **Note 1**).
6. Interaction buffer: 100 mM KCl, 20 mM HEPES pH 7.5, 1 mM EDTA, 0.1 mM DTT, and 10 % glycerol.
7. Nunc plates with four wells.

2.2 Antibodies and Detection

1. Primary reagent: Any histone modification-specific antibody or purified reading domain. Best results were obtained with GST-tagged reading domains.
2. Secondary antibody: horseradish conjugated anti-rabbit antibody, or anti-GST antibody for the analysis of GST-tagged reading domains, or another adequate antibody.
3. Tertiary horseradish-conjugated anti-goat antibody in the case of GST-tagged reading domains.
4. ECL-developing solution.

3 Methods

All experiments should be carried out at room temperature unless specified otherwise.

3.1 Incubation of the Peptide Array with Antibodies

1. Insert the peptide array in a Nunc plate with clean forceps and block it with 5 mL blocking solution for 1 h at room temperature with mild shaking or overnight at +4 °C (*see* **Note 2**).
2. Wash the peptide array with 3.5 mL TTBS for 5 min with mild shaking (*see* **Note 3**).
3. Repeat **step 2** twice.
4. Dissolve the primary anti-histone modification antibody in 2–5 mL blocking solution (use dilution recommended by the manufacturer for western blot) and incubate for 1 h (*see* **Note 3**).
5. Wash the peptide array with 3.5 mL TTBS for 5 min with mild shaking.
6. Repeat **step 5** twice.
7. Dissolve the secondary horseradish peroxidase-conjugated antibody in 2–5 mL blocking solution (use dilution recommended by the manufacturer) and incubate for 1 h.
8. Wash the peptide array with 3.5 mL TTBS for 5 min with mild shaking.
9. Repeat **step 8** three times (*see* **Note 4**).
10. Add 0.4 mL final volume of ECL developing solution per array (*see* **Note 5**).
11. Detect chemiluminescence either on X-ray film or with chemiluminescence imaging instrument.
12. Use different exposition times and retain images where the signal is not saturated (*see* **Note 6**).

3.2 Incubation of the Peptide Array with Recombinant Reading Domains

1. Insert the peptide array in a Nunc plate with clean forceps and block it with 5 mL blocking solution for 1 h at room temperature with mild shaking or overnight at +4 °C (*see* **Note 2**).
2. Wash the peptide array with 3.5 mL TTBS for 5 min with mild shaking (*see* **Note 3**).
3. Repeat **step 2** once.
4. Wash the peptide array with 3.5 mL interaction buffer for 5 min with mild shaking.
5. Add 1–1000 nM recombinant protein dissolved in 2–4 ml interaction buffer and incubate for at least 2 h at room temperature (*see* **Note 3**).
6. Wash the peptide array with 3.5 mL TTBS for 5 min with mild shaking.
7. Repeat **step 6** twice.
8. Dissolve the secondary anti-TAG antibody in 2–5 mL blocking solution (use dilution recommended by the manufacturer for western blot) and incubate for 1 h (*see* **Note 3**).
9. Wash the peptide array with 3.5 mL TTBS for 5 min with mild shaking.
10. Repeat **step 9** twice.
11. Dissolve the tertiary horseradish peroxidase-conjugated antibody in 2–5 mL blocking solution (use dilution recommended by the manufacturer) and incubate for 1 h.
12. Wash the peptide array with 3.5 mL TTBS for 5 min with mild shaking.
13. Repeat **step 12** three times (*see* **Note 4**).
14. Add 0.4 mL final volume of ECL developing solution per array (*see* **Note 5**).
15. Detect chemiluminescence either on X-ray film or with chemiluminescence imaging instrument.
16. Use different exposition times and retain images where the signal is not saturated (*see* **Note 6**).

3.3 Bioinformatic Analysis of the Peptide Array

1. Click on *start analysis* and open your file with primary data (*tiff* format) (*see* **Note 7**).
2. Assign spots on the right- and left-hand side. To do this, click on a random bound spot on the left hand side (in this case C1), select its position (you may refer to the excel file for help), and press *assign left spot*. Repeat the same for the right-hand side. After assigning the right and left spots, press *assign spots*. This gives the coordinates for gridline positioning (Fig. 2). If the gridline is not positioned correctly, press *reset analysis* and try once more.

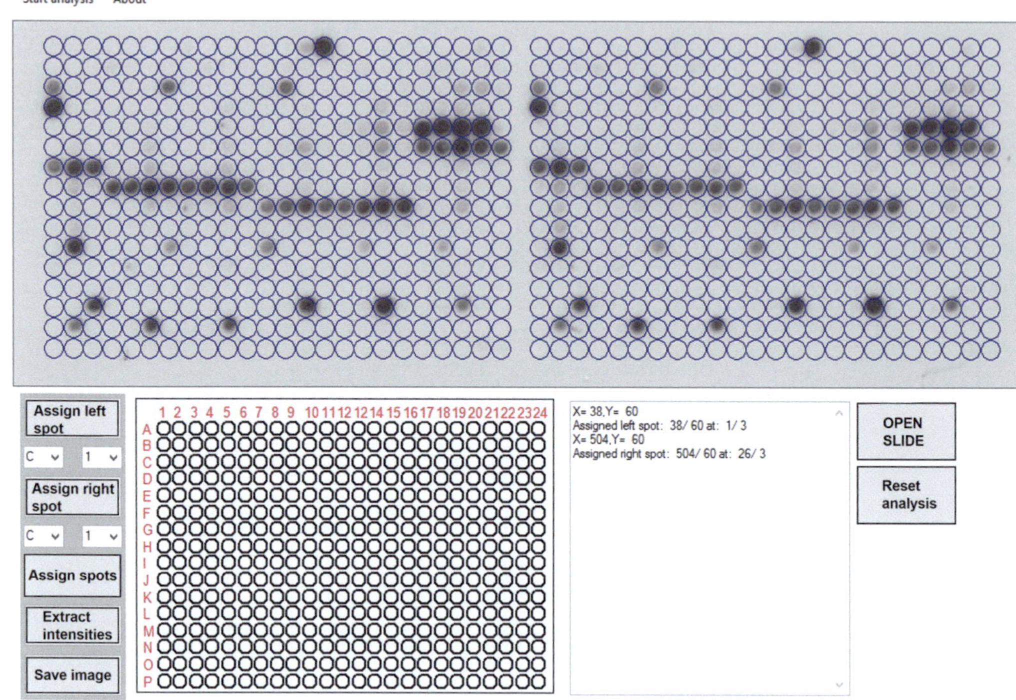

Fig. 2 Correct spot assignment and grid positioning

3. Now you are ready to analyze the array. Press *extract intensities.* A window with spot statistics will appear. The scatter plot shows you the reproducibility of spot intensity on both sides of the peptide array (Fig. 3a). This provides a first assessment of the technical reproducibility of the results. Errors between 5 and 30 % are acceptable.
4. Press *continue* to set the background level. This can be modulated manually, but most of the time the default setting is sufficient (Fig. 3b).
5. Now you can proceed to specificity analysis by pressing *calculate specificity.* When you click on the *specificity analysis* bar, you see a bar diagram showing the specificity factor for different histone modifications (Fig. 3c).
6. For manual inspection of the peptide array one can click on the *annotated array* bar and assign different colors to different histone PTMs. Also, one can map either one or both sides of the array (Fig. 4). This is what we typically do for data presentation in scientific publications and internal lab seminars.
7. All successive files (scatter plot, background subtraction, specificity bar diagram images, and spreadsheet files with ranking intensities) are automatically saved in the folder of origin.

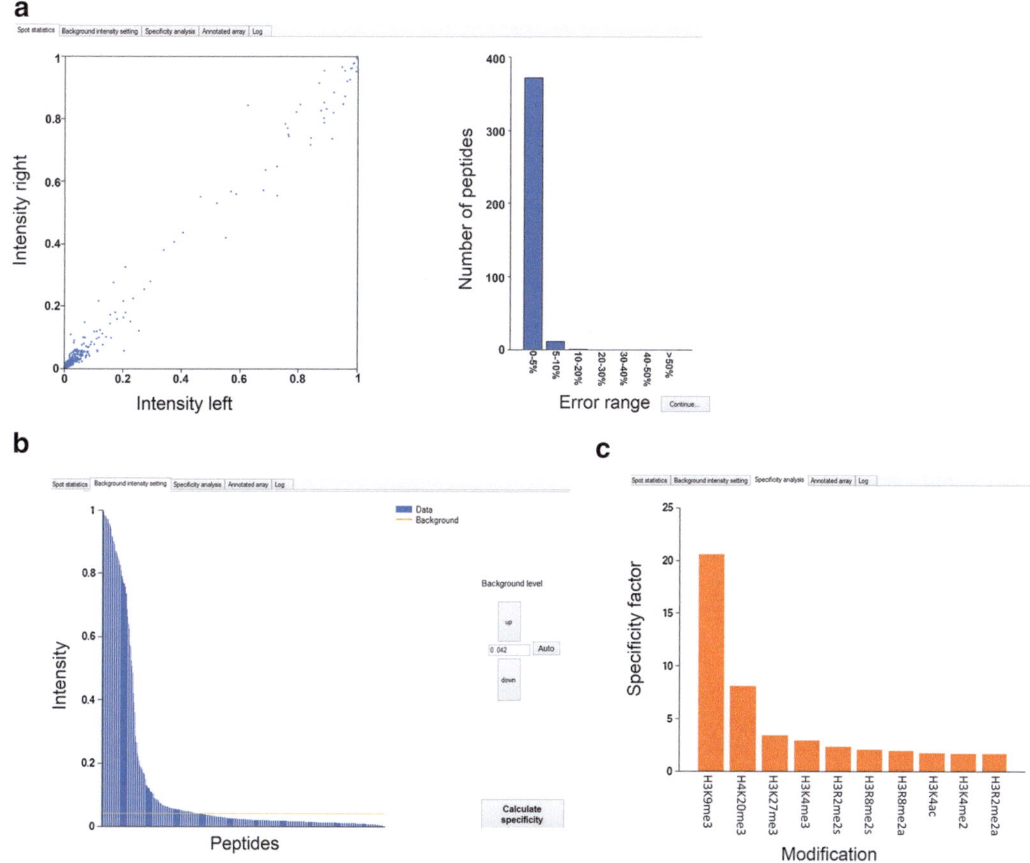

Fig. 3 Initial data analysis. (**a**) *Scatter plot* showing the reproducibility of spot intensity on both sides of the peptide array (*left panel*) and *bar diagram* showing the error range (*right panel*). (**b**) *Diagram* showing the background levels. (**c**) *Bar diagram* showing the specificity factor for different histone PTMs

4 Notes

1. When preparing the 10× TTBS solution, cut the end of a 1 ml tip, so you can aspirate Tween-20 easily and accurately. To prepare 1× TTBS, add 900 mL Milli-Q water to a 1 L graduated cylinder and fill it up with 100 mL 10× TTBS. To prepare the blocking solution, add 5 g of nonfat skim milk, fill up to 100 ml with 1× TTBS in a reagent bottle and mix with a stir bar until the powder is fully dissolved. We typically keep the blocking solution at +4 °C for not more than 5 days.
2. Keep the peptide arrays at +4 °C. Avoid touching the side of the array where the peptides are spotted and always use clean gloves and forceps when you transfer the array into the Nunc plate.
3. After washing the array, one has to tilt the Nunc plate for ~30° in order to completely remove the washing (TTBS) buffer. Different dilutions of the antibody or concentrations of the

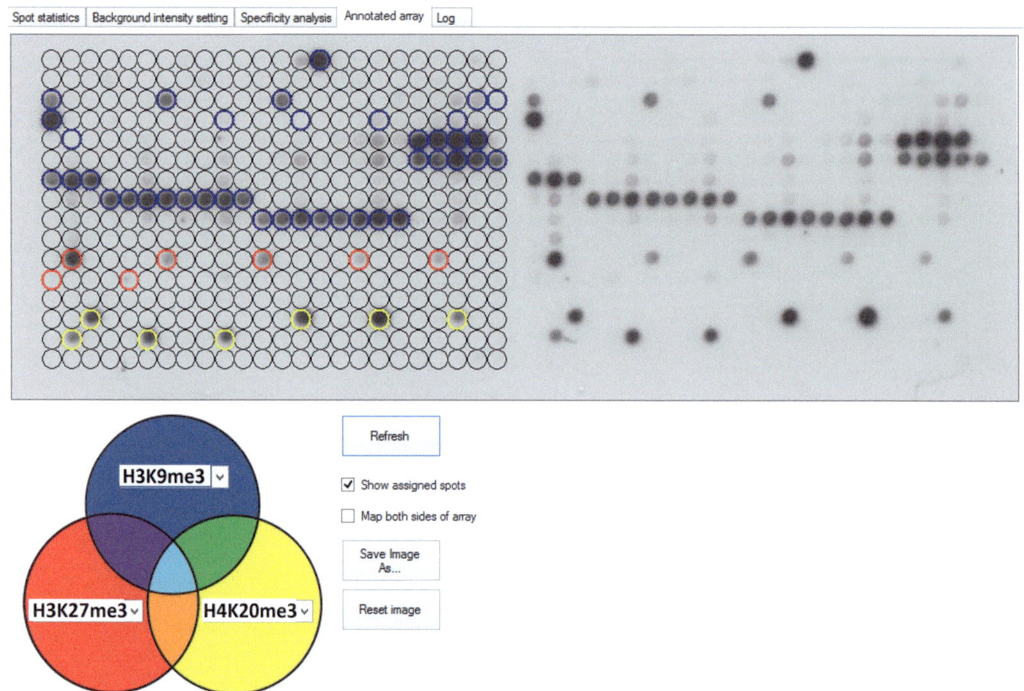

Fig. 4 Image of a peptide array with custom color annotation (H3K9me3—*blue*, H3K27me3—*red*, H4K20me3—*yellow*)

recombinant protein may be used, which can provide additional information about the relative binding affinity at each spot. When adding the antibody/recombinant protein solution, pipette 25 % of the total volume in each corner of the well to have equal distribution of solution and more uniform spot intensity on both sides of the array.

4. After the final wash, right before detection, lift the array up with forceps or a 200 μl or 1 ml tip and carefully put it on a pre-cut transparent foil, add the ECL solution and cover it up with precut transparent foil (if using X-ray films for detection). Remove all bubbles before proceeding to detection. It usually takes between 0.5- and 30-min exposition time. In case no signal is detected after 30 min, either wash three times with 1× TTBS and add ultrasensitive ECL developing solution or incubate with lower antibody dilution or higher recombinant domain concentration. Upon detection, discard the peptide array in a waste bin designated for glassware.

5. Regardless of whether X-ray or imaging detection is carried out, always mark the angles/borders of the peptide array on the film or overlay the obtained chemiluminescence signal with white-light image of the array, respectively. This will be helpful in the bioinformatic annotation of spots.

6. Save the primary data in *tiff* format. This is used for bioinformatic analysis with the ArrayAnalyze software. You can use *jpeg*- or *png*-formatted pictures for your lab book or report presentations.
7. Always try to use straight image encompassing only the peptide array, it will be easier to annotate the peptide spots and it will add aesthetic value to your image.

Acknowledgements

Work in the authors' lab has been supported by the BMBF (0315886B).

References

1. Hubner MR, Spector DL (2010) Chromatin dynamics. Annu Rev Biophys 39:471–489
2. Luger K, Mader AW, Richmond RK et al (1997) Crystal structure of the nucleosome core particle at 2.8 A resolution. Nature 389: 251–260
3. Kouzarides T (2007) Chromatin modifications and their function. Cell 128:693–705
4. Margueron R, Reinberg D (2010) Chromatin structure and the inheritance of epigenetic information. Nat Rev Genet 11:285–296
5. Bannister AJ, Kouzarides T (2011) Regulation of chromatin by histone modifications. Cell Res 21:381–395
6. Tan M, Luo H, Lee S et al (2011) Identification of 67 histone marks and histone lysine crotonylation as a new type of histone modification. Cell 146:1016–1028
7. Suva ML, Riggi N, Bernstein BE (2013) Epigenetic reprogramming in cancer. Science 339:1567–1570
8. Beck S, Rakyan VK (2008) The methylome: approaches for global DNA methylation profiling. Trends Genet 24:231–237
9. Huttenhofer A, Vogel J (2006) Experimental approaches to identify non-coding RNAs. Nucleic Acids Res 34:635–646
10. Bock I, Dhayalan A, Kudithipudi S et al (2011) Detailed specificity analysis of antibodies binding to modified histone tails with peptide arrays. Epigenetics 6:256–263
11. Egelhofer TA, Minoda A, Klugman S et al (2011) An assessment of histone-modification antibody quality. Nat Struct Mol Biol 18: 91–93
12. Fuchs SM, Strahl BD (2011) Antibody recognition of histone post-translational modifications: emerging issues and future prospects. Epigenomics 3:247–249
13. Nishikori S, Hattori T, Fuchs SM et al (2012) Broad ranges of affinity and specificity of anti-histone antibodies revealed by a quantitative peptide immunoprecipitation assay. J Mol Biol 424:391–399
14. Peach SE, Rudomin EL, Udeshi ND et al (2012) Quantitative assessment of chromatin immunoprecipitation grade antibodies directed against histone modifications reveals patterns of co-occurring marks on histone protein molecules. Mol Cell Proteomics 11:128–137
15. Hattori T, Taft JM, Swist KM et al (2013) Recombinant antibodies to histone post-translational modifications. Nat Methods 10: 992–995
16. Heubach Y, Planatscher H, Sommersdorf C et al (2013) From spots to beads-PTM-peptide bead arrays for the characterization of anti-histone antibodies. Proteomics 13:1010–1015
17. Taverna SD, Li H, Ruthenburg AJ et al (2007) How chromatin-binding modules interpret histone modifications: lessons from professional pocket pickers. Nat Struct Mol Biol 14:1025–1040
18. Patel DJ, Wang Z (2013) Readout of epigenetic modifications. Annu Rev Biochem 82: 81–118
19. Winkler DF, Hilpert K, Brandt O et al (2009) Synthesis of peptide arrays using SPOT-technology and the CelluSpots-method. Methods Mol Biol 570:157–174
20. Bock I, Kudithipudi S, Tamas R et al (2011) Application of Celluspots peptide arrays for the analysis of the binding specificity of epigenetic reading domains to modified histone tails. BMC Biochem 12:48

21. Dhayalan A, Tamas R, Bock I et al (2011) The ATRX-ADD domain binds to H3 tail peptides and reads the combined methylation state of K4 and K9. Hum Mol Genet 20:2195–2203
22. Kycia I, Kudithipudi S, Tamas R et al (2014) The Tudor domain of the PHD finger protein 1 is a dual reader of lysine trimethylation at lysine 36 of histone H3 and lysine 27 of histone variant H3t. J Mol Biol 426:1651–1660
23. Pradeepa MM, Sutherland HG, Ule J et al (2012) Psip1/Ledgf p52 binds methylated histone H3K36 and splicing factors and contributes to the regulation of alternative splicing. PLoS Genet 8, e1002717
24. Du J, Zhong X, Bernatavichute YV et al (2012) Dual binding of chromomethylase domains to H3K9me2-containing nucleosomes directs DNA methylation in plants. Cell 151: 167–180
25. Darvekar S, Johnsen SS, Eriksen AB et al (2012) Identification of two independent nucleosome-binding domains in the transcriptional co-activator SPBP. Biochem J 442:65–75
26. Kimura H (2013) Histone modifications for human epigenome analysis. J Hum Genet 58: 439–445
27. Park S, Martinez-Yamout MA, Dyson HJ et al (2013) The CH2 domain of CBP/p300 is a novel zinc finger. FEBS Lett 587:2506–2511
28. Pestell RG, Yu Z (2014) Long and noncoding RNAs (lnc-RNAs) determine androgen receptor dependent gene expression in prostate cancer growth in vivo. Asian J Androl 16: 268–269
29. Zucchelli C, Tamburri S, Quilici G et al (2014) Structure of human Sp140 PHD finger: an atypical fold interacting with Pin1. FEBS J 281:216–231
30. Alsarraj J, Faraji F, Geiger TR et al (2013) BRD4 short isoform interacts with RRP1B, SIPA1 and components of the LINC complex at the inner face of the nuclear membrane. PLoS One 8, e80746
31. Widiez T, Symeonidi A, Luo C et al (2014) The chromatin landscape of the moss Physcomitrella patens and its dynamics during development and drought stress. Plant J 79: 67–81
32. Du J, Johnson LM, Groth M et al (2014) Mechanism of DNA methylation-directed histone methylation by KRYPTONITE. Mol Cell 55(3):495–504

Chapter 25

Prion-Specific Antibodies Produced in Wild-Type Mice

Peter M.H. Heegaard, Ann-Louise Bergström, Heidi Gertz Andersen, and Henriette Cordes

Abstract

Peptide-specific antibodies produced against synthetic peptides are of high value in probing protein structure and function, especially when working with challenging proteins, including not readily available, non-immunogenic, toxic, and/or pathogenic proteins. Here, we present a straightforward method for production of mouse monoclonal antibodies (MAbs) against peptides representing two sites of interest in the bovine prion protein (boPrP), the causative agent of bovine spongiform encephalopathy ("mad cow disease") and new variant Creutzfeldt-Jakob's disease (CJD) in humans, as well as a thorough characterization of their reactivity with a range of normal and pathogenic (misfolded) prion proteins. It is demonstrated that immunization of wild-type mice with ovalbumin-conjugated peptides formulated with Freund's adjuvant induces a good immune response, including high levels of specific anti-peptide antibodies, even against peptides very homologous to murine protein sequences. In general, using the strategies described here for selecting, synthesizing, and conjugating peptides and immunizing 4–5 mice with 2–3 different peptides, high-titered antibodies reacting with the target protein are routinely obtained with at least one of the peptides after three to four immunizations with incomplete Freund's adjuvant.

Key words Prion, Monoclonal antibodies, Synthetic peptide, Peptide-specific antibodies, Peptide-carrier protein conjugates

1 Introduction

Prion proteins (PrP), specifically their misfolded forms, are intimately linked with the pathogenesis of prion diseases, a group of rare, invariably fatal neurodegenerative diseases including Creutzfeldt-Jakob's disease in humans, scrapie in sheep and goats, and bovine spongiform encephalopathy (BSE, mad cow disease) in cattle [1]. Importantly, misfolded PrP is the single available molecular biomarker of these diseases, and its detection in brain material by PrP-specific antibodies forms the basis of postmortem tests for these diseases on brain material. For example, in immunoblotting-based rapid tests, brain homogenates are subjected to protease treatment before SDS-PAGE followed by transfer to a blotting membrane and development with PrP-specific antibodies, reacting

Gunnar Houen (ed.), *Peptide Antibodies: Methods and Protocols*, Methods in Molecular Biology, vol. 1348, DOI 10.1007/978-1-4939-2999-3_25,

with the misfolded, protease-resistant fragment of PrP [2]. This is often combined with immunohistochemistry, using the PrP-specific antibodies to reveal characteristic patterns of PrP depositions ("plaques") in the brain tissue. Specific antibodies reacting with high affinity with PrP are therefore crucial for postmortem diagnosis of prion diseases.

The PrP protein is highly homologous between species [3]. Overall homology between bovine (bo), human (hu), and murine (mu) PrP is 83.8 %, and as shown in the alignment in Fig. 1a, the few nonhomologous positions are evenly distributed throughout the mature sequence with the exception of an additional so-called octarepeat insertion in the bovine sequence (aa95–103). None of the hydrophilic stretches correspond to regions of nonhomology (Fig. 1b, Kyte and Doolittle plot [4]). The 3-D structure of PrP is unusual as more than half the polypeptide chain (the N-terminal half) is natively unfolded [5]. Thus, NMR spectroscopy does not provide any structure for aa25–125 of boPrP (amino acids 1–25 not being present in the mature protein), while the C-terminal half of the molecule is globular and revealed by NMR to contain three α-helices and two β-strands (Fig. 1c) [5]. Although the exact 3-D structure of the misfolded form has not been solved, it is known to have quite a high protease resistance, to have an increased content of β-structure and to involve the polypeptide chain starting from around amino acid 80–95 and thus comprising the globular part as well as a minor part of the unfolded N-terminal half of the normally folded protein [6]. Due to the high homology of PrP proteins between species, the protein is not very immunogenic *per se*. Even so, in an early work, Kascsak et al. [7] described the production of PrP-specific antibodies obtained by immunizing wild-type mice with PrP fibrils isolated from the brains of scrapie-inoculated hamsters. In contrast, PrP fibrils from scrapie-infected mouse brains induced very low levels of antibodies in these mice, even after a vigorous immunization scheme, involving eight injections over several months using complete and incomplete Freund's adjuvant and intradermal, subcutaneous, intramuscular, and intraperitoneal injections; it was also observed that mice inoculated with scrapie and developing disease did not develop an antibody response against the prion protein. These experiments were cumbersome due to the paucity of material for immunization; however, it was demonstrated that an immune response against hamster scrapie PrP fibrils could indeed be obtained in normal mice, and a widely used mouse monoclonal antibody (3F4) resulted from this work.

Many subsequent studies have reported using $PrP^{0/0}$ mice for production of antibodies against PrP proteins and/or peptides. For example, Korth et al. [8] immunized $PrP^{0/0}$ mice with recombinant bovine PrP (rboPrP) administering the protein subcutaneously with Freund's complete (first injection) and incomplete (subsequent injections) adjuvant in 3-week intervals

a

```
CLUSTAL O(1.2.1) multiple sequence alignment

SP|P10279|PRIO_BOVIN MVKSHIGSWILVLFVAMWSDVGLCKKRPKPGGGWNTGGSRYPGQGSPGGNRYPPQGGGGW 60
SP|P04156|PRIO_HUMAN --MANLGCWMLVLFVATWSDLGLCKKRPKPG-GWNTGGSRYPGQGSPGGNRYPPQGGGGW 57
SP|P04925|PRIO_MOUSE --MANLGYWLLALFVTMWTDVGLCKKRPKPG-GWNTGGSRYPGQGSPGGNRYPPQG-GTW 56
                       :.:*  *:*.***:  *:*:********** ************************ * *

SP|P10279|PRIO_BOVIN GQPHGGGWGQPHGGGWGQPHGGGWGQPHGGGWGQPHGGGGWGQGGTHGQWNKPSKPKTNM 120
SP|P04156|PRIO_HUMAN GQPHGGGWGQPHGGGWGQPHGGGWGQPHGGGWGQG--------GGTHSQWNKPSKPKTNM 109
SP|P04925|PRIO_MOUSE GQPHGGGWGQPHGGSWGQPHGGSWGQPHGGGWGQG--------GGTHNQWNKPSKPKTNL 108
                     **************.*******.***********          **** ***********:

SP|P10279|PRIO_BOVIN KHVAGAAAAGAVVGGLGGYMLGSAMSRPLIHFGSDYEDRYYRENMHRYPNQVYYRPVDQY 180
SP|P04156|PRIO_HUMAN KHMAGAAAAGAVVGGLGGYMLGSAMSRPIIHFGSDYEDRYYRENMHRYPNQVYYRPMDEY 169
SP|P04925|PRIO_MOUSE KHVAGAAAAGAVVGGLGGYMLGSAMSRPMIHFGNDWEDRYYRENMYRYPNQVYYRPVDQY 168
                     **:*************************:****.*:*********:***********:*:*

SP|P10279|PRIO_BOVIN SNQNNFVHDCVNITVKEHTVTTTTKGENFTETDIKMMERVVEQMCITQYQRESQAYYQ-- 238
SP|P04156|PRIO_HUMAN SNQNNFVHDCVNITIKQHTVTTTTKGENFTETDVKMMERVVEQMCITQYERESQAYYQ-- 227
SP|P04925|PRIO_MOUSE SNQNNFVHDCVNITIKQHTVTTTTKGENFTETDVKMMERVVEQMCVTQYQKESQAYYDGR 228
                     **************:*:****************:***********:***::******:

SP|P10279|PRIO_BOVIN RGASVILFSSPPVILLISFLIFLIVG 264
SP|P04156|PRIO_HUMAN RGSSMVLFSSPPVILLISFLIFLIVG 253
SP|P04925|PRIO_MOUSE RSSSTVLFSSPPVILLISFLIFLIVG 254
                     *.:*  :********************
```

b

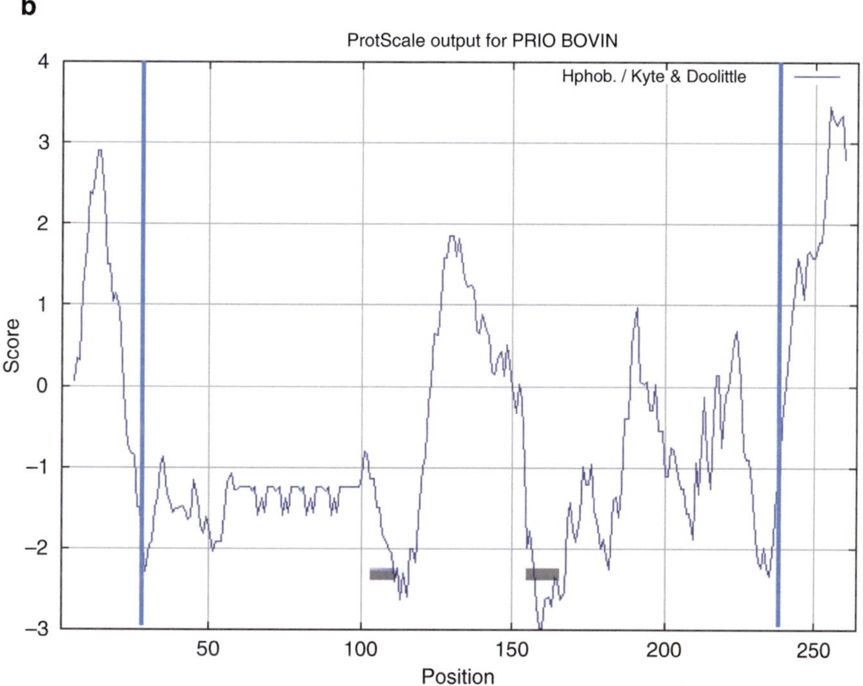

Fig. 1 (**a**) Multiple alignment of bovine, human, and mouse PrP sequences, indicating identical amino acids (*), conservative (:) or semiconservative (.) substitutions. Non-marked residues belong to N- and C-terminal pro-peptide regions; *blue line* indicates globular domain and *black lines* indicate locations of peptides used in this work: boPrP 102–113 and boPrP153–165 (bovine sequence). (**b**) Bovine PrP (P10279) hydropathy plot (Kyte and Doolittle 1982, http://www.expasy.org/), moving window $n=9$; *blue vertical lines* showing boundaries of mature protein (25–241) and *grey horizontal lines* showing position of the two peptides used in this work. (**c**) Representative structure of boPrP globular domain (125–227, PDB 1DX1, RasMol representation), location of peptide comprising the minimal 6H4 sequence (resulting MAb: 1.6F4) indicated in *yellow* and the peptide preceding the N-terminal (G125) of the globular domain (resulting MAb: 2.12) indicated in *blue* as a hypothetical random coil structure

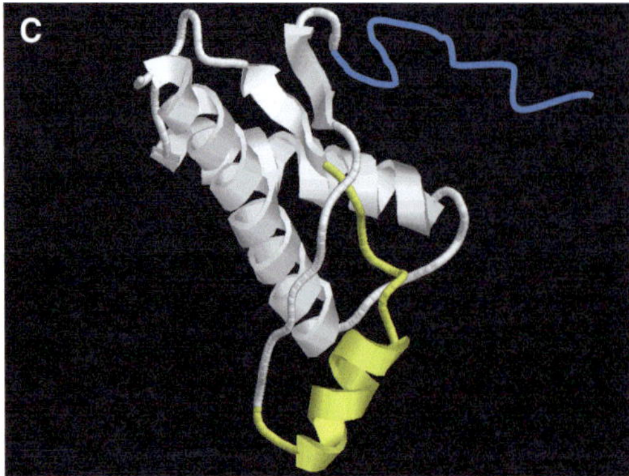

Fig. 1 (continued)

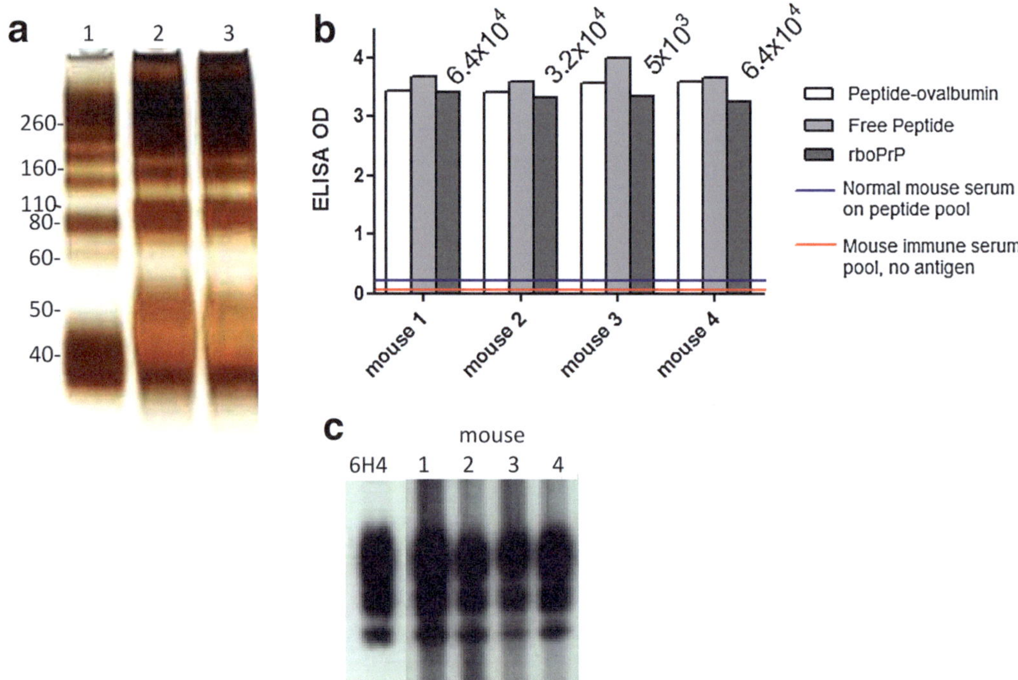

Fig. 2 (**a**) Example of analysis of peptide-ovalbumin conjugates by silver-stained SDS-PAGE. NuPage 10 % Bis-Tris. The composition of the 1× NuPAGE® LDS Sample Buffer is as follows: 141 mM Tris base, 2 % LDS, 10 % Glycerol, 0.51 mM EDTA, 0.22 mM SERVA Blue G, 0.175 mM Phenol Red pH 8.5. (*1*) BMPS-activated ovalbumin, (*2*) peptide-ovalbumin conjugate (peptide: HADGSFSDEC), (*3*) peptide-ovalbumin conjugate (peptide: HAEGTFTGGGC). (**b**) Example of mouse antiserum reactivity obtained after immunization with a peptide-ovalbumin conjugate. In this example, boPrP102–120 was used for immunization of four mice. Each of the four mouse antisera were tested against the immunizing peptide-ovalbumin conjugate, the free (uncoupled) peptide itself and recombinant bovine prion protein (reboPrP). Titers against reboPrP were determined in an independent ELISA and are indicated above the columns for each of the four mouse antisera. (**c**) Immunoblotting of BSE (bovine spongiform encephalopathy, mad cow disease) brain homogenates, visualized with mouse antisera (from the same four mice as in **b**), i.e., raised against boPrP102–120) compared to 6H4 (commercial Mab having as its minimal epitope boPrP153–165)

and boosting mice twice, 6 and 7 days after the last immunization (i.p. and i.v., respectively), followed by killing and preparation of spleen cell fusions at day 8. This led to the production of another mouse monoclonal antibody becoming widely used (6H4) and used as part of the Prionics rapid test kit for immunoblotting-based detection of misfolded, protease-resistant PrP. The minimal epitope of this monoclonal antibody was mapped to boPrP 155–163 (DYEDRYYRE; *see* Fig. 1a) [8, 9]. The same overall strategy was followed by Jones et al. [10]; however, using purified natural human, normally folded PrP. Also in this study Freund's adjuvant (complete and incomplete) was used to increase the immune response in $PrP^{0/0}$ mice. In yet another approach, Handisurya et al. [11] used virus-like particles and immunization of rabbits and rats to produce antibodies against the 6H4-minimal epitope-like peptide DWEDRYYRE of the murine and rat PrP protein (amino acids 144–152; *see* Fig. 1a), resulting in antibodies reacting both with the peptide and the corresponding human peptide (differing in one amino acid). The peptide was incorporated into an immunogenic surface loop of the L1 major capsid protein of bovine papillomavirus type 1 (BPV-1), expressed by recombinant baculovirus technology. Non-transgenic white rabbits and rats received a total of four injections with the CpG oligonucleotide adjuvant ODN 2006 and incomplete Freund's adjuvant (rabbits only).

Synthetic peptides are clearly relevant for producing antibodies with predetermined epitope specificity against elusive proteins such as misfolded PrP, which in addition is a highly pathogenic protein. We previously published the characterization of antibodies specific for a single bovine PrP epitope (boPrP153–165, an extended version of the 6H4 epitope) produced by the simple approach of immunizing a standard inbred mouse type (NMRI) with a carrier protein-coupled synthetic peptide representing the epitope [12, 13]. We present here the detailed protocol used to obtain these antibodies and also include previously unpublished data on antibodies raised using the same approach against another boPrP peptide, namely, boPrP 102–113 (*see* Fig. 1a for the position of these peptides in the bovine, human, and mouse PrP sequences).

Here, a high-affinity MAb (2.12) against a 12 amino acid bovine PrP peptide differing in only two positions (a missing G as amino acid no. 3 and G replacing N at position 7 (G108)) from the murine sequence was obtained in wild-type mice after three immunizations with the peptide-ovalbumin conjugate mixed with Freund's adjuvant. This MAb also reacted with recombinant huPrP, also having only two differing amino acid-positions, although to a lesser degree than was seen with boPrP (Table 1B). In addition, and as reported previously [13], a monoclonal antibody (1.6F4) was produced by the same strategy against a 13 amino acid bovine PrP peptide, also having only two amino acids differing from the murine sequence, namely, an S in the place on an N as the second amino acid and a conservative change from W to Y in the fourth position.

Table 1
A: Affinity constants with rboPrP and peptide competitors (indicated to the left) of the developed MAbs compared to the 6H4b MAb, as determined by ELISA. Boxed numbers indicate affinity with immunizing peptide. 6H4 was produced against whole rboPrP; however, the affinity for peptide containing the 6H4 minimal epitope is indicated (dashed box). N.I.: No inhibition was obtained at or below the competitor concentration indicated. B: Reactivities of the two developed MAbs compared to 6H4 MAb with bovine, murine, and human prion proteins, either recombinant, normally folded (by ELISA), or brain tissue-derived misfolded PrP (PrPRES) from natural BSE, mouse brains affected by experimental infection with mouse-adapted scrapie (RML 79A), and human patients (spontaneous CJD), respectively (by WB). *n.d.* not done. C: Reactivities of the two developed MAbs compared to 6H4 MAb in immunohistochemistry with misfolded brain-derived prion proteins from cow (natural BSE), sheep (natural scrapie), and hamster (experimental infection with hamster-adapted scrapie, 263 K), respectively. *n.d.* not done

A: Antigen-binding affinity of monoclonal antibodies

	2.12C4	1.6F4	6H4
PrP102–113	1.4×10^{12} M^{-1}	N.I.(1.2×10^{7} M)	N.I.(1.2×10^{7} M)
PrP153–165	N.I.(1.4×10^{7} M)	9.6×10^{9} M^{-1}	N.I.(1.4×10^{7} M)
rboPrP	N.I.(1.0×10^{6}M)	N.I.(1.0×10^{6}M)	3.5×10^{7} M^{-1}

B: Antigen reactivity (ELISA and immunoblotting)

	Species	rPrP ELISA	PrP^{RES} WB
2.12C4 (IgG_1)	Bovine	+++	–
	Murine	n.d.	–
	Human	+	–
1.6F4 (IgG_1)	Bovine	++	++
	Murine	n.d.	–
	Human	n.d.	+++
6H4 (IgG_1)	Bovine	++	+++
	Murine	n.d.	+++
	Human	++	+++

C: Antigen reactivity (immunohistochemistry)

	BSE	Bovine control	Scrapie	Ovine control	263 K	Hamster control
2.12C4	++	+	++	–	+	–
1.6F4	+++	++	++++	+	n.d.	n.d.
6H4	+	–	+	–	++	–

Concerning MAb 1.6F4, as also noted by Cordes et al. [13], its reaction with the various forms of PrP was quite different from that of 6H4, although nominally sharing the same epitope. In short, although the affinity for the peptide comprising the 6H4 epitope was low, and lower than that of 1.6F4 for the same peptide (Table 1A), MAb 6H4 showed better reactivity with protease-cropped, denatured disease-derived PrP, while 1.6F4 was superior in reacting with minimally denatured, non-proteinase-treated disease-derived PrP (Table 1C). This demonstrates that two different antibodies raised against the same epitope through different immunization methods may show substantially different reactivity with a protein comprising the peptide sequence in question.

Interestingly, MAb 2.12 did not react with the protease-resistant core of misfolded boPrP, as seen by its total absence of reactivity with BSE, murine scrapie, and human spontaneous CJD PrP in western blotting (Table 1B and Fig. 3a); still, by immunohistochemistry, reactivity with BSE, sheep, and murine scrapie PrP was seen (Table 1C). Finally, as seen in Fig. 3b, c MAb 2.12 reacted quite selectively with human PrP derived from a human GSS patient, showing no or very little reaction with human PrP from sCJD and fCJD patients as well as with BSE and sheep scrapie PrP.

The reactivity of MAb 2.12 with the protease-resistant, misfolded form of PrP in GSS brain samples unequivocally identifies the location of the pre-globular domain peptide corresponding to boPrP102–113 (huPrP91–102) as part of the protease-resistant misfolded domain of GSS-derived human PrP and suggests that this region may be absent in the protease-resistant cores of the other forms of PrP investigated (from BSE, scrapie, CJD, hamster, and murine scrapie). Whenever this region is not removed by protease treatment of disease-derived PrP (immunohistochemistry having no protease K digestion step) or when using recombinant PrP (ELISA), Mab 2.12 reacted with the protein, suggesting the accessibility of the region in normal as well as misfolded PrP.

This is an example of peptide-specific antibodies being powerful tools for probing protein structures, here pointing to the presence of specific, fine structure-related variations in the spatial structure of the misfolded PrP, constituting the difference between different "prion types," in this case GSS PrP versus a range of other disease-related PrPs; the 2.12 epitope seems to be included in the protease-resistant core of GSS prions but not in the protease-resistant cores of PrP from the other prion diseases investigated here.

2 Materials

Bovine PrP peptides boPrP102–113, GQGGTHGQWNKP and boPrP153–165, GSDYEDRYYRENM both having an additional N-terminal cysteine residue were prepared by solid-phase peptide

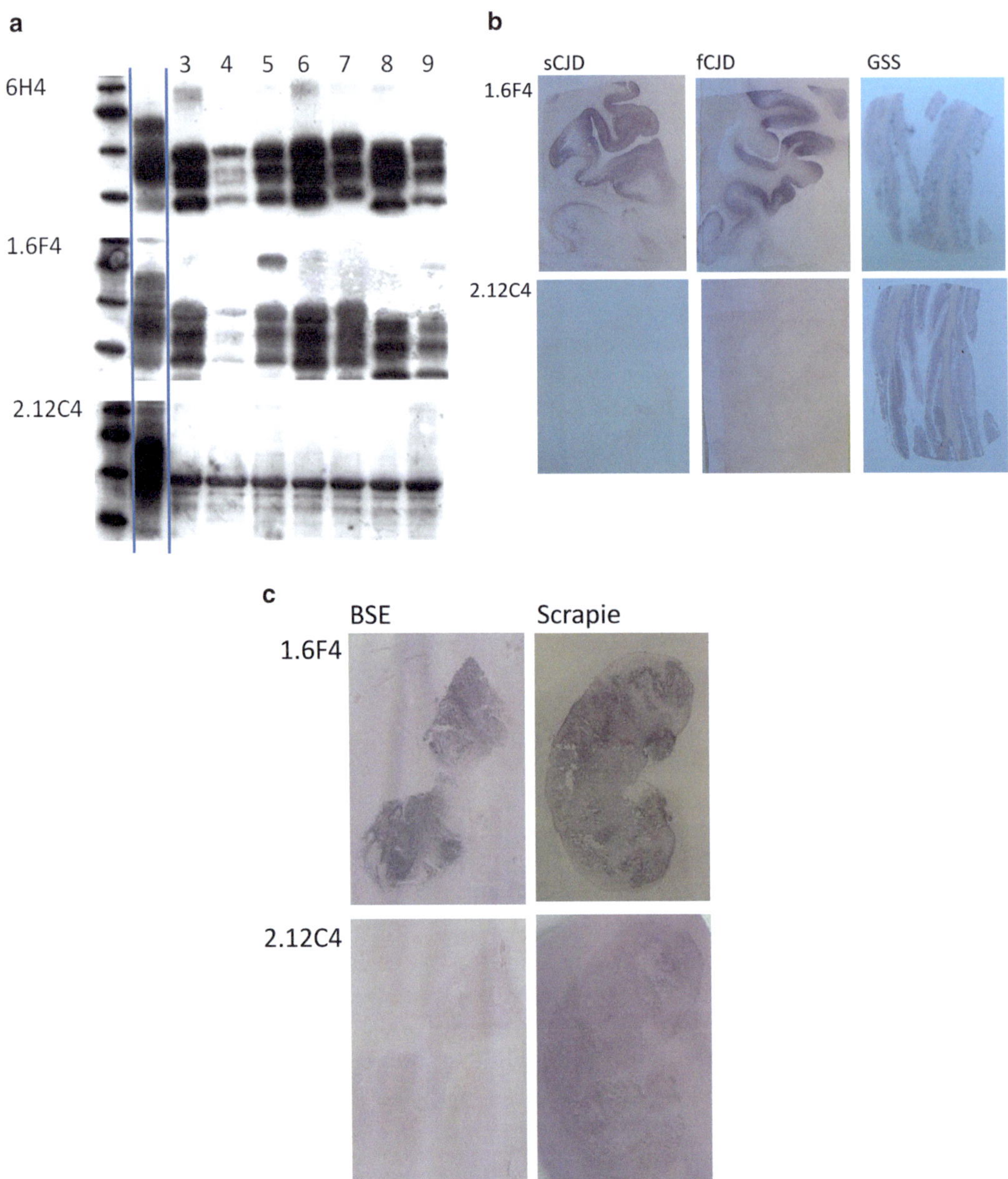

Fig. 3 (**a**) Immunoblotting with human brain tissue (frontal cortex) samples subjected to proteinase K digestion as described in ref. 13: *First lane:* Molecular weight marker (Magic Mark, InVitrogen), 20, 30, 40, 50 kDa, bands are visible. *Second lane*: Undigested control brain sample *Lanes 3–9*: Brain tissue from spontaneous Creutzfeldt-Jakob's disease patients. Antibodies: 6H4 (Prionics, 0.5 μg/ml), 1.6F4 (Mab antibovine PrP153–165, 1 μg/ml), 2.12C4 (Mab antibovine PrP102–113, 1 μg/ml). 6H4 and 1.6F4 results were published previously by Cordes et al. [13]. (**b**) PET blots, human brain samples (cerebellum region); *sCJD* spontaneous Creutzfeldt-Jakob Disease; *fCJD* familial Creutzfeldt-Jakob Disease; *GSS* Gerstmann-Sträussler-Scheinker syndrome. Antibodies are used at 0.4 μg/ml. 1.6F4 blots were originally published by Cordes et al. [13]. (**c**) PET blots, bovine and ovine brain samples, respectively (both from brainstem (obex) region); *BSE* bovine spongiform encephalopathy. Antibodies are used at 0.4 μg/ml. 1.6F4 blots were originally published by Cordes et al. [13]

synthesis and documented by HPLC mass spectrometry as previously described [14].

PrP102–113 is located in the unstructured N-terminal part of the protein, close to the beginning of the globular part (G125), and PrP153–165 covers the minimal 6H4 epitope and comprises α-helix one and a part of the stretch of polypeptide chain following the helix (*see* Fig. 1c).

PBS

ELISA plates and reagents

Recombinant human (rhu), bovine (rbo), and ovine (rov) PrP expressed in *E. coli* as His-tagged full-length prion proteins (human: 23–230, bovine: 25–241, ovine: 25–234) (supplied as purified histidine-tagged full-length fusion proteins).

Commercial antibodies: monoclonal mouse antibody 6H4, raised against bovine PrP and having the minimal epitope DYEDRYYRE as determined by epitope mapping (bovine PrP155–163, *see* Fig. 1) [8, 9].

Mouse immunoglobulin IgG1 and IgG2a isotype controls, respectively (e.g., X0931 and X0944).

Postmortem human and animal (mouse, hamster, sheep, and cow) brain samples were obtained as described in Cordes et al. [13].

3 Methods

3.1 Preparation of Peptide-Carrier Protein Conjugate

1. Make a 10 mg/ml solution of ovalbumin in 0.1 M sodium carbonate pH 8.2. Make a 5 mg/ml solution of BMPS (*N*-[β-maleimidopropyloxy]succinimide ester) in *N*-methyl pyrrolidone and use immediately: 200 μL BMPS/NMP solution is mixed with 500 μL of the ovalbumin solution and left for 1 h at 20 °C. Then, the mixture is desalted on a PD10 desalting column equilibrated in 0.05 M sodium acetate pH 5.5 following the manufacturer's instruction (*see* **Note 1**).
2. 2–5 mg cysteine-containing peptide is then mixed with the desalted BMPS-activated ovalbumin and left for 2 h at 20 °C and then dialyzed thoroughly against PBS (*see* **Note 2**).

Analyze resulting peptide-ovalbumin conjugates by SDS-PAGE (*see* **Note 3** and Fig. 2).

3.2 Immunization

1. Adjust peptide-ovalbumin conjugate to 1 mg/ml in PBS and emulsify thoroughly with equal volume of Freund's complete adjuvant (*see* **Note 4**).
2. Emulsify in an Eppendorf tube by vortexing at high speed for a minimum of 30 min. Total volume should be less than 1.5 ml (*see* **Note 5**).

3. Test emulsion stability by droplet test: A droplet of the formulation is placed on a water surface in a beaker. If a stable emulsion has been obtained, the drop must not disperse on the surface, but should retain its shape. If not, the emulsion is not stable and the mixing will have to be repeated (*see* **Note 6**).
4. Inject 100 μL immunogen emulsion, corresponding to 50 μg ovalbumin-peptide conjugate subcutaneously in the skin of the back of the neck of a NMRI mouse (*see* **Note 7**) using a 21G hypodermic, sterile needle, and a 1 mL syringe (*see* **Note 8**). Use at least three mice for each immunogen. This is repeated after 14 and 28 days.
5. Ten days after the third immunization, a blood sample is taken from each mouse by tail, eye, or chin bleeding (5–20 μL is enough), and sera are tested for reactivity against rboPrP by indirect ELISA (*see* **Note 9**).

3.3 Fusion and Cloning

1. After testing bleedings for reactivity against rboPrP, the mouse having the highest titer of specific antibody reactivity (*see* **Note 10**) is given an intraperitoneal booster injection of 100 μg ovalbumin-peptide conjugate in saline (at least 1 mg/ml) at least 14 days after the final subcutaneous immunization.
2. Four days after the intraperitoneal booster injection, the mouse is killed by decapitation, cervical dislocation, or another instant, nondrug-dependent method. Immediately thereafter, the spleen is dissected out and immediately transported to a flow bench and disintegrated, and the splenocytes fused with myeloma cells (P3X63 Ag.8653 murine myeloma cells) and set up for culture, essentially as described by 15 for production of monoclonal antibodies a.m. Köhler and Milstein [16].
3. The screening of hybridomas for reactivity towards rboPrP is done by testing cell culture supernatants by ELISA (*see* **Note 11**), selecting, and re-cloning positive cell cultures [15] (*see* **Note 12**).
4. After cloning and expansion of stable, positive hybridomas, monoclonal antibody in the form of mouse immunoglobulin G is purified from cell culture supernatants by protein A agarose chromatography [17]. Antibodies are isotyped using any mouse Ig subclass-specific kit (e.g., Mouse MonoAb-ID Kit (HRP)) (*see* **Note 13**).

3.4 Immunochemical Characterization of Antibodies

1. ELISA (*see* **Note 9**) is employed for characterization of antibody specificity. ELISA plates are coated with dilution series of rboPrP, rovPrP, and rhuPrP, respectively. Antibodies are incubated for 1 h at optimal dilution and compared to positive controls in the form of commercially available antibodies 6H4 and 3F4, directed against bovine and hamster PrP epitopes, respectively.

2. Antibody affinity is determined by competition ELISA, coating the wells with rboPrP followed by incubation of the antibody preparation to be analyzed in the presence of a dilution series of free peptide or protein (PrP peptide or rboPrP) (*see* **Note 14**; for ELISA, *see* **Note 9**). 50 μL anti-peptide antibody and 50 μL of a twofold dilution series of either rboPrP or PrP peptide are applied. Data were fitted to a one-site competition model using nonlinear regression, and the reciprocal inhibitory concentration giving half maximal OD was taken as equivalent to the affinity constant (GraphPad Prism vers. 3.0) (*see* **Note 15**).
3. For both human and animal brain samples, homogenates are prepared from frozen brain tissue as 10 % homogenates. Animal brain samples can be homogenized in the homogenization buffer of the Prionics kit (Check Western Kit, Prionics, Zürich) following the instructions of the manufacturer. Human brain samples can be homogenized in 100 mM Tris, 100 mM NaCl, 10 mM EDTA, 0.5 % Nonidet P-40, and 0.5 % sodium deoxycholate (pH adjusted to 6.9).
4. Immunoblotting: Briefly, 10 % brain homogenate (bovine brain stem and human frontal cortex samples, respectively [13]) is treated with Prionics kit proteinase K for 30 min. at 37 °C or with 100 μg/ml proteinase K at 37 °C for 2 h with slight agitation. Stop protease digestion by adding phenylmethylsulfonyl fluoride (PMSF) to 5 mM (*see* **Note 16**). Run samples with reducing sample buffer (e.g., from Prionics) in 12–15 well NuPAGE gels and blot onto PVDF membranes by electrotransfer [18]. After blocking, incubate membranes with antibodies to be tested in appropriate dilution, and use alkaline phosphatase-conjugated anti-mouse Ig for detection by chemiluminescent ECL WB detection reagent (AP Biotech, Denmark) (*see* **Note 17**).
5. PET blot for detection of PrPRES in prion-infected and noninfected brain tissue. This is done according to Schulz-Schaeffer et al. [19] with a few modifications as described by 13. After collection of paraffin-embedded tissue sections (2–4 μm) on nitrocellulose membranes attached to glass slides, and further processing as described previously [13], blots are incubated with primary antibody (0.4 μg/ml) diluted in blocking buffer overnight at 4 °C and, after washing with alkaline phosphatase-conjugated goat-anti-mouse immunoglobulin, developed in NBT/BCIP (*see* **Note 18**).

4 Notes

1. A number of proteins can be used as carrier protein for peptides for the purpose of producing anti-peptide antibodies by immunization of animals provided that the protein is foreign to the

animal species used and has a number of accessible functional groups for derivatization. Also, generally, monomeric proteins are preferred. Ovalbumin (egg albumin) is commercially available, relatively inexpensive, has 20 surface amino groups accessible for peptide coupling [20], and, being of avian origin, shows good immunogenicity in lab rodents. Similarly, any of a big number of coupling reagents may be used. The heterobifunctional coupling agent BMPS [21] reacts readily with primary amines through its succinimide ester. It is important to ensure a high protein concentration (above 2 mg/ml) as hydrolysis of the ester by the aqueous solvent is competing with the acylation of primary amines. Then, through its maleimide group, BMPS forms a stable thioether linkage with a thiol group. This reaction is highly specific at slightly acidic pH and allows coupling through the thiol group of cysteine-containing peptides. Cysteine residues may be incorporated into synthetic peptides at any position that does not interfere with the intended epitope. For a number of model peptides, we have found that an intermediary cysteine residue, placed in the middle of the peptide sequence, can lead to higher antibody titers; however, this is also correlated proportionally with the peptide-carrier protein ratio [22]. Generally, the most preferred position of the cysteine residue is at one of the peptide termini. The enhanced peptide immunogenicity generally observed upon conjugation to a carrier protein method is presumed to rely on the presence of T-cell epitopes in the carrier protein giving rise to activation of T helper cells. Additionally, the multimeric display of the peptide may result in high-avidity interactions with the specific B-cell receptors in question, also increasing peptide immunogenicity. Finally, it might be speculated that, at least for some peptides, carrier protein coupling may stabilize peptide conformations making the structure of the peptide resemble the stabilized conformation of the same peptide when part of a protein sequence.

2. Instead of dialysis, PD10 desalting may also be used for the final step; however, as the conjugates are quite stable, it is not a demand that buffer exchange is performed quickly. This is in contrast to the maleimide-coupled ovalbumin which should be further processed quickly (buffer exchanged and coupled with cysteine peptide) due to the relative hydrolytic reactivity of the maleimide group caused by the neighboring aromatic ring in BMPS [23]. If the pH 5.5 acetate buffer does not cause adverse effects in the immunized animal, the final dialysis step may be skipped altogether.

3. We use SDS-PAGE as a relatively quick and easy qualitative analysis of resulting peptide-ovalbumin conjugates. An example of such an analysis (silver-stained SDS-PAGE) with two peptides not related to PrP is shown Fig. 2. This preparation of ovalbumin is quite heterogeneous containing in addition to

the monomer protein at approximately 40 kDa dimers and higher oligomers. For BMPS-derivatized ovalbumin which has a similar band pattern to non-derivatized ovalbumin, the 80 and 120 kDa bands as well as higher molecular weight oligomers are clearly seen (lane 1, Fig. 2). In lane 2 and 3, the results of coupling with two different peptides, respectively, are shown. Peptide coupling leads to an increase in molecular weight and a broadening of the protein band, indicative of population of peptide-protein conjugates having different peptide to protein ratios. This is seen for the monomer as well as for the dimer and tetramer bands, while the higher molecular weight bands although clearly affected are not well separated.

4. Any type of emulsion or aluminum-based adjuvant can be used. In our experience, both Freund's complete and incomplete adjuvant and a number of other water-in-oil emulsion forming adjuvants are very efficient at augmenting immune responses leading to high-titered antiserum by subcutaneous or intramuscular administration. Due to adverse effects, the use of Freund's complete adjuvant should be limited to the first immunization or avoided completely.

5. The following two methods can be used for immunogen preparation by emulsifying the antigen/adjuvant mixture with an oil-based adjuvant, leading to the water phase forming stable microscopic water droplets inside the oil phase. Larger volume batches of emulsified antigen may be prepared for immediate or later use. If to be used later, the immunogen emulsion may be stored at minus 20 °C (turns white upon freezing).

   For 1–1.5 ml total volumes: Mix in an Eppendorf tube and vortex at high speed for a minimum of 30 min. Specific adaptors (e.g., TurboMix Attachment from Scientific Industries) are available for this. They are quite noisy but also very efficient.

   For 5–20 ml volumes: Use the syringe extrusion technique in which adjuvant and antigen are drawn into two different syringes. The two syringes are connected with an I-connector, and then the two liquids are emulsified by pushing the liquids back and forth between the two syringes, completely displacing the total volume of liquid into one of the syringes and then back to the other one. Accordingly, the syringes should each be large enough to be able to hold the total volume of the adjuvant and the antigen solution. The first 20 cycles are performed slowly, followed by 60 cycles performed as fast as possible.

6. If the drop diffuses on the surface of the water, the emulsion is not stable and the antigen will disperse from the adjuvant in the aqueous environment after injection. This will severely compromise the immunogenicity of the immunogen formulation which is thought to rely on the combined effects of slow release ("depot" effect) and concurrent immune cell stimulation by the oil phase in the same location.

7. Subcutaneous administration is preferred due to its ease of use under the skin at the back of the neck of the mouse. This site also causes relatively little discomfort to the mouse.
8. The emulsion is quite viscous and must be delivered steadily and slowly taking great care to avoid accidents with the needle as Freund's adjuvants may lead to unpleasant local reactions (inflammation with granuloma formation).
9. All ELISAs are performed using standard microwell plates, e.g., Maxisorp plates from Nunc coated with antigen (peptide or rboPrP, both typically at 1 μg/ml overnight at 4 °C in PBS). Then, blocking is performed for 30 min using PBS/T/BSA (PBS with 0.05 % Tween 20, 1 % BSA). This buffer is also used as the dilution buffer. This is followed by washing (filling/emptying wells (at least 300 μl/well)) 5 times with PBS/T (PBS with 0.05 % Tween 20) and incubation with sample, i.e., serum samples, hybridoma cell culture samples, or samples of protein A-purified mouse antibody using PBS/T/BSA for dilution. After 1 h of incubation at room temperature, plates are washed as above, and then incubation with peroxidase-conjugated goat anti-mouse IgG 1/2000 in PBS/T/BSA is performed for 1 h (detection layer). After a final wash as above, the plate is emptied and developed with TMB plus one component solution, 100 μl/well, stopping color development by adding 100 μl 1 N H_2SO_4 to each well. Optical densities of wells are read at 450 nm, subtracting 650 nm (background).
10. The titer against the immunizing peptide can be used as a measure of the activation of the immune system. A high serum antibody titer is thought to indicate a high number of B cells being activated to become antibody producing plasma cells and therefore giving a higher chance of obtaining high numbers of antibody producing spleen cells for fusion into hybridoma cells. However, in order to ensure that relevant antibodies are being produced as a result of immunization with a peptide-carrier protein conjugate, preferably titers against the target protein should be determined as anti-peptide antibodies will not always react with the target peptide sequence in a protein, even if the anti-peptide titer might be high. Titers will increase upon boosting and are therefore normally determined after the final immunization only. Titers in the 10^4–10^5 range are not unusual after three immunizations; if not higher than $2–3 \times 10^3$, a couple of additional immunizations should be done in order to increase the titer. If still not satisfactory, a new peptide and/or peptide conjugate should be prepared.
11. Selection of hybridomas is done by cloning [15] and screening cell culture supernatants from clones coating with the antigen version most closely resembling that targeted in the intended use of the final MAb.

12. If needed, and capacity allows, more than one screening can be performed pr. cloning step, for example, with the purpose of detecting clones producing IgM instead of IgG. Antibodies of the IgM class are most often not optimal as they tend to show a high level of unspecific binding and therefore may generally be considered "false-positives" when turning up as hits in the ELISA.
13. Isotyping is typically done using commercial ELISA-based kits that will yield both heavy chain isotype and light chain isotype.
14. When rboPrP is used as competitor, it has proven necessary to reduce and alkylate rboPrP in the coating layer. This prevents free rboPrP from binding to the rboPrP in the coating layer by S-S reshuffling forming intermolecular S-S bonds between coating layer and competitor rboPrP. Monomeric bovine PrP can be reduced and S-alkylated in the following way: dissolve rBoPrP at 1 mg/ml in 25 mM Tris, 0.075 M NaCl, and 2.5 mM EDTA pH 8.5, and add dithiothreitol to 5 mM. After incubation for 2 h at 37 °C, cool to room temperature, add 0.5 M iodoacetamide, continue incubation for 2 h at room temperature, and then dialyze extensively against 10 mM sodium acetate pH 4.0.
15. The competition ELISA employs a coating with recombinant bovine prion protein, evidently comprising the two linear epitopes towards which the MAb in this work are directed (boPrP102–113 and boPrP153–165, respectively). To this is added the MAb together with the free antigen ligand (in this case a peptide or a recombinant protein) in a dilution series, competing with the coated antigen for binding to the Mab. After a suitable incubation time, this is then followed by a detection antibody, leading to color development in direct proportion to the amount of MAb present. The more free, competing antigen ligand present during the MAb incubation step, the less MAb will be available for binding to the coating layer and the less color signal is generated. In this setup, the affinity of the antibody in question towards the added antigen ligand can be determined as the inverse concentration of the free added antigen ligand at which the signal is at half maximum, as in this case the concentration of free MAb must be equal to the concentration of the MAb-antigen complex, i.e., the affinity constant equation, $Kaff = [MAb\text{-}antigen\ complex]/([MAb] \times [antigen])$ reduces to $Kaff = 1/[antigen]$.
16. PMSF is less toxic than diisopropylfluorophosphate (DFP).
17. Any other type of secondary antibody conjugate suitable for signal generation on PVDF membranes may be used; however, the ECL (enhanced chemiluminescent) detection offers a high sensitivity and immediate "recording" of the signal in the form of either film-based detection or by digital imaging.

It also potentially allows stripping and reprobing of the blot as no color is deposited on the PVDF membrane.

18. IHC and PET blotting differ in several aspects: PET blotting includes an extensive PK-digestion step, leaving only PK-resistant PrPSc (PrPRES) bound to the membrane. Therefore, the PET blot, in contrast to IHC, shows the presence of PrPRES only. Accordingly, all non-diseased control tissues were completely blank in PET blotting with all antibodies tested. Also, the PET procedure includes a denaturation step unfolding PrPRES. In contrast, although PrPSc might be denatured to some degree by the fixation and the mild formic acid steps of IHC, leaving out the denaturation step in PET blotting completely abolishes all antibody reactivity, possibly because denaturation step is needed to make PrPRES bind the membrane with adequate efficiency. The PET blot therefore resembles immunoblotting with respect to the trimming of PrPSc by PK followed by denaturation and IHC with respect to the preserved localization and aggregation state of PrPSc.

Acknowledgments

We would like to thank Regina Lund, Karin Larsen, Panchale Olsen, Annie Ravn Pedersen, and Anne-Marie Petersen for their excellent technical assistance.

References

1. Chesebro B (2003) Introduction to the transmissible spongiform encephalopathies or prion diseases. Br Med Bull 66:1–20
2. Kübler E, Oesch B, Raeber AJ (2003) Diagnosis of prion diseases. Br Med Bull 66:267–279
3. Wopfner F, Weidenhofer G, Schneider R et al (1999) Analysis of 27 mammalian and 9 avian PrPs reveals high conservation of flexible regions of the prion protein. J Mol Biol 289: 1163–1178
4. Kyte J, Doolittle RF (1982) A simple method for displaying the hydropathic character of a protein. J Mol Biol 157:105–132
5. Wüthrich K, Riek R (2001) Three-dimensional structures of prion proteins. Adv Protein Chem 57:55–82
6. Govaerts C, Wille H, Prusiner SB et al (2004) Evidence for assembly of prions with left-handed beta-helices into trimers. Proc Natl Acad Sci U S A 101:8342–8347
7. Kascsak RJ, Rubenstein R, Merz PA et al (1987) Mouse polyclonal and monoclonal antibody to scrapie-associated fibril proteins. J Virol 61:3688–3693
8. Korth C, Stierli B, Streit P et al (1997) Prion (PrPSc)-specific epitope defined by a monoclonal antibody. Nature 390:74–77
9. Korth C, Streit P, Oesch B (1999) Monoclonal antibodies specific for the native, disease-associated isoform of the prion protein. Methods Enzymol 309:106–122
10. Jones M, McLoughlin V, Connolly JG (2009) Production and characterization of a panel of monoclonal antibodies against native human cellular prion protein. Hybridoma 28:13–20
11. Handisurya A, Gilch S, Winter D et al (2007) Vaccination with prion peptide-displaying papillomavirus-like particles induces autoantibodies to normal prion protein that interfere with pathologic prion protein production in infected cells. FEBS J 274:1747–1758
12. Bergstrom AL, Jensen TK, Heegaard PM et al (2006) Short-term study of the uptake of PrP(Sc) by the Peyer's patches in hamsters after

oral exposure to scrapie. J Comp Pathol 134:126–133

13. Cordes H, Bergström AL, Ohm J, Laursen H et al (2008) Characterisation of new monoclonal antibodies reacting with prions from both human and animal brain tissues. J Immunol Methods 337:106–120
14. Heegaard PM, Pedersen HG, Flink J et al (2004) Amyloid aggregates of the prion peptide PrP106–126 are destabilised by oxidation and by the action of dendrimers. FEBS Lett 577:127–133
15. Goding JW (1980) Antibody production by hybridomas. J Immunol Methods 39:285–308
16. Köhler G, Milstein C (1976) Derivation of specific antibody-producing tissue culture and tumor lines by cell fusion. Eur J Immunol 6:511–519
17. Manil L, Motte P, Pernas P et al (1986) Evaluation of protocols for purification of mouse monoclonal antibodies. Yield and purity in two-dimensional gel electrophoresis. J Immunol Methods 90:25–37
18. Towbin H, Staehelin T, Gordon J (1979) Electrophoretic transfer of proteins from polyacrylamide gels to NC sheets: procedure and applications. Proc Natl Acad Sci U S A 76: 4350–4354
19. Schulz-Schaeffer WJ, Tschoke S, Kranefuss N et al (2000) The paraffin-embedded tissue blot detects PrP(Sc) early in the incubation time in prion diseases. Am J Pathol 156:5156
20. Muller S (1999) Peptide-carrier conjugation. In: Regenmortel MHV, Muller S (eds) Synthetic peptides as antigens. Elsevier, Amsterdam, pp 79–132
21. Kitagawa T, Aikawa T (1976) Enzyme coupled immunoassay of insulin using a novel coupling reagent. J Biochem 79:233–236
22. Pedersen MK, Sorensen NS, Heegaard PM et al (2006) Effect of different hapten-carrier conjugation ratios and molecular orientations on antibody affinity against a peptide antigen. J Immunol Methods 311:198–206
23. Hermanson GT (1996) Bioconjugate techniques. Academic Press, San Diego

Chapter 26

Immunoblotting with Peptide Antibodies: Differential Immunoreactivities Caused by Certain Amino Acid Substitutions in a Short Peptide and Possible Effects of Differential Refolding of the Peptide on a Nitrocellulose or PVDF Membrane

Takenori Yamamoto, Taisuke Matsuo, Atsushi Yamamoto, Ryohei Yamagoshi, Kazuto Ohkura, Masatoshi Kataoka, and Yasuo Shinohara

Abstract

Immunodetection using antibodies, e.g., Western blotting, is generally utilized to measure the amount of a certain protein in a protein mixture. For valid interpretation of results observed by immunodetection, strict attention must be paid to the factors affecting the immunoreactivities of the antibodies. We here describe the step-by-step procedures to demonstrate that substitution of certain amino acids in a peptide can cause remarkable differences in its immunoreactivity with antibodies against epitope tags in the immobilized peptide. Refolding of the peptide on the membrane in a way that masks the epitope to different degrees was the possible reason for their distinct immunoreactivities with the antibodies. The results in this chapter suggest that we need to interpret carefully the experimental results involving immunodetection.

Key words Antibody, Immunodetection, Nitrocellulose membrane, PVDF membrane, Refolding, Western blotting

1 Introduction

Immunodetection is a powerful technique used for specific detection of certain proteins [1]. However, for valid interpretation of results observed by this method, strict attention must be paid to the factors affecting the immunoreactivities of the antibodies, because these immunoreactivities may be significantly different depending on how the antibodies are prepared. In the case of monoclonal antibodies, these are expected to react only with a certain epitope region of the protein antigen. In the case of polyclonal antibodies raised against a short peptide corresponding to a certain

Gunnar Houen (ed.), *Peptide Antibodies: Methods and Protocols*, Methods in Molecular Biology, vol. 1348, DOI 10.1007/978-1-4939-2999-3_26, © Springer Science+Business Media New York 2015

region of the target protein, antibody molecules present in the antiserum would be expected to react only with the epitope region of this short sequence in the target protein. The reactivities of either monoclonal or polyclonal antibodies with a protein antigen would be significantly reduced if a conformational change in the protein antigen causes masking of the epitope region. On the contrary, in the case of polyclonal antibodies raised by using "whole protein" as antigen, various species of antibodies recognizing various regions of the antigen as epitopes are present in the antiserum. Therefore, the observed reactivities of these antibodies reflect the "average" of the immunoreactivities of individual antibody species. For this reason, changes in the reactivities of a polyclonal antibody raised against a whole protein as the immunogen due to a conformational change in the protein antigen would be more moderate than in the case of a monoclonal antibody or polyclonal antibodies raised against a short peptide as the immunogen. In the case of SDS-PAGE separation and subsequent Western blotting, proteins are generally thought to be denatured by SDS. Thus, the differences in immunoreactions observed by Western blotting are often simply interpreted as being due to the different amounts of protein antigen present. The possible effects of a conformational change in the protein antigen on the observed immunoreactivities have not been previously taken into serious consideration. However, refolding of a denatured protein on the nitrocellulose membrane can occur [2, 3]. In such cases, interpretation of the observed immunoreactivities of antibodies would be difficult. We describe herein the step-by-step procedures that provide clear examples of peptides showing significantly distinct reactivities with antibodies, possibly reflecting the refolding of the immobilized peptides on the membranes.

2 Materials

2.1 Preparation of Peptides and Their Purification

1. Expression vector pColdIV.
2. *E. coli* host strain BL21(DE).
3. Ni-Sepharose resin.

2.2 SDS-PAGE

1. Resolving gel solution: 1.5 M Tris–HCl, pH 8.8.
2. Stacking gel solution: 0.5 M Tris–HCl, pH 6.8.
3. Thirty percent acrylamide/Bis solution (29.2:0.8 acrylamide: Bis).
4. Ammonium persulfate: 10 % solution in water.
5. *N,N,N′,N′*-tetramethyl-ethylenediamine (TEMED).
6. SDS-PAGE running buffer: 0.025 M Tris–HCl, 0.192 M glycine, 0.1 % SDS.

7. SDS lysis buffer (2×): 25 mM Tris–HCl (pH 6.8), 2 % SDS, 20 % glycerol, 2 % dithiothreitol, and 0.05 % Coomassie Brilliant Blue (CBB).
8. CBB staining solution: 0.1 % CBB R-250, methanol 50 %, acetic acid 7.5 %.
9. Destaining solution: 30 % methanol, acetic acid 7.5 %.

2.3 Immunoblotting

1. Nitrocellulose membrane Hybond ECL (*see* **Note 1**).
2. PVDF membrane Hybond P (*see* **Note 1**).
3. Transfer buffer: 100 mM Tris, 192 mM glycine, 0.02 % SDS, and 5 % methanol.
4. Tween solution: 0.05 % Tween 20, 150 mM NaCl, and 20 mM sodium phosphate, pH 7.4.
5. Blocking solution: Tween solution including 0.3 % skim milk.

3 Methods

3.1 Preparation of Peptides and Their Purification

In this study, we prepared various mutants of Pf3 coat protein (Fig. 1). We mutated four amino acids at the positions of X1–X4. In this manuscript, the peptides thus prepared will be referred to just simply by the amino acid sequence of X1–X4. For instance, DDDD represents a peptide having four Asp residues at X1–X4. We designed the following nine peptides: DDDD, DDNN, DDRK, NNDD, NNNN, NNRK, KRDD, KRNN, and KRRK.

1. cDNAs encoding individual peptides are prepared by overlap extension PCR [4].
2. Their expression vectors are constructed by using pColdIV.

MQSVIT N VTGQLTAVQA N ITTIGGAILLLIVLAAVVLGI N WI N AQFF

<u>Myc</u>KQSVITX$_1$VTGQLTAVQAX$_2$ITTIGGAILLLIVLAAVVLGIX$_3$WIX$_4$AQFFK<u>T7His</u>

Myc; MEQKLISEE
T7; MASMTGGQQMG
His; HHHHHH

Fig. 1 Amino acid sequences of the peptides used in the present study. The *letters on the first line* indicated the amino acid sequence of the parental peptide, Pf3(3L-4N) [9]; and those on the *second line*, the core amino acid sequence of the peptides used in the present study. The latter is made by incorporation of Myc, T7, and His tags into the former. The amino acid sequences of these tags are MEQKLISEE, MASMTGGQQMG, and HHHHHH, respectively. For preparation of various mutant peptides, the four Asn (N) amino acids at positions X1–X4 are varied; and the peptides thus prepared are simply referred to as X1X2X3X4; for instance, in the case of KRNN, Lys, Arg, Asn, and Asn are used at positions of X1–X4, respectively

3. Peptides are expressed in *E. coli* host strain BL21(DE3) by the procedure recommended by the supplier (*see* **Note 2**).
4. Expressed peptides are dissolved (*see* **Note 3**) and homogeneously purified by Ni-Sepharose chromatography (*see* **Notes 4** and **5**).

3.2 16.5 % Sodium Dodecyl Sulfate Polyacrylamide Gel Electrophoresis

1. Mix 3.6 ml of resolving buffer, 5.94 ml of acrylamide mixture, and 5.0 ml of water. Add 144 μl of 10 % SDS, 120 μl of ammonium persulfate, and 10 μl of TEMED, and case gel within a 7 cm × 10 cm × 1 mm gel cassette. Allow space for stacking the gel and gently overlay with isopropanol. Incubate at room temperature for 1 h.
2. Prepare the stacking gel by mixing 1.5 ml of stacking gel solution, 0.67 ml of acrylamide mixture, and 3.4 ml water. Add 60 μl of 10 % SDS, 40 μl of ammonium persulfate, and 10 μl of TEMED. After removing the isopropanol, gently overlay the stacking gel solution mixture over the resolving gel. Insert a 12-well gel comb. Incubate at room temperature for 1 h.
3. Dissolve purified peptides in 10 μl of 1× sample buffer (*see* **Note 6**). Then, load the peptide samples (10 μl/lane) into the gel. Electrophorese at 30 mA/gel until the dye front (from the BPB dye in the samples) has reached the bottom of the gel.
4. Place the gel in a plastic container containing CBB staining solution, and slowly shake the container for 1 h. Then, remove the staining solution, add destaining solution, and slowly shake. As shown in the photo described as "CBB" in Fig. 2a, the peptides will show relatively distinct migration regardless of the similarity of their molecular mass (from 8032.3 for DDDD to 8140.7 for KRRK).

3.3 Transfer of Proteins to Membranes

1. Immediately following SDS-PAGE, separate the gel plates and remove the stacking gel.
2. Transfer the gel into a container containing transfer buffer and keep it there for 5 min.
3. Rinse the gel carefully with deionized water.
4. Place three filter papers presoaked in transfer buffer on the apparatus of a semidry transfer apparatus. Place the immersed gel on the filter papers, and then carefully place a PVDF or nitrocellulose membrane on the gel.
5. Place another three papers presoaked with transfer buffer on the membranes, and then place the lid on the transfer apparatus.
6. Connect the power supply and perform electrophoresis at a constant voltage (*see* **Note 7**).

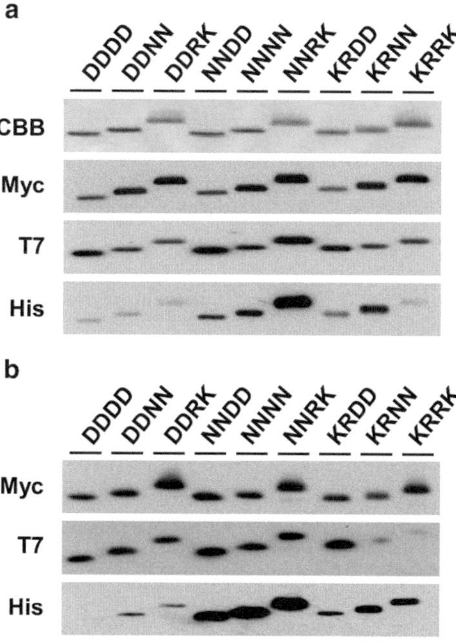

Fig. 2 Immunoreacitivities of nine peptides transferred onto nitrocellulose (**a**) or PVDF (**b**) membranes and detected with antibodies against the Myc, T7, or His tag. Aliquots (0.2 μg) of individual peptides were subjected to SDS-PAGE and visualized by CBB. For immunodetection with antibodies against the Myc-tag, 0.02-μg aliquots of individual peptides were subjected to SDS-PAGE; whereas for that with antibodies against the T7 or His tag, 0.05-μg aliquots of peptides were used. After SDS-PAGE, the separated peptides were transferred onto nitrocellulose (**a**) of PVDF (**b**) membranes

3.4 Immuno detection

1. Block the membrane with blocking solution for 1 h.
2. Incubate with blocking solution including antibody against Myc tag (1:10,000 dilution, for 1 h), T7 tag (1:5,000 dilution, overnight), or His tag (1:5,000 dilution, overnight).
3. Wash the membrane with Tween solution twice, 5 min each time.
4. Incubate with Tween solution including antibody against anti-mouse IgG conjugated with horseradish peroxidase (1:10,000 dilution for Myc tag, 1:5000 dilution for T7 tag or His tag) for 1 h.
5. Wash the membrane with Tween solution, five times, for 5 min each time.
6. Immunoreactive protein bands are visualized by using an ECL kit. As a result, remarkable differences are observed in the reactivity of antibodies against Myc, T7, or His tags with individual mutant peptides immobilized on a nitrocellulose

membrane (Fig. 2a, *see* **Note 8**) or PVDF membrane (Fig. 2b, *see* **Note 9**) [5]. The observed changes in the reactivities of peptides with antibodies are considered to be caused by a conformational change due to the refolding of the immobilized peptides on the membrane (*see* **Note 10**). That is, possible refolding of proteins and peptides on a nitrocellulose or PVDF membrane might cause the epitopes recognized by the antibodies to be masked, thus leading to their differential immunoreactivities with the antibodies. Therefore, for proper interpretation of the experimental results involving immunodetection, we need to pay attention to the effect of the refolding of proteins and peptides on their immunoreactivities with antibodies.

4 Notes

1. Immerse a nitrocellulose membrane in transfer buffer or a PVDF membrane in methanol. Rinse twice in distilled water and once with transfer buffer.
2. The expression of peptides is induced by shifting the culture temperature from 37 to 15 °C and by adding IPTG to a final concentration of 0.4 mM.
3. The bacterial cells are disrupted by sonication in PN medium (500 mM NaCl, 20 mM NaPi buffer; pH 7.4). The bacterial pellet is resuspended in PN medium containing 1 % Triton X-100. After centrifugation at 15,000 × *g* for 15 min, the supernatant is recovered.
4. The above supernatant containing the expressed peptide is loaded onto a Ni-Sepharose column pre-equilibrated with PN medium. After washing of the column with PN buffer containing 0.05 % Triton X-100 and 100 mM imidazole, the peptide is eluted with PN buffer with 0.05 % Triton X-100 and 500 mM imidazole.
5. Removal of Triton X-100 from the purified protein samples by acetone precipitation is useful to clearly detect the protein band in the SDS-PAGE and Western blots.
6. The peptide solutions need to be heated at 95 °C for 10 min for clear detection.
7. One of the major factors affecting the results of Western analysis of proteins is the difference in the efficiency of transfer of proteins to the membrane [6–8]. However, when we examined the efficiency of transfer of peptides to the nitrocellulose membrane, it was not markedly different among the peptides (data not shown).

8. First, antibody against the Myc tag gives similar high immunoreactivities with six peptides, i.e., DDNN, DDRK, NNNN, NNRK, KRNN, and KRRK, and slightly weaker ones with DDDD, NNDD, and KRDD. Antibody against the T7 tag gives strong protein bands for DDDD, NNDD, NNNN, NNRK, and KRDD, but much weaker ones for the remaining four peptides. Differences in immunoreactivities among these peptides with antibody against the His tag are quite significant. This antibody reacts very strongly with NNRK; moderately with NNDD, NNNN, and KRNN; but weakly with the remaining five peptides. Although these five peptides show markedly weak reactivities with the antibody against the His tag, this weakness is not due to truncation or degradation of the C-terminal His-tag sequence, because all nine peptides are successfully purified by Ni-Sepharose column chromatography.
9. Antibody against the Myc tag yields similar signal intensities with all nine peptides. The antibody against the T7 tag gives similar signal intensities with all peptides except KRNN and KRRK, whose bands are almost negligible. Again, the antibody against the His tag shows markedly distinct signals with these peptides. It gives strong signals with three peptides, i.e., NNDD, NNNN, and NNRK, and moderate ones with the others except for DDDD, which was not detectable with the antibody against the His tag under the conditions used.
10. When we tested the effects of heating (121 °C, 20 min) nitrocellulose filters on the reactivities of immobilized peptides with antibodies, we observed remarkable changes in the reactivities of immobilized peptides with antibodies (data not shown). This result also strongly supports the validity of our interpretation.

References

1. Gershoni JM, Palade GE (1983) Protein blotting: principles and applications. Anal Biochem 131:1–15
2. Klinz FJ (1994) GTP-blot analysis of small GTP-binding proteins. The C-terminus is involved in renaturation of blotted proteins. Eur J Biochem 225:99–105
3. Karlsson-Borgå A, Rolfsen W (1991) Methodological considerations when using nitrocellulose immunoblotting from polyacrylamide gels to study the mould allergens Aspergillus fumigatus and Alternaria alternata. J Immunol Methods 136:91–102
4. Ho SN, Hunt HD, Horton RM, Pullen JK, Pease LR (1989) Site-directed mutagenesis by overlap extension using the polymerase chain reaction. Gene 77:51–59
5. Matsuo T, Yamamoto T, Katsuda C, Niiyama K, Yamamoto A, Yamazaki N, Ohkura K, Kataoka M, Shinohara Y (2009) Substitution of certain amino acids in a short peptide causes a significant difference in their immunoreactivities with antibodies against different epitopes: evidence for possible folding of the peptide on a nitrocellulose or PVDF membrane. Biologicals 37:44–47
6. Dunn SD (1986) Effects of the modification of transfer buffer composition and the renaturation of proteins in gels on the recognition of proteins on Western blots by monoclonal antibodies. Anal Biochem 157:144–153
7. Van Oss CJ, Good RJ, Chaudhury MK (1987) Mechanism of DNA (Southern) and protein (Western) blotting on cellulose nitrate and other membranes. J Chromatogr 391:53–65

8. Hoffman WL, Jump AA, Kelly PJ, Ruggles AO (1991) Binding of antibodies and other proteins to nitrocellulose in acidic, basic, and chaotropic buffers. Anal Biochem 198:112–118

9. Ridder AN, Kuhn A, Killian JA, de Kruijff B (2001) Anionic lipids stimulate Sec-independent insertion of a membrane protein lacking charged amino acid side chains. EMBO Rep 2:403–408

Chapter 27

Immunocytochemical and Immunohistochemical Staining with Peptide Antibodies

Tina Friis, Klaus Boberg Pedersen, David Hougaard, and Gunnar Houen

Abstract

Peptide antibodies are particularly useful for immunocytochemistry (ICC) and immunohistochemistry (IHC), where antigens may denature due to fixation of tissues and cells. Peptide antibodies can be made to any defined sequence, including unknown putative proteins and posttranslationally modified sequences. Moreover, the availability of large amounts of the antigen (peptide) allows inhibition/adsorption controls, which are important in ICC/IHC, due to the many possibilities for false-positive reactions caused by immunoglobulin Fc receptors, nonspecific reactions, and cross-reactivity of primary and secondary antibodies with other antigens and endogenous immunoglobulins, respectively. Here, simple protocols for ICC and IHC are described together with recommendations for appropriate controls.

Key words Adsorption control, Immunocytochemistry, Immunohistochemistry, Peptide antibodies, Specific staining

1 Introduction

Antibodies allow precise and specific cellular and tissue localization of antigens by immunocytochemistry (ICC) and immunohistochemistry (IHC) [1, 2]. Peptide antibodies, in particular, are useful for localization of many peptides and proteins including unknown/putative peptides/proteins predicted to exist from DNA sequences. Another advantage of peptide antibodies is that they can be made to posttranslationally modified sequences, e.g., phosphorylated sequences or citrullinated sequences. Also, in ICC and IHC, denaturing conditions are often used (e.g., heat treatment or ethanol fixation), and the epitopes of peptide antibodies will generally be less sensitive to such treatments, compared to three-dimensional epitopes. Finally, since peptides usually can be produced in large quantities, they allow control experiments in the form of inhibition/adsorption controls (Fig. 1).

Peptide antibodies may be produced following procedures described in this volume. Incubation of cells and tissues with

Gunnar Houen (ed.), *Peptide Antibodies: Methods and Protocols*, Methods in Molecular Biology, vol. 1348, DOI 10.1007/978-1-4939-2999-3_27, © Springer Science+Business Media New York 2015

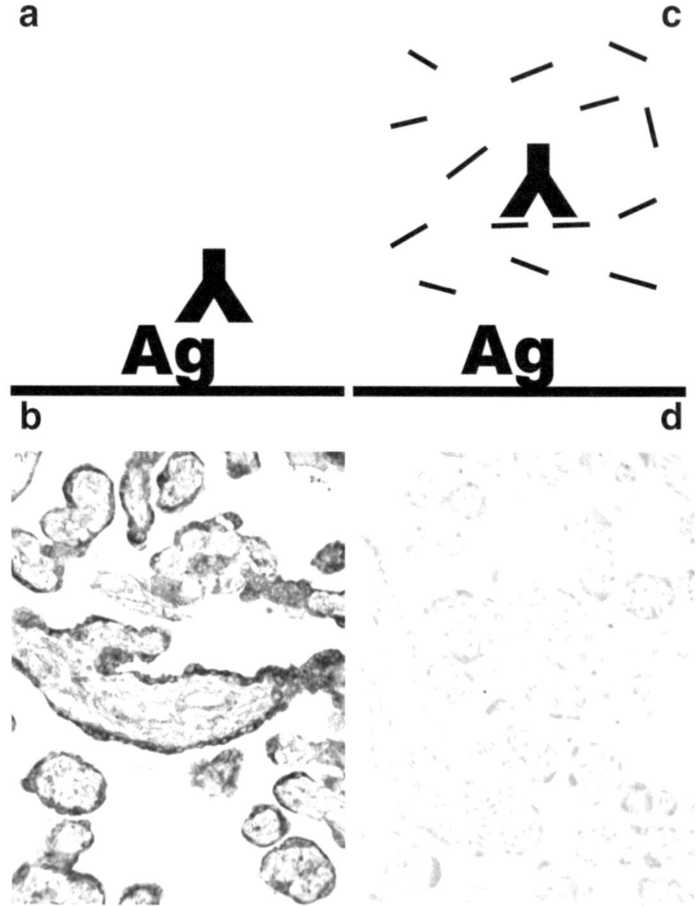

Fig. 1 Principle of ICC/IHC and adsorption control with peptide antibody (Ab). (**a**) Binding of peptide Ab to epitope in cell/tissue antigen (Ag). (**b**) Adsorption control with excess peptide containing the epitope for the Ab. (**c**) Localization of calreticulin in human placenta by IHC with enzyme activity staining (peroxidase) and light microscopy. The tissue section was incubated with a sequence-specific rabbit antiserum against a synthetic peptide from the C-terminus of calreticulin [16] followed by incubation with horseradish peroxidase-conjugated goat immunoglobulins against rabbit IgG. Note the strong staining of the syncytiotrophoblast layer. (**d**) Adsorption control with excess peptide. Note the essentially complete absence of staining

peptide antibodies is not different from the use of other antibodies, and bound antibodies may be visualized by a number of well-known techniques (i.e., enzyme activity staining or fluorescence) [1, 2].

For ICC and IHC cells or tissues may be obtained commercially, by cell culture or from clinical departments and animal facilities, taking appropriate ethical precautions.

Cells may be stained directly with antibodies or prepared by fixation with cross-linking agents (e.g., formaldehyde) or precipitating agents (e.g., ethanol). In the case of cell surface antigens, living cells can be stained directly. With intracellular antigens, cells have to be fixated and permeabilized to make the antigen accessible. Depending on the antigen and the method of fixation, various methods for permeabilization (i.e., detergent treatment) and epitope/antigen retrieval may have to be attempted (e.g., heat treatment, protease treatment, incubation with different detergents, buffers at low or high pH, etc.) [2]. Tissues may also be sectioned directly for subsequent staining, but usually tissue samples are frozen before sectioning/fixation/staining or fixated before freezing/sectioning/staining (Fig. 2).

Visualization of bound antibodies is achieved by labeling of antibodies (primary or secondary) with fluorophores or enzymes. Staining patterns can then be inspected by fluorescence microscopy or by light microscopy after enzyme activity staining (e.g., phosphatase or peroxidase). Staining of abundant components of cells and tissues (e.g., nuclei) with conventional dyes (contrast staining) may be used to facilitate microscopical analysis [1–3].

Validation of specific antibody staining in ICC/IHC requires performance of a number of positive and negative control experiments, and a staining pattern is only specific if all controls exclude nonspecific staining [2–5]. Special for peptide antibodies is the usually high specificity and the availability of large amounts of the antigen (peptide) at a reasonable cost, thus providing the possibility to perform well-defined inhibition/adsorption controls (Fig. 1).

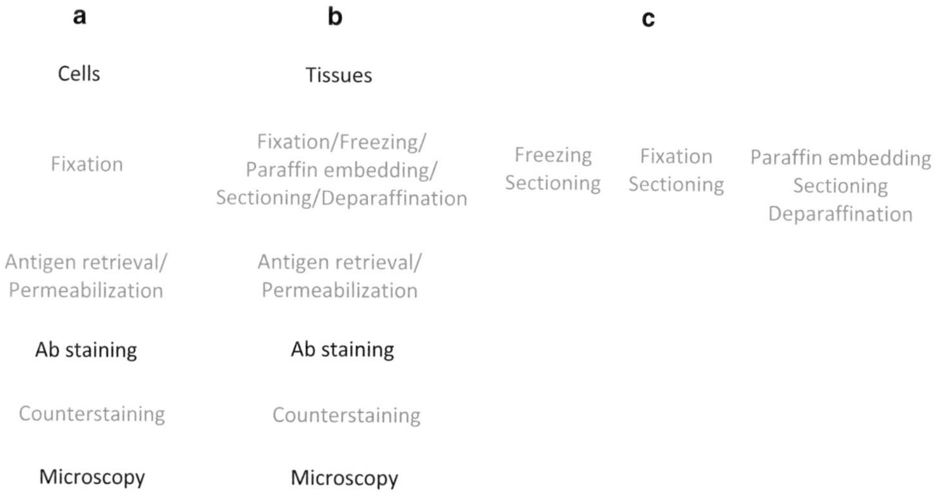

Fig. 2 (**a**) Work flow for immunocytochemical staining with antibodies. (**b**) Work flow for immunohistochemical staining with antibodies. (**c**) Options for tissue processing before Ab staining. *Black*: obligatory. *Gray*: optional

These are particularly important when using newly developed or uncharacterized antibodies in ICC/IHC due to the heterogenous nature of the substrate and the possibility of false-positive reactions caused by endogenous enzyme activity or autofluorescence, Fc receptors binding primary or secondary antibodies, or cross-reactivity of primary or secondary antibodies with other antigens or endogenous immunoglobulins, respectively. For ideal adsorption controls, staining is abolished by inhibition/preadsorption of the primary antibody with purified antigen but not with carrier protein or irrelevant antigen. Unfortunately, simple preincubation of antibody with antigen does not always eliminate antibody staining completely, since antigen may dissociate from antibody during incubations, thus allowing unbound antibody to bind antigens in the section. To circumvent this, it may be advisable to use antigens coupled to beads for preadsorption of specific antibodies and to use more rounds of adsorption.

Reagent substitution controls may be used to identify false-positive staining reactions (e.g., by omitting primary or secondary antibodies or by replacing the primary antibody with an irrelevant species- and isotype-matched antibody known not to react with the cells/tissues under study).

Positive and negative controls may be performed on other relevant cells/sections prepared as those being tested. A negative control with tissue known not to contain the antigen of interest may exclude false-positive staining, and a positive tissue control may exclude the possibility of false-negative staining. The positive tissue control is most valuable when performed on a section containing low concentrations of the antigen of interest and thus displaying weak staining, since it will be more sensitive to impacts that influence the antibody staining than a section containing high antigen levels.

This chapter describes simple protocols for ICC/IHC together with suggestions for appropriate controls to verify the specificity of the observed reactions (*see* **Note 1**).

2 Materials

2.1 Chemicals

Tween 20.

Triton X-100.

EtOH.

Glycerol.

Xylene.

Hexane.

Paraffin.

Tissue-Tek.

Hematoxylin.

Eosin.

4′, 6-Diamidino-2-phenylindole (DAPI).

Hoechst 33342.

Propidium iodide (PI).

3, 3′-Diaminobenzidine (DAB).

5-Bromo-4-chloro-3′-indolyl-phosphate (BCIP).

Nitroblue tetrazolium (NBT).

Proteinase K.

Peptides may be obtained commercially or can be synthesized and characterized according to procedures described in this volume.

Peptides may be coupled to various matrices (e.g., CNBr-Sepharose) or beads (e.g., paramagnetic beads or beads for immunoassays) following the instructions of the manufacturer for preadsorption controls.

2.2 Equipment

Scalpels, forceps, and scissors.

Plate for cutting.

Metal plate for freezing of tissue samples.

Metal tissue "holders."

Paper towels.

Gloves.

Plastic (polystyrene) tubes/containers.

Syringes.

Disposable 0.45 μm filters.

Glass slides.

Glass cover slips.

Microtome.

Microscope (light, fluorescence).

Beakers.

Plastic (polystyrene) ware.

Pipettes and tips.

Boxes for incubation of slides (humidity chambers).

2.3 Buffers and Fixatives

PBS (50 mM NaP_i, 0.15 M NaCl).

TBS (50 mM Tris, pH 7.5, 0.15 M NaCl).

TTN (50 mM Tris, pH 7.5, 1 % Tween 20, 0.3 M NaCl).

Saline (0.15 M NaCl).

Formaldehyde (HCHO) (37 %).

2.4 Antibodies

Primary antibodies are obtained commercially or produced following procedures described in this volume. The species of origin and the isotype must be known. Optimally, control antibodies from the same species and of the same isotype but another specificity should also be obtained.

Secondary antibodies are obtained commercially, usually as horseradish peroxidase- or alkaline phosphatase-conjugated rabbit or goat immunoglobulins against mouse immunoglobulins or rabbit immunoglobulins, respectively (for light microscopy), or as FITC- or rhodamine-conjugated rabbit or goat immunoglobulins against mouse immunoglobulins or rabbit immunoglobulins, respectively (for fluorescence microscopy) (*see* **Note 2**).

Animal sera should be available from the same species as the secondary antibodies and optionally also for the primary antibodies, in the case of polyclonal antibodies.

2.5 Cells

Cells are obtained commercially, are grown in culture, or are obtained from blood samples or other biological fluids following current ethical guidelines.

Cells may be stored at −80 °C, −135 °C, −150 °C, or −180 °C, or in liquid N_2, with or without prior fixation.

2.6 Tissues

Tissue samples are obtained from clinical departments, animal departments, or slaughter houses following current ethical guidelines and, for animal tissue, current guidelines for animal welfare (e.g., for sedation and sacrifice of animals).

Tissues may be stored at −80 °C, −135 °C, −150 °C, or −180 °C, or in liquid N_2, with or without prior fixation, or at room temperature for paraffin-embedded tissues.

3 Methods

3.1 Immunocyto chemistry

3.1.1 Cells

Premade cell preparations may be obtained from a number of commercial suppliers (e.g., HEp-2 cells for immunofluorescence cytochemistry).

Hemopoietic/lymphopoietic cells and mature leukocytes are prepared from heparinized blood samples or other biological fluids using differential (gradient) centrifugation or Ab-coupled (paramagnetic) beads (*see* **Note 3**). Cells from such samples may be stained directly or prepared for ICC by centrifugation (cytospins) or smeared on glass slides, where they may be stained directly, air dried, and/or fixated before staining.

Cell lines may be cultured in suspension or on glass slides, coverslips, or plastic ware designed for ICC (*see* **Note 4**).

When cells have reached an appropriate density and have been treated according to experimental protocols (e.g., stimulation with growth factors or treatment with chemicals), they are washed briefly with medium/PBS/TBS/saline on the support or in suspension with intermittent centrifugations ($1000 \times g$) followed by resuspension to 10^6 cells/ml.

Cells can then either be stained directly with antibodies (cell surface antigens), fixated in suspension or on the support (adherent cells), or smeared on a glass slide (suspension cells) and then air dried and/or treated with an appropriate fixative before further processing.

3.1.2 Direct Immunofluorescent Staining of Living Cells

Wash cells briefly with medium at 5–37 °C depending on the antigen under study (5 °C for cell surface localization, 37 °C for internalization studies) (*see* **Note 5**).

Incubate 20–30 min (in a humidity chamber) with fluorescence-labeled primary antibody diluted 1:100, 1:1000, and 1:10,000 in cell growth medium or an appropriate buffer (with or without 1 % serum albumin) at the chosen temperature (*see* **Note 6**).

If the primary antibody is unlabeled or biotinylated, wash cells briefly 2–3 times with medium/PBS/saline and incubate 20–30 min (in a humidity chamber) with fluorescence-labeled secondary antibody or streptavidin diluted 1:1000, 1:2000, or 1:5000.

Wash cells briefly with medium or buffer at 5–37 °C and proceed with fluorescence microscopy (and/or flow cytometry for suspension cells).

3.1.3 Fixation of Cells

Cells may be fixated to varying degrees depending on the antigen(s) under study (*see* **Note 7**).

Formaldehyde fixation

Incubate the cells in 0.1–3.7 % formaldehyde in PBS for 10–30 min (or overnight) at 5 °C or for 10–30 min at room temperature.

Wash 3 times with PBS, TBS, or saline.

Ethanol fixation

Incubate the cells with cold EtOH (96–99 %, 5 °C) for 10–15 min. Then wash briefly with buffer and proceed with antibody staining.

3.1.4 Permeabilization of Cells (Optional)

If cells have been fixated with EtOH, this may be sufficient to allow access to the antigen(s). For aldehyde fixation, cells are treated with a detergent to dissolve membrane lipids and permit access to internal structures.

Wash 2–3 times with TTN buffer for 10–30 min at 5 °C or 5–10 min at room temperature (*see* **Note 8**).

The degree of permeabilization may be tested with a nucleic acid stain (*see* **Note 9**).

3.1.5 Antigen Retrieval (Optional)

This is mainly used in case of prolonged formaldehyde fixation. *See* Subheading 3.2.6 for methods.

3.1.6 Pretreatment (Optional)

If necessary (depending on the cells and the antibodies), cells are incubated with buffers containing blocking agents (e.g., H_2O_2 for blocking of endogenous peroxidase, levamisole for blocking of endogenous alkaline phosphatase, and avidin for blocking of endogenous biotin).

Autofluorescence due to endogenous compounds or aldehyde groups may be a problem and is most easily relieved by switching to another fluorophore (*see* **Note 10**).

Irreversible inhibition of endogenous peroxidase activity

Incubate 15 min with 0.5 % H_2O_2 in methanol or PBS, depending on the cells and antigens.

Wash 3 × 5–10 min in medium/saline/PBS/TBS.

Competitive inhibition of endogenous phosphatase activity

Include 1 mM levamisole in washing buffer before activity staining and in buffer for activity staining.

Inhibition of endogenous biotin

Incubate 30 min with streptavidin (1 μM) in TTN/PBS/TBS.

Wash 3 times 10 min in TTN/PBS/TBS.

Incubate 30 min with 1 mM biotin in TTN/PBS/TBS.

Wash 3 times 10 min in TTN/PBS/TBS.

3.1.7 Antibody Staining and Inhibition/Adsorption Controls

Wash 3 times 10 min in buffer (PBS/TBS for non-permeabilized cells, TTN, or PBS/TBS with 0.1–1 % detergent/Triton X-100, Tween 20, or others). Optionally add 1 % BSA or another abundant protein for blocking of nonspecific binding. Another means of reducing nonspecific binding is to include immunoglobulins from the species in which the secondary antibody is produced (e.g., inclusion of 0.1–1 % rabbit serum).

Incubate with primary antibody diluted 1:1000 or another predetermined dilution (*see* **Note 11**) in buffer.

NB. Perform all incubations with antibodies in a humidity chamber.

Inhibition/adsorption controls

Inhibition control: Include at least a 1000-fold molar excess of peptide together with primary antibody (*see* **Note 12**) and preincubate 1 h at room temperature or overnight at 5 °C before incubation on the cells. Adsorption control: Preincubate the primary antibody with peptide coupled to a solid support (*see* **Note 13**) before incubation on the cells.

Reagent control: Substitute the primary antibody with a species- and isotype-matched antibody of irrelevant specificity (*see* **Note 14**).

Wash 3 times 10 min in buffer.

Incubate with secondary antibody (conjugate) diluted 1:1000, 1:2000, or 1:5000 (as determined by preliminary titration experiments) in the same buffer as used for the primary antibody.

Wash 3 times 10 min in buffer.

3.1.8 Development (Only for Light Microscopy)

For these procedures, it is optimal to use commercially available substrate tablets.

Peroxidase

Incubate with 0.01–0.03 % H_2O_2 (3–9 mM) and a suitable chromogenic substrate, e.g., DAB, 0.05 % (1 mM) in TBS/PBS for 10–30 min at room temperature.

Wash 3 times with buffer or water.

Alkaline phosphatase

Incubate with 0.02 % BCIP (0.5 mM) and 0.03 % NBT (0.04 mM) in TBS for 30–60 min (or overnight if required).

Wash 3 times with buffer or water.

3.1.9 Contrast Staining (Optional)

Double staining with different sets of primary and secondary antibodies may be used for localization of other antigens (*see* **Note 15**). Otherwise, staining of, e.g., DNA or protein can aid in microscopical analysis.

Fluorescence ICC

Incubate with, e.g., 4′,6-diamidino-2-phenylindole (DAPI, 1 μg/mL), PBS/TBS for 5 min (*see* **Note 16**).

Wash 3 times with PBS/TBS for 1–2 min.

Enzyme activity ICC (light microscopy)

Stain for a few min with a suitable cytochemical stain, e.g., Wright-Giemsa stain, May-Grünwald stain, or Papanicolaou stain (*see* **Note 17**)

Wash 3 times with PBS/TBS for 1–2 min.

3.1.10 Microscopy

Mount slides with coverslips using a drop of buffer (e.g., PBS) or buffer with glycerol (e.g., PBS with 90 % glycerol). A commercial fainting-reducing mounting medium may also be used.

Inspect with fluorescence microscope or light microscope using magnifications from 10 to 100. Reactions may be semiquantified using a scale from 0 (no staining) to 3 (strong staining). Slides with positive staining verified by stringent controls may be studied using more advanced techniques (*see* **Note 18**).

3.2 Immuno histochemistry

IHC procedures are generally equivalent to ICC procedures, but longer times and higher concentrations of Abs may be required for some applications.

3.2.1 Tissue Samples

Tissue slides may be obtained commercially as single tissues or as tissue arrays. Alternatively, tissues are obtained from clinical departments and animal facilities with appropriate ethical consent and following current ethical and animal welfare guidelines.

3.2.2 Tissue Preparation for Immunohistochemical Staining

As shown if Fig. 2, tissues may be processed in various ways before staining with antibodies, depending on the antigens in question and the purpose of the studies. Tissues may be frozen first, then sectioned, fixated, and prepared for immunohistochemical staining. Alternatively, tissues may be fixated and sectioned directly or frozen and then sectioned. Finally, fixated tissue samples may be dehydrated, embedded in paraffin, sectioned, deparaffinated, and then prepared for immunochemical staining. Each procedure has advantages and disadvantages, particularly with respect to antigen preservation and storage of specimens (*see* **Note 19**).

In general, more modifications are introduced to antigens the more and the harsher treatments are used in sample processing following the order: fresh, frozen, and alcohol-, aldehyde-, and paraffin-exposed tissue. However, some treatments may diminish or abolish some forms of nonspecific binding (e.g., FcRs), thus improving detection of specific antigens.

3.2.3 Freezing, Sectioning, and Fixation of Tissue Samples

Cut tissue in appropriate pieces of approximately 1–10 mm^3 or a height of a few mm for larger pieces.

Place the pieces on a precooled (dry ice) metal tissue holder for subsequent sectioning or on a metal plate cooled with dry ice and allow to freeze (*see* **Note 20**).

Record the samples in a logbook and store in plastic containers at −80 °C or lower (*see* **Note 21**).

Disinfect equipment and workplace with EtOH after use (*see* **Note 22**).

3.2.4 Sectioning

For mounting, tissue pieces may be firmly attached to the tissue holders using drops of PBS or with, e.g., Tissue-Tek.

Place the holders with tissue in the (cryo)microtome and cut sections of 2–4 μm. Gently transfer the sections to (optionally precoated, e.g., with aminoalkylsilane) glass slides optionally with the aid of a fine brush and allow to attach

The sections may be dried voluntarily or in a stream of cold air.

The location of the tissue can be marked with a waterproof pen for ease of subsequent incubations.

3.2.5 Fixation, Paraffin Embedding, Sectioning, and Deparaffination of Tissue Samples

Fix the tissue sample for 12–48 h in 3.5 % paraformaldehyde in 0.1 M phosphate buffer.

The tissue may now be sectioned directly, frozen for later sectioning (see above), stored in 70 % EtOH at 5 °C, or embedded in paraffin or a suitable polymer.

Before embedding in paraffin, dehydrate the tissues [(1) 30 min in 70 % ethanol, (2) 30 min in 95 % ethanol, (3) 30 min in 99.9 % ethanol, (4) 1 h in 99.9 % ethanol, (5) 1 h in 99.9 % ethanol, (6) 1 h in 99.9 % ethanol, (7) 30 min in 99.9 % ethanol/xylene, (8) 1 h in xylene, (9) 2 h in xylene, (10) 2 h 30 min in paraffin wax, (11) 4 h or longer in paraffin wax].

Slice the tissue in 4–5 μm sections and transfer sections to a waterbath (~45 °C)

Transfer sections to (optionally precoated) glass slides and allow to attach freely (*see* **Note 23**).

Dry slides freely in air (or in a stream of cold air, or for 20 min at 60 °C).

For paraffin-embedded sections, melt paraffin in oven to firmly adhere tissue sections to glass slides and subsequently deparaffinate and rehydrate by 4 washes in xylene, 3 washes in 99.9 % ethanol, and 3 washes in 96 % ethanol (and then buffer).

3.2.6 Pretreatment/Epitope Retrieval

Tissue sections may be treated with detergents, proteases, or other enzymes to facilitate tissue penetration and allow access to individual antigens (*see* Subheading 3.1.4). Moreover, several methods for epitope/antigen retrieval may be used (*see* **Note 24**). Most often, heat treatment, which reverses some of the chemical modifications from fixation, is used.

Detergen t treatment

Incubate in TTN or PBS/TBS with 1 % Tween-20/Triton X-100 for 30–60 min.

Protease treatment

Incubate with proteinase K (10–20 μg/ml) in PBS/TBS/TTN for 10–20 min at room temperature.

Heat-induced epitope retrieval

Treat sections in buffer (e.g., TTN/PBS/TBS/0.5 M Tris buffer, pH 10.0, or other buffer, e.g., 10 mM sodium citrate, pH 6.0) for 10 min in a microwave oven at 800 W.

Cool for approximately 20 min.

Wash 5 min in buffer or water.

Dry the sections by using a cold air blower.

Use a Pap pen to draw a circle around the tissue.

Dry the sections by using a cold air blower.

Rehydrate the sections for 5 min in TBS.

Block endogenous enzyme activity (peroxidase, alkaline phosphatase) or biotin as described under Subheading 3.1.6.

Wash shortly in buffer or water, followed by washing for 2 × 5 min in TBS

3.2.7 Antibody Staining and Inhibition/Adsorption Controls

Incubate sections for 1 h at room temperature or overnight at 4 °C with primary antibody diluted appropriately in TBS + 1 % BSA.

Inhibition/adsorption control: Primary antibody preincubated with excess antigen (*see* Subheading 3.1.7).

Reagent control: isotype-matched antibody from the same species.

Negative control: TBS + 1 % BSA.

Tissue controls: Positive and negative tissue samples with known concentration of antigen of interest (e.g., using tissue arrays).

Wash the slides 3 × 5 min in TBS.

Incubate all sections (except control without secondary antibody) for 30 min with secondary antibody at appropriate dilution (*see* **Note 25**).

Wash the slides 3 × 5 min in TBS.

3.2.8 Enzyme Activity Staining (Only Light Microscopy)

Peroxidase

Incubate with 0.01–0.03 % H_2O_2 (3–9 mM) and a suitable chromogenic substrate, e.g., DAB, 0.05 % (1 mM) in TBS/PBS for 10–30 min at room temperature.

Alkaline phosphatase

Incubate with 0.02 % BCIP (0.5 mM) and 0.03 % NBT (0.04 mM) in TBS for 30–60 min (or overnight if required).

Wash the slides for 3–5 min in buffer or running water.

3.2.9 Contrast Staining

Fluorescence histochemistry

Incubate with DAPI or another fluorescent dye (*see* Subheading 3.1.9).

Enzyme histochemistry

Incubate slides with a suitable histochemical stain (*see* **Note 26**).

Wash the slides for 5 min in running water.

3.2.10 Microscopy

Mount coverglass with buffer (optionally add glycerol to 90 %) or an aqueous mountain medium. A commercial fainting-reducing mounting medium may also be used.

Inspect with fluorescence microscope or light microscope using magnifications from 10 to 100. Reactions may be semiquantified using a scale from 0 (no staining) to 3 (strong staining). Slides with positive staining verified by stringent controls may be studied using more advanced techniques (*see* **Note 18**).

4 Notes

1. The protocols mainly relate to initial use and characterization of peptide antibodies for research purposes. For diagnostic and routine uses, consult recent textbooks, reviews, and articles [2, 3, 6–8].
2. A variety of secondary antibodies with other fluorescence labels are available, e.g., Alexa Fluor, CyDye, and quantum dots.
3. *See*, e.g., refs. 9–11.
4. *See*, e.g., refs. 11, 12.
5. With intermittent centrifugation (e.g., 1000 × *g*, 10 min) for suspension cells.
6. Depending on the antibody, other dilutions may have to be used. In this case, perform a twofold titration from the concentration yielding a satisfactory staining.
7. Mild non-permeabilizing fixation can be done with 0.1 % formaldehyde for 10–30 min at 5 °C. Higher concentrations, longer times, and higher temperatures will result in stronger fixation. Ethanol fixation at 5–37 °C will produce an almost instantaneous membrane-disrupting fixation.
8. The optimal detergent and buffer may have to be determined individually for each antigen, especially pH, ionic strength, and presence/absence of Ca^{2+}; other ions or cofactors may have to be taken into consideration. Triton X-100 (0.1–1 %), saponin, or another detergent in PBS 5–10 min at room temperature is often used.
9. Several DNA intercalating stains are useful (e.g., DAPI, propidium iodide, Hoechst 33342).
10. Autofluorescence due to residual aldehyde groups may be reduced by incubation with amine-containing buffers (e.g., 50 mM Tris/hydroxylamine/glycine, pH 7–8) or by $NaBH_4$ treatment (1 % in PBS).
11. A titration experiment with primary antibodies diluted 1:100, 1:1000, and 1:10,000 (and later on a twofold titration from one of these) will determine a suitable dilution.
12. Since the concentration of specific antibodies can only be accurately determined for monoclonal antibodies, a molar excess cannot be calculated for polyclonal antibodies. In this case, use peptide concentrations of 0.1–1 mg/mL.
13. Use CNBr-Sepharose, magnetic beads, or other matrices following the instructions of the manufacturer.
14. For sera obtained by immunization, a preimmunization bleed may also be included as an extra control.

15. Double staining with different sets of primary and secondary antibodies.
16. Other dyes may be used, e.g., Hoechst 33342 (10 μg/mL for 10–20 min).
17. *See*, e.g., refs. 3, 13.
18. Confocal laser scanning microscopy or more advanced techniques (e.g., photon correlation microscopy) [11, 14, 15].
19. Since a number of antibodies are not suitable for the use on both frozen and paraffin-embedded tissue, it might be valuable to divide the tissue samples in two parts and freeze one part and use the other part for paraffin embedding.
20. Alternatively, embed the tissue in Tissue-Tek and snap freeze in n-hexane.
21. The tissue should be kept at −80 °C for at least 24 h before use.
22. Disposal of tissues must follow local guidelines for disposal of biological waste.
23. Coating of slides with aminoalkylsilane or polylysine may facilitate attachment.
24. The optimal antigen retrieval protocol depends on the antigen and must be determined individually. Other suitable treatments for antigen retrieval include incubation for 10 min in a microwave oven in 0.01 M citrate buffer pH 6.0 or by enzymatic treatment, e.g., treatment for 15 min at 37 °C with 0.1 % trypsin in 0.1 % $CaCl_2$, pH 7.8 (dissolve the trypsin in 37 °C warm $CaCl_2$ solution).
25. For biotinylated Abs, incubate sections for 30 min with preincubated StreptABComplex/HRP prepared according to the instructions of the manufacturer.
26. For example, HE (hematoxylin for DNA/nuclei and eosin for protein/cytoplasm) [3, 13].

References

1. Van Noorden S (2002) Advances in immunocytochemistry. Folia Histochem Cytobiol 40: 121–124
2. Ramos-Vara JA, Miller MA (2014) When tissue antigens and antibodies get along: revisiting the technical aspects of immunohistochemistry–the red, brown, and blue technique. Vet Pathol 51:42–87
3. Bancroft JC, Gamble M (eds) (2002) Theory and practice of histological techniques. Har court publishers, London
4. Bordeaux J, Welsh A, Agarwal S, Killiam E, Baquero M, Hanna J, Anagnostou V, Rimm D (2010) Antibody validation. Biotechniques 48: 197–209
5. Burry RW (2011) Controls for immunocytochemistry: an update. J Histochem Cytochem 59:6–12
6. Skoog L, Tani E (2011) Immunocytochemistry: an indispensable technique in routine cytology. Cytopathology 22:215–229
7. Taylor CR (2011) New revised Clinical and Laboratory Standards Institute Guidelines for Immunohistochemistry and Immunocyto chemistry. Appl Immunohistochem Mol Morp hol 19:289–290

8. Neumeister VM (2014) Tools to assess tissue quality. Clin Biochem 47:280–287
9. Strasser EF, Eckstein R (2010) Optimization of leukocyte collection and monocyte isolation for dendritic cell culture. Transfus Med Rev 24:130–139
10. Neurauter AA, Bonyhadi M, Lien E, Nøkleby L, Ruud E, Camacho S, Aarvak T (2007) Cell isolation and expansion using Dyna beads. Adv Biochem Eng Biotechnol 106: 41–73
11. Celis JE (1994) Cell biology – a laboratory handbook. Academic Prsss, San Diego
12. Pollard JW (1997) Basic cell culture. Methods Mol Biol 75:1–11
13. Exbrayat J-M (2013) Histochemical and cytochemical methods of visualization. CRC Press, London
14. Miyashita T (2004) Confocal microscopy for intracellular co-localization of proteins. Methods Mol Biol 261:399–410
15. Diaspro A, Chirico G, Collini M (2005) Two-photon fluorescence excitation and related techniques in biological microscopy. Q Rev Biophys 38:97–166
16. Houen G, Jakobsen MH, Svaerke C, Koch C, Barkholt V (1997) Conjugation to preadsorbed preactivated proteins and efficient generation of anti peptide antibodies. J Immunol Methods 206:125–134

Chapter 28

Designing B-Cell Epitopes for Immunotherapy and Subunit Vaccines

Harinder Singh, Sudheer Gupta, Ankur Gautam, and Gajendra P.S. Raghava

Abstract

Rationally designed subunit vaccines mainly consist of small peptides or B-cell epitopes, which can stimulate the body's immune response. Development of subunit vaccines is a very tedious and costly process. One of the imperative and crucial steps of vaccine development is the identification of highly competent B-cell epitopes as most of the proteins and fragments of proteins are immunologically irrelevant. With the advances in bioinformatics tools, it can be possible to precisely narrow down potential B-cell epitopes from the whole proteome of any pathogen. This chapter sheds light on prediction and designing of B-cell epitopes using two in silico tools LBtope and IgPred.

Key words B-cell epitopes, Bioinformatics, In silico, LBtope, IgPred, Vaccine

1 Introduction

Traditional vaccine formulations consist of inactivated or killed pathogens. Once injected, the body's immune system recognizes these inactivated pathogens and develops an immune response against them. These traditional vaccine formulations have shown effectiveness against some types of infectious diseases. However, many times these vaccine formulations are not very safe and have shown some unwanted adverse effects like development of inflammation and autoimmune response [1]. Nowadays, scientific attention has been focused to develop novel subunit vaccines [2], which consist of smaller noninfectious fragments of pathogens instead of the whole organism [3]. Subunit vaccines are much safer and more efficient than the traditional vaccine formulations. In addition, multiple epitopes deriving from different antigens can be used in a single vaccine.

Gunnar Houen (ed.), *Peptide Antibodies: Methods and Protocols*, Methods in Molecular Biology, vol. 1348, DOI 10.1007/978-1-4939-2999-3_28,

With the advances in peptide synthesis technologies and comparatively low production cost, peptides have become the preferred choice for subunit vaccine candidates [4, 5]. The success of a subunit vaccine highly depends on the immunogenicity of the fragments that are part of the formulations. Most of the proteins and fragments of proteins are immunologically irrelevant. Therefore, identifying the minimum fragments of these proteins that can generate a strong immune response is paramount and a crucial task of vaccine development.

In an era of next-generation sequencing, a considerable amount of genomes from different pathogens have been sequenced [6], and proteomics data of these pathogens are now available. With the advances in bioinformatics, it can be possible to identify all such peptide fragments which are immunologically relevant. Therefore, over the last few years, much attention has been paid to develop bioinformatic tools [7–13], which can predict the immunogenicity of peptides. In this chapter, we discuss two B-cell epitope prediction in silico tools LBtope [9] and IgPred [10], which have been developed recently. LBtope is useful to predict B-cell epitopes in a protein sequence and also help in rationally designing of B-cell epitopes. IgPred allows prediction of antibody class-specific B-cell epitopes.

2 Prediction of B-Cell Epitopes Using LBtope

LBtope allows accurate prediction of linear B-cell epitopes in a protein sequence. It is the first method developed using both positive and negative B-cell epitopes and allows prediction in variable length peptide sequence. The web server consists of three different modules for prediction of B-cell epitope in protein, peptides sequence and mutants of a given peptide sequence. The result page provides a graphical visualization of the probable regions having B-cell epitopes.

2.1 Antigen Sequence Module

The antigen sequence module allows prediction of B-cell epitopes in multiple proteins using five different models. These five models can be classified into two categories: fixed length and variable length of epitopes. The default is LBtope_Variable model, consisting of information of thousands of experimentally positive and negative B-cell epitopes. The LBtope_Variable_non_redundant model is based on 80 % non-redundant data of LBtope_Variable. The LBtope_confirm model is based on the epitopes that are experimentally validated by two or more studies. The result page shows the query sequence in a graphical box, with color code for identification of strong B-cell epitopic region. The amino acids in black color have 0–20 %, orange color has 21–40 %, sky blue color has 41–60 %, green color has 61–80 %, and red color has 81–100 % probability to form B-cell epitope. A B-cell epitope consists of a linear stretch of residues. Thus, an area having a stretch of residues in red color has the highest probability to form B-cell epitope. A single residue colored in the red region and had neighboring residues in black color is a false prediction. Any region in the sequence having continuous red-colored amino acids has the highest probability to form B-cell epitope as shown in Fig. 1. Users can also download the result in text format using the Download link. A pop-up will show when the

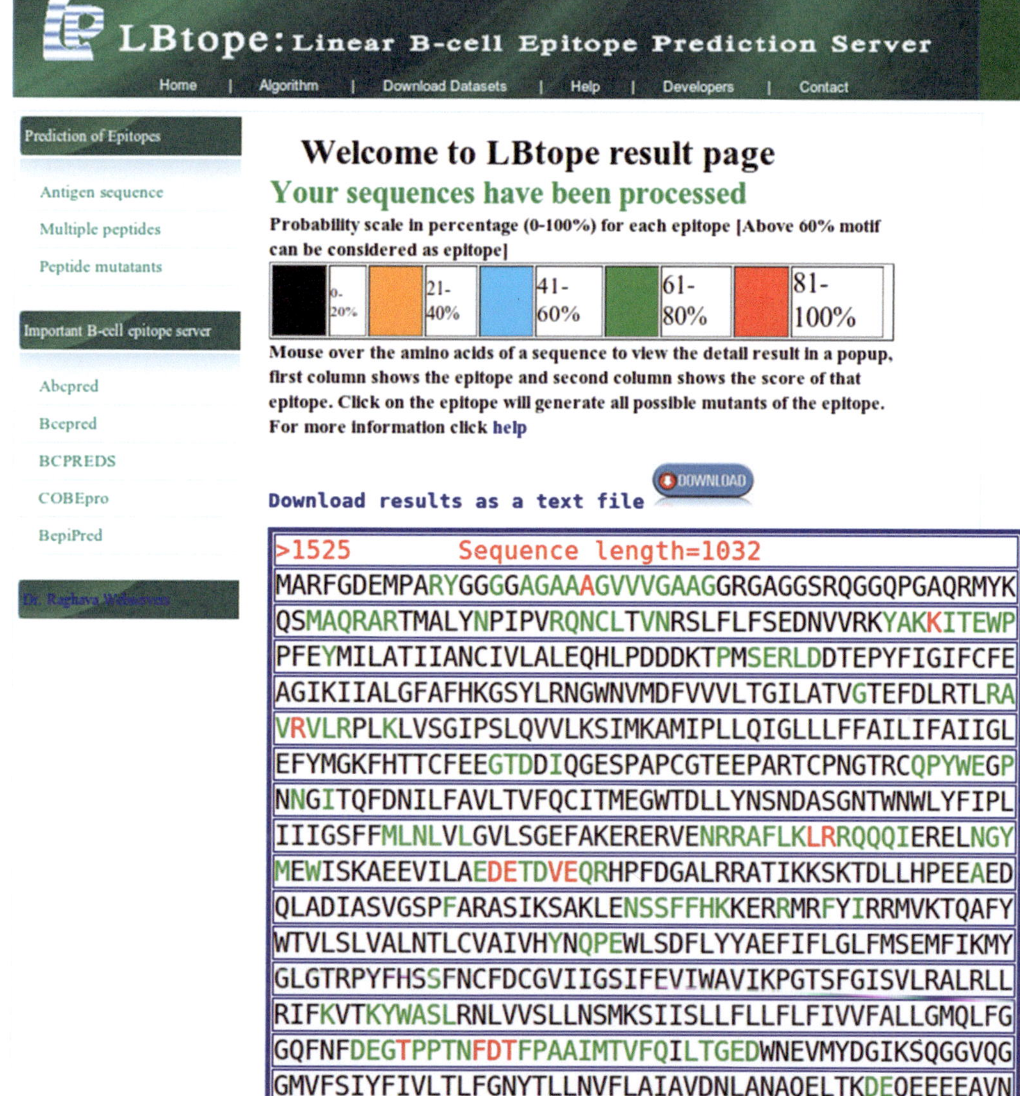

Fig. 1 The output page of antigen sequence module, displaying the probable epitopes in different color coding

user mouse over an amino acid. It displays the epitope having the selected amino acid as the center amino acid of the epitope, along with the probability score of epitope and the position of the selected amino acid in the protein sequence. This module of LBtope allows users to increase any region of protein epitopic efficiency. Clicking on the epitope sequence in pop-up will generate all the possible mutants of the epitope and display the B-cell probability score of each mutant as shown in Fig. 2. A careful analysis of the result page allows users to understand which mutation increased or decreased the B-cell probability score.

2.2 Peptide Mutant Module

The third and last module of LBtope is peptide mutant, which makes all possible mutants of any given peptide and predicts the potential B-cell epitopes. The module first generates all possible mutants, based upon point mutation in peptide. Next, it predicts the B-cell epitope probability score of each mutant along with

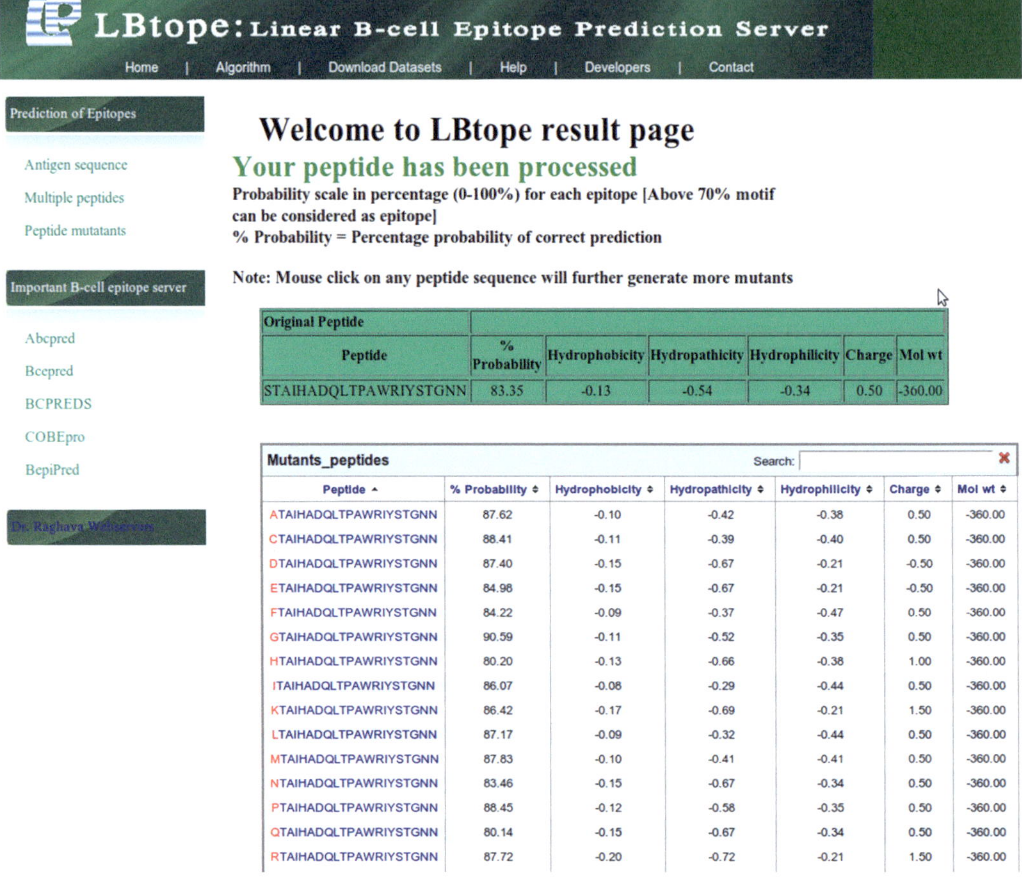

Original Peptide						
Peptide	% Probability	Hydrophobicity	Hydropathicity	Hydrophilicity	Charge	Mol wt
STAIHADQLTPAWRIYSTGNN	83.35	-0.13	-0.54	-0.34	0.50	-360.00

Mutants_peptides

Peptide	% Probability	Hydrophobicity	Hydropathicity	Hydrophilicity	Charge	Mol wt
ATAIHADQLTPAWRIYSTGNN	87.62	-0.10	-0.42	-0.38	0.50	-360.00
CTAIHADQLTPAWRIYSTGNN	88.41	-0.11	-0.39	-0.40	0.50	-360.00
DTAIHADQLTPAWRIYSTGNN	87.40	-0.15	-0.67	-0.21	-0.50	-360.00
ETAIHADQLTPAWRIYSTGNN	84.98	-0.15	-0.67	-0.21	-0.50	-360.00
FTAIHADQLTPAWRIYSTGNN	84.22	-0.09	-0.37	-0.47	0.50	-360.00
GTAIHADQLTPAWRIYSTGNN	90.59	-0.11	-0.52	-0.35	0.50	-360.00
HTAIHADQLTPAWRIYSTGNN	80.20	-0.13	-0.66	-0.38	1.00	-360.00
ITAIHADQLTPAWRIYSTGNN	86.07	-0.08	-0.29	-0.44	0.50	-360.00
KTAIHADQLTPAWRIYSTGNN	86.42	-0.17	-0.69	-0.21	1.50	-360.00
LTAIHADQLTPAWRIYSTGNN	87.17	-0.09	-0.32	-0.44	0.50	-360.00
MTAIHADQLTPAWRIYSTGNN	87.83	-0.10	-0.41	-0.41	0.50	-360.00
NTAIHADQLTPAWRIYSTGNN	83.46	-0.15	-0.67	-0.34	0.50	-360.00
PTAIHADQLTPAWRIYSTGNN	88.45	-0.12	-0.58	-0.35	0.50	-360.00
QTAIHADQLTPAWRIYSTGNN	80.14	-0.15	-0.67	-0.34	0.50	-360.00
RTAIHADQLTPAWRIYSTGNN	87.72	-0.20	-0.72	-0.21	1.50	-360.00

Fig. 2 The mutant module displays the mutants of the given peptide along with a probability score and other properties

hydrophobicity, hydropathicity, hydrophilicity, charge, and molecular weight. The result page also displays the B-cell epitope probability score of the original peptide as reference. The mutant result page consists of a sortable table, with a search bar for searching the best B-cell epitope mutant.

As shown in Fig. 2, the module generates all possible mutants of peptide STAIHADQLTPAWRIYSTGNN; the mutated residues are displayed in red color. A point mutation in the first residue serine (S) into glycine (G) increases the B-cell epitope probability score from 83.35 to 90.59 % (Fig. 2). The table is sortable; clicking on the % probability will sort the table with highest scoring mutant at the top. The top 20 mutants having the highest probability to form B-cell epitope are shown in Fig. 3. A point mutation in second residue into glutamate (SEAIHADQLTPAWRIYSTGNN) increases the probability score from 85.35 to 95.70 %. Similarly, the table can also be sorted based upon other properties of the peptide. The mutant module helps the user in predicting the possible location and amino acid to increase the epitopic efficiency of any peptide.

Mutants_peptides Search:

Peptide	% Probability	Hydrophobicity	Hydropathicity	Hydrophilicity	Charge	Mol wt
SEAIHADQLTPAWRIYSTGNN	95.70	-0.15	-0.68	-0.18	-0.50	-360.00
STAIHADQLTPAWRIYSYGNN	95.03	-0.12	-0.57	-0.43	0.50	-360.00
STAIHADQLTPAWRIYSTNNN	92.84	-0.17	-0.69	-0.33	0.50	-360.00
STAIHADQLTPAWRIYSTGTN	92.14	-0.11	-0.41	-0.37	0.50	-360.00
STAVHADQLTPAWRIYSTGNN	92.04	-0.14	-0.56	-0.32	0.50	-360.00
STSIHADQLTPAWRIYSTGNN	91.79	-0.15	-0.67	-0.30	0.50	-360.00
STAIHADQLTPAWRIYSEGNN	91.33	-0.15	-0.68	-0.18	-0.50	-360.00
SIAIHADQLTPAWRIYSTGNN	90.60	-0.08	-0.30	-0.40	0.50	-360.00
GTAIHADQLTPAWRIYSTGNN	90.59	-0.11	-0.52	-0.35	0.50	-360.00
SYAIHADQLTPAWRIYSTGNN	90.46	-0.12	-0.57	-0.43	0.50	-360.00
STAIHADQLTPAWRIYSTTNN	89.90	-0.14	-0.56	-0.36	0.50	-360.00
STAWHADQLTPAWRIYSTGNN	88.59	-0.15	-0.80	-0.41	0.50	-360.00
PTAIHADQLTPAWRIYSTGNN	88.45	-0.12	-0.58	-0.35	0.50	-300.00
STAYHADQLTPAWRIYSTGNN	88.44	-0.16	-0.82	-0.36	0.50	-360.00
CTAIHADQLTPAWRIYSTGNN	88.41	-0.11	-0.39	-0.40	0.50	-360.00
STAIHADQLTPAWRIYSMGNN	88.35	-0.11	-0.42	-0.38	0.50	-360.00
STAIHADQLTPAWRIYSTGNQ	88.18	-0.13	-0.54	-0.34	0.50	-360.00
STAFHADQLTPAWRIYSTGNN	88.14	-0.13	-0.62	-0.37	0.50	-360.00
STLIHADQLTPAWRIYSTGNN	88.09	-0.11	-0.45	-0.40	0.50	-360.00
SAAIHADQLTPAWRIYSTGNN	87.87	-0.11	-0.42	-0.34	0.50	-360.00
MTAIHADQLTPAWRIYSTGNN	87.83	-0.10	-0.41	-0.41	0.50	-360.00

Fig. 3 Top 20 mutants having the highest probability to form a B-cell epitope as compared to the original peptide

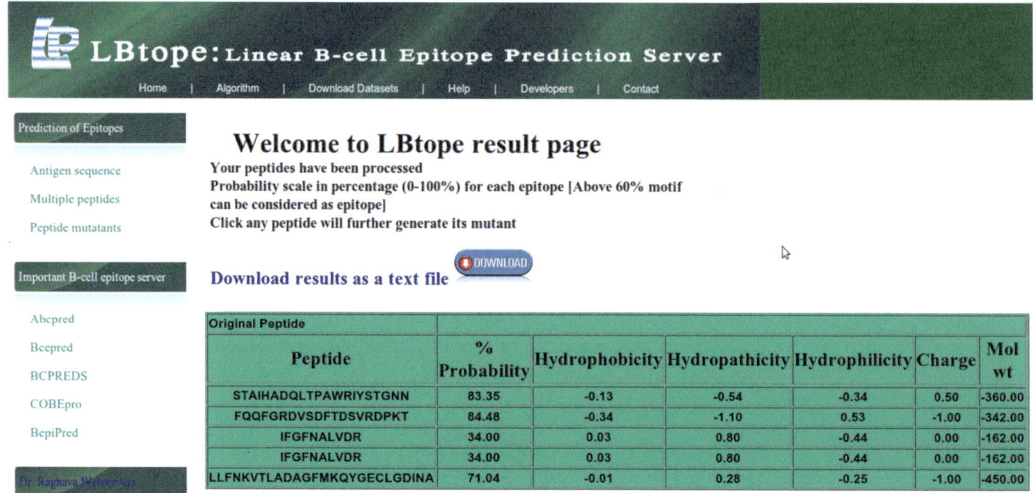

Fig. 4 Result page is displaying the B-cell probability score of multiple peptides along with various other properties

2.3 Multiple Peptide Module

The second module of LBtope allows prediction of B-cell epitopes in a given list of peptides. This module allows prediction of variable length of multiple peptides. The result page displays every peptide with the probability score to form B-cell epitope. Peptides having probability score above >60 % can be considered as potential epitopes. Along with a probability score, other important physico-chemical properties of peptides like hydrophobicity, hydropathicity, hydrophilicity, charge, and molecular weight are also displayed (Fig. 4). The result can also be downloaded in text format using the Download link.

3 Antibody Class-Specific B Cell Epitope Prediction

Until now, we have discussed the regions in an antigen, which can bind with the antibody or BCEs. Apart from just binding to an antibody, it has also now become imperative to know the class of antibody with which the BCE binds. There are different effector functions which Abs can perform for example; neutralization, cell mediated cytotoxicity [14], phagocytosis [15], complement activation [16], and mast cell binding [17]. Based on the effector functions the antibodies are classified into different classes like IgG, IgE, IgA, IgD, and IgM. The IgM and IgG class of Abs are mostly involved in the systemic immunity, IgA participates in the mucosal immunity, and the IgE class is responsible for the allergic and inflammatory reaction in the body.

To date, the designing of subunit vaccines required the induction of B cells, T cells, and other antigen-presenting cells and different in silico tools have served well for this purpose.

Fig. 5 Home page of IgPred web server having different modules for analysis and prediction of class-specific BCEs

However, the appropriate effector function or class of Abs secreted by activated B cells could not be inspected till date, by any computational method except for IgE. Here we describe a Web-based in silico tool "IgPred," which predicts the tendencies of antigenic regions to induce a particular class of antibody. The web server (http://crdd.osdd.net/raghava/igpred/) is composed of different modules/sections, which enable the user to analyze and predict Ab class-specific BCEs (Fig. 5). The modules are developed for a different function as described below:

3.1 Prediction of Epitopes/Peptides to Envisage Its Ability to Induce Ab Class-Specific BCEs

In the direction of subunit vaccine design, if we have to envisage the Ab response of a particular class, for the epitope/peptide in question, we can use this module. In this module, the user can submit multiple query sequences. The module has two submodules, which use different type of input sequences; for example, the variable length submodule can be selected if the length of queries vary from 4 to 50, or else a fixed length submodule can be preferred where only 20-mer peptides/epitopes can be submitted. The thresholds can be adjusted as per stringency of prediction. The queries can simultaneously be predicted for their ability to

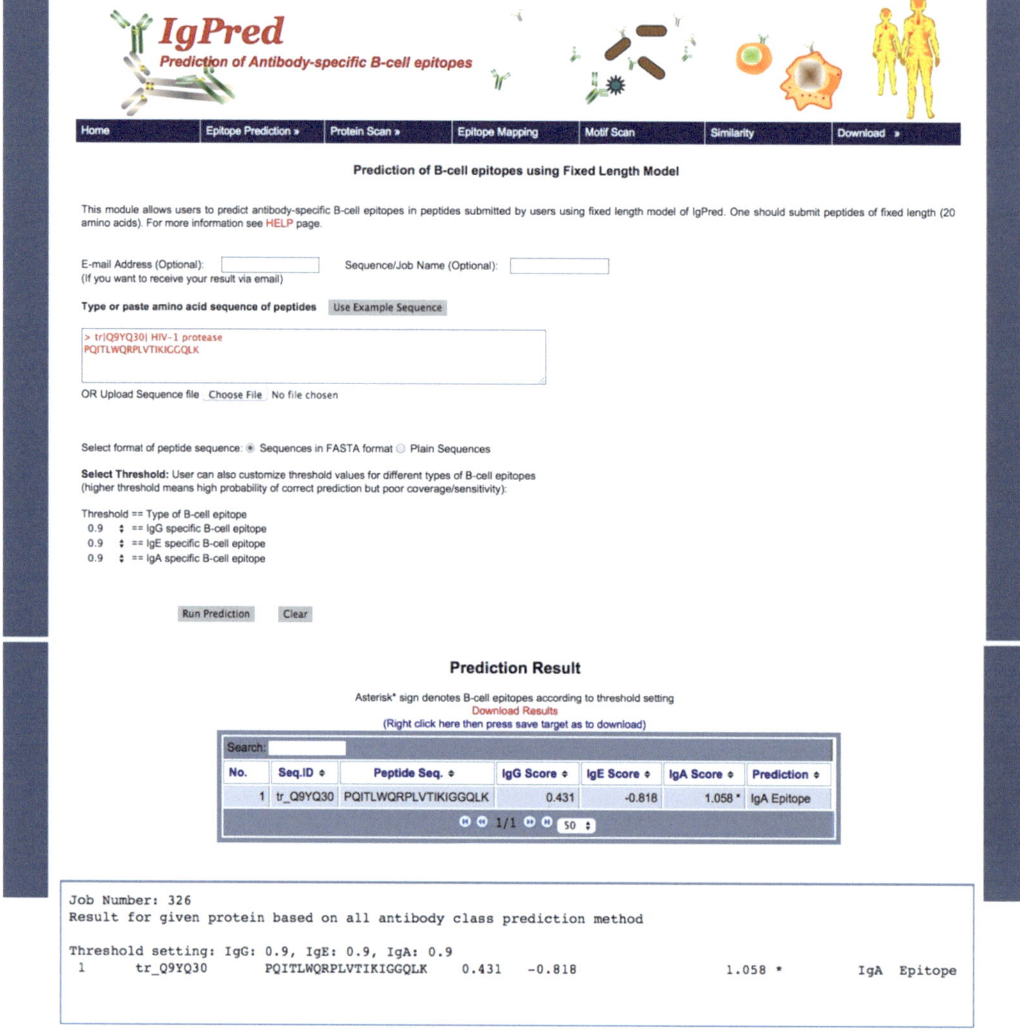

Fig. 6 Epitope prediction module of IgPred with fixed-length submodule in selection

induce all classes of antibodies and the SVM score for each class will be displayed. A higher score shows the higher chances for the induction of a particular class of Ab. For example in Fig. 6, if we submit a peptide fragment of HIV-I protease to understand if it has Ab class-specific property. The results show that this fragment is an IgA inducing BCE, and it also has chances of inducing IgG, but there is no allergenic property by generating IgE Abs. The result tables are downloadable in the form of a text file in Download link.

3.2 Identification of Class-Specific Ab-Inducing Regions in an Antigen

Apart from the vaccine candidate epitopes/peptides, if one has to identify antigenic regions in a full-length protein, this module can be very useful. Here the full-length protein sequence is scanned for the presence of Ab-class-specific BCEs. The threshold and window size can be selected as per stringency and size of epitopes respectively. The module can take multiple sequences as input, and the output comes in the form of graphical presentation where query sequences are aligned with predicted sequences having Ab-class specific BCEs as immunogenic regions. Each BCE starts with a red residue in the sequence and continues as blue. For example, in Figs. 7 and 8 we have shown submission of VP40 protein of Zaire Ebola virus and CH1 allergen of *Felis catus*, in variable and fixed length submodules. The submission results in the immunogenic regions in VP40 protein, which might induce IgA along with IgG but not IgE. Similarly, the CH1 protein has immunogenic regions after the 40th residue, which may be responsible for its allergenicity.

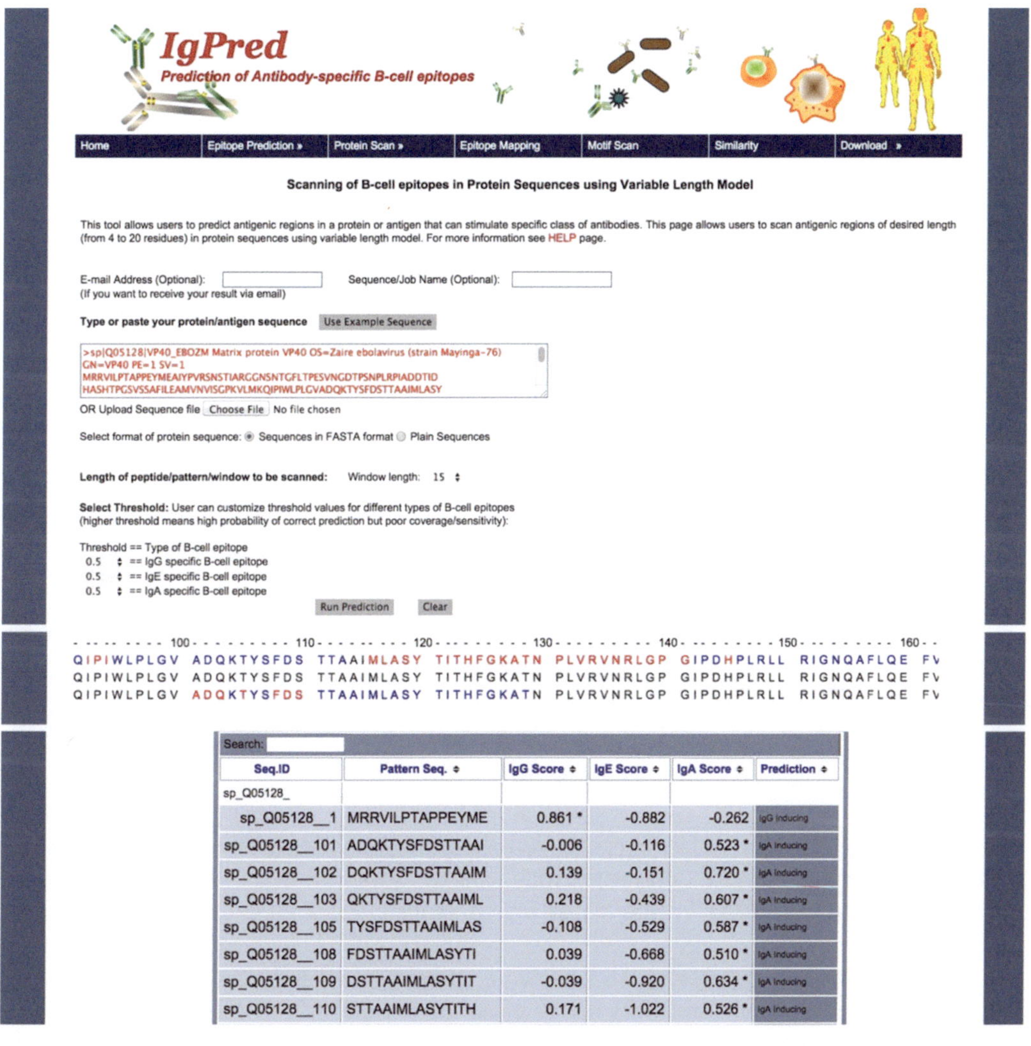

Fig. 7 Protein Scan module of IgPred showing scanning with variable length submodule

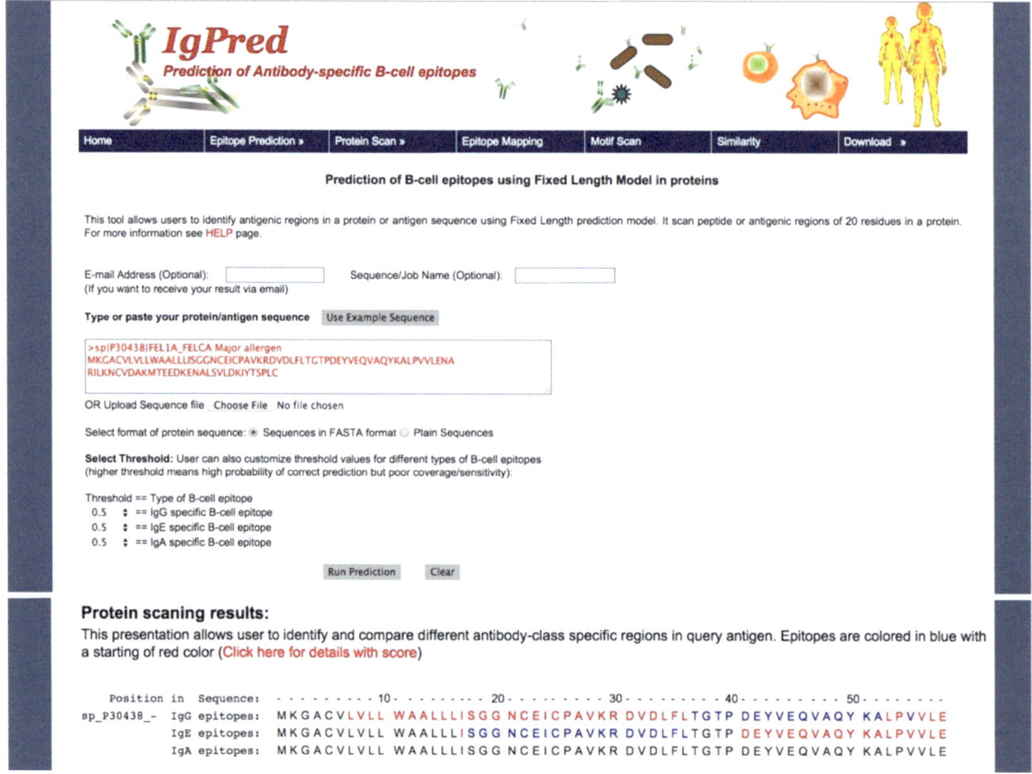

Fig. 8 Protein Scan module of IgPred showing scanning with fixed length submodule

3.3 Mapping of Experimentally Validated Epitopes in an Antigen

This module is one of the most important modules since here we can search for experimentally validated epitopes, which are present in our query sequence. The query is searched against the Ab-class specific BCEs from IEDB database, and the exactly matching epitopes are presented as aligned sequences. The results can be further explored for immune information by clicking links to IEDB database, which leads to the experimental assays for the epitope. For example, here we have submitted 2S albumin protein from *Ricinus communis* (castor bean). We found several experimental epitopes mapped onto the query sequences (Fig. 9), which means that if these regions are used in any therapeutics, they may cause allergy.

3.4 Scanning for the Presence of Ab Class-Specific Motifs

As it has already been shown in previous studies that the presence of certain motifs may lead to induction of a particular class of Abs [18], it may be of worth, if we check our query sequence for the presence of Ab class-specific motifs. The module is based on the MEME-MAST package where the MEME was applied for discovery of motifs in the class-specific BCEs and MAST is used for the searching of class-specific motifs in the query protein. For example, we submitted GP (Envelop Glycoprotein) of Sudan Ebola virus in

Fig. 9 Epitope mapping module of IgPred showing mapping of experimentally validated BCEs

this module for detection of any IgA-specific motif. This resulted in the graphical presentation of alignment where the query was aligned with an IgA motif with a p-value of 3.21e-04 (Fig. 10). Thus, it might be a region in the GP protein, which may induce IgA Abs.

3.5 Search Similarity Against Experimentally Validated B-Cell Epitopes

Since the "mapping of experimentally validated epitopes" module, as described above, is based on the exact match of the sequence, even a single residue mismatch may exclude the experimental epitope from being mapped onto the query. For similarity search, we have developed a module, which can do a similarity search with the experimentally validated class-specific BCEs, using Smith-Waterman algorithm (SSEARCH). The query is searched for similarity against selected Ab class BCEs, and the result comes in the form of alignment of epitopes with the query (Fig. 11). The aligned epitopes are also provided with links to the IEDB database for further

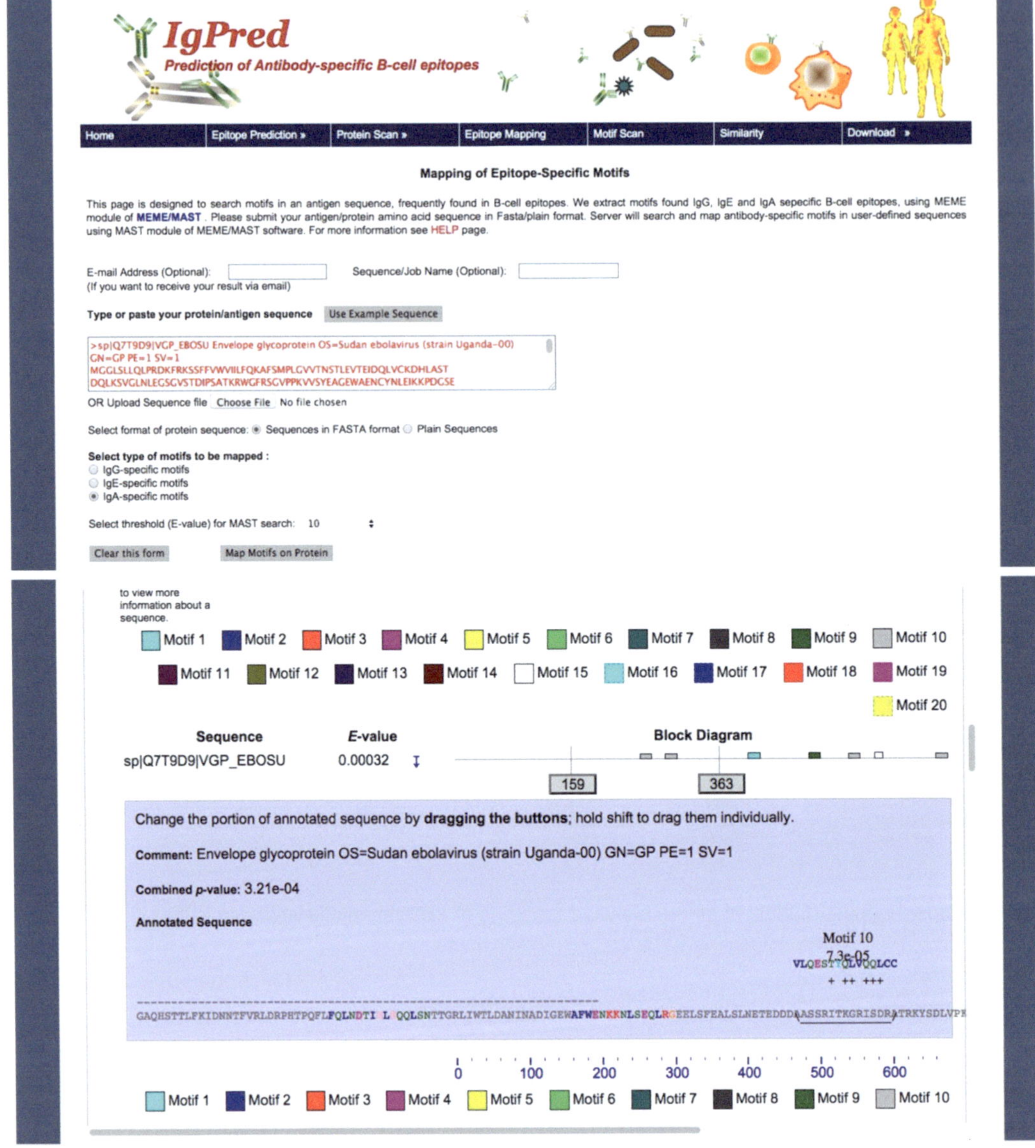

Fig. 10 Motif scan module of IgPred showing search for Ab class-specific motifs

immunological evidence. As an example, when we submit EBNA6 protein of Epstein-Barr virus for similarity search with experimental class-specific BCEs, we get several highly similar BCEs in result.

4 Limitations and Future Prospects

One of the major problems in developing accurate B-cell epitope prediction tools is the limitation of the experimental data, particularly the data for non-BCEs. Earlier methods were developed

Fig. 11 Similarity search module of IgPred showing similarity search of experimentally validated class-specific BCEs in the query sequences

using a small dataset (~1000 epitopes as positive examples). Also, in these methods, random peptides from Swissprot were considered as negative examples. Though this problem was addressed in LBtope, which was based upon thousands of experimentally validated positive and negative epitopes. Still the validity of an epitope to act as positive or negative is controversial. Many epitopes in IEDB are reported both as positive and negative epitopes. There is a need for a highly curated dataset of thousands of experimentally validated positive and negative epitopes and to the development of a better method for prediction of autoimmune disease. The current best methods are based upon machine learning and understanding how an epitope is predicted using machine learning is difficult for biologists. There is a need to develop more statistical methods for prediction of BCEs for better understanding, as well as an accurate prediction of BCEs.

Furthermore, in the case of class-specific BCE prediction, IgPred includes prediction of only the IgG, IgE and IgA inducing epitopes. The epitopic data is very limited for other classes, including IgM, IgD, and subclasses like IgG1, IgG2, IgG3, IgG4, IgA1, and IgA2, which are clinically more relevant. This limitation leads to the scope for the future, where with the availability of data for these classes and subclasses, we can develop prediction models. We anticipate in the future; the class and subclass specific BCE prediction tools will be developed, which will be of great use in the identification and designing of immunotherapeutics against different pathogenic organism and diseases like cancer.

References

1. Black M, Trent A, Tirrell M, Olive C (2010) Advances in the design and delivery of peptide subunit vaccines with a focus on toll-like receptor agonists. Expert Rev Vaccines 9:157–173
2. Purcell AW, McCluskey J, Rossjohn J (2007) More than one reason to rethink the use of peptides in vaccine design. Nat Rev Drug Discov 6:404–414
3. BenMohamed L, Wechsler SL, Nesburn AB (2002) Lipopeptide vaccines – yesterday, today, and tomorrow. Lancet Infect Dis 2:425–431
4. Vlieghe P, Lisowski V, Martinez J, Khrest chatisky M (2010) Synthetic therapeutic peptides: science and market. Drug Discov Today 15:40–56
5. Kaspar AA, Reichert JM (2013) Future directions for peptide therapeutics development. Drug Discov Today 18:807–817
6. Weinstock GM, Peacock SJ (2014) Nextgeneration pathogen genomics. Genome Biol 15:528
7. Dhanda SK, Gupta S, Vir P, Raghava GP (2013) Prediction of IL4 inducing peptides. Clin Dev Immunol 2013:263952
8. Dhanda SK, Vir P, Raghava GP (2013) Designing of interferon-gamma inducing MHC class-II binders. Biol Direct 8:30
9. Singh H, Ansari HR, Raghava GP (2013) Improved method for linear B-cell epitope prediction using antigen's primary sequence. PLoS One 8, e62216
10. Gupta S, Ansari HR, Gautam A, Raghava GP (2013) Identification of B-cell epitopes in an antigen for inducing specific class of antibodies. Biol Direct 8:27
11. Ansari HR, Raghava GP (2010) Identification of conformational B-cell Epitopes in an antigen from its primary sequence. Immunome Res 6:6
12. Hu YJ, Lin SC, Lin YL, Lin KH, You SN (2014) A meta-learning approach for B-cell conformational epitope prediction. BMC Bioin formatics 15:378
13. Krawczyk K, Liu X, Baker T, Shi J, Deane CM (2014) Improving B-cell epitope prediction and its application to global antibody-antigen docking. Bioinformatics 30:2288–2294
14. Roos A, Bouwman LH, van Gijlswijk-Janssen DJ, Faber-Krol MC, Stahl GL et al (2001) Human IgA activates the complement system via the mannan-binding lectin pathway. J Immunol 167:2861–2868
15. Webster SD, Galvan MD, Ferran E, Garzon-Rodriguez W, Glabe CG et al (2001) Antibody-mediated phagocytosis of the amyloid beta-peptide in microglia is differentially modulated by C1q. J Immunol 166:7496–7503
16. Mitchell TJ, Andrew PW, Saunders FK, Smith AN, Boulnois GJ (1991) Complement activation and antibody binding by pneumolysin via a region of the toxin homologous to a human acute-phase protein. Mol Microbiol 5: 1883–1888
17. Galli SJ, Tsai M (2012) IgE and mast cells in allergic disease. Nat Med 18:693–704
18. Jain S, Rosenthal KL (2011) The gp41 epitope, QARVLAVERY, is highly conserved and a potent inducer of IgA that neutralizes HIV-1 and inhibits viral transcytosis. Mucosal Immunol 4:539–553

Chapter 29

Enterovirus-Specific Anti-peptide Antibodies

Chit Laa Poh, Katherine Kirk, Hui Na Chua, and Lara Grollo

Abstract

Enterovirus 71 (EV-71) is the main causative agent of hand, foot, and mouth disease (HFMD) which is generally regarded as a mild childhood disease. In recent years, EV71 has emerged as a significant pathogen capable of causing high mortalities and severe neurological complications in large outbreaks in Asia. A formalin-inactivated EV71 whole virus vaccine has completed phase III trial in China but is currently unavailable clinically. The high cost of manufacturing and supply problems may limit practical implementations in developing countries. Synthetic peptides representing the native primary structure of the viral immunogen which is able to elicit neutralizing antibodies can be made readily and is cost effective. However, it is necessary to conjugate short synthetic peptides to carrier proteins to enhance their immunogenicity. This review describes the production of cross-neutralizing anti-peptide antibodies in response to immunization with synthetic peptides selected from in silico analysis, generation of B-cell epitopes of EV71 conjugated to a promiscuous T-cell epitope from Poliovirus, and evaluation of the neutralizing activities of the anti-peptide antibodies. Besides neutralizing EV71 in vitro, the neutralizing antibodies were cross-reactive against several Enteroviruses including CVA16, CVB4, CVB6, and ECHO13.

Key words Enterovirus 71, Enterovirus-specific Anti-peptide antibodies, Cross-reactive neutralizing antibody epitopes, In silico analysis of B-cell epitopes

1 Introduction

As the number of polio infections decreases, Asia is experiencing an increasing number of epidemics caused by Enteroviruses (EVs) such as Enterovirus 71(EV71) and Coxsackie virus 16(CA16). Hand, foot, and mouth disease (HFMD) infections are generally mild and are endemic in countries like China, Singapore, Taiwan and Malaysia. However, in recent years, EV71 has caused severe HFMD with associated neurological complications such as acute flaccid paralysis and brain stem encephalitis, leading to hundreds of deaths in several countries in Asia. Large HFMD outbreaks involving over 1.3 million children in China in 2013 were associated with high fatalities of 243 cases [1]. There are currently no vaccine or antiviral agent to prevent or treat serious HFMD caused by EV71. The whole virus formalin-inactivated vaccine candidate has

Gunnar Houen (ed.), *Peptide Antibodies: Methods and Protocols*, Methods in Molecular Biology, vol. 1348,
DOI 10.1007/978-1-4939-2999-3_29, © Springer Science+Business Media New York 2015

gone through phase III clinical trial in 10,000 healthy children in China and appears to be promising [2]. However, practical implementation of the formalin-inactivated EV71 vaccine for clinical application will require a carefully regulated manufacturing process and the establishment of good global vaccine standards as different companies in different countries will use different vaccine strains, different cell substrates and different production processes. Although the immunogenicity of the formalin-inactivated EV71 vaccine is good, it is not long lasting and needs at least a booster dose after 6 months to remain effective. Due to antigenic variability of RNA viruses, there is also the added problem of vaccine coverage of different EV71 genotype/subgenotype strains which requires further investigations. Further difficulties involved the necessity to grow large amounts of pathogenic viruses for inactivation and the possibility of inducing inappropriate allergic and/or reactogenic responses in the host [3].

An alternative immunization approach is to identify peptide epitopes that can elicit the required immune response and to use the synthetic versions of the peptides as vaccines. Synthetic peptide vaccines are cost effective and do not carry the risk of reversion. Unlike the inactivated vaccine, it does not have a safety risk of incomplete inactivation or has components that can contribute to unwanted side effects. The use of synthetic peptide vaccines is desirable in situations when the natural protein antigen is unavailable or is difficult to prepare in large quantities. Synthetic peptide vaccines can be prepared to high purity and stored freeze-dried. This avoids the "cold-chain" requirement for storage, transport, and distribution. Rational design of synthetic peptide vaccines to include multiple epitopes from the same virus or multiple determinants from several viruses will greatly expand the usefulness of the vaccine. For example, an inactivated vaccine may not have broad protection against all the serotypes, genotypes/subgenotypes and there is a need to include several serotypes, genotypes/subgenotypes in the formulation. The inactivated Poliovirus vaccine had to include all three serotypes—*viz.* Sabin 1, 2, and 3. In several HFMD epidemics, the less virulent CA 16 virus was found to be co-circulating with EV71. A bivalent synthetic peptide vaccine can be designed to include virus-neutralizing epitopes from both EV71 and CA16 viruses. Although promising, synthetic peptide vaccines have been shown to be less immunogenic than the traditional vaccines. The challenge is to identify a synthetic peptide vaccine that can optimally stimulate both B-cell and T-cell immune responses. The added limitation to the use of a synthetic peptide vaccine is that it is mainly restricted to representing linear epitopes. If conformational B-cell epitopes are required to elicit neutralizing antibodies, peptides representing the B-cell epitopes could be assembled on a suitable backbone which represents the three-dimensional structure [4]. Despite the shortcomings of low immunogenicity, several synthetic peptide vaccines are under development against human immunodeficiency virus (HIV1), hepatitis C virus (HCV), malaria, influenza, cytomegalovirus (CMV), and human papilloma virus (HPV) [5]. Most of the candidate peptide vaccines under development against infectious pathogens are currently in phase I and II clinical studies but some peptide vaccines against cancer have reached phase III studies. However, no human peptide-based vaccine has reached the market but strategies to improve the immunogenicity, stability, and delivery will enable some peptide vaccines to enter the human therapeutics market in the near future.

1.1 Design of Synthetic Peptide Vaccines

The development of an effective peptide vaccine will involve the identification of the immunodominant epitopes which are capable of inducing both humoral and cell-mediated immunity against the viral pathogen. Synthetic peptide vaccines can be designed to target a humoral response which is mediated by specific neutralizing antibodies or a cytotoxic immune response mediated by cytotoxic T lymphocytes ($CD8^+$ T cells) or a combination of both. Both arms of immunity are further dependent on the induction of a helper T cell response. Vaccines that induce antibody formation should contain the B-cell epitope and the T-helper epitope whereas vaccines that are designed to generate the cytotoxic response should carry the T-cell epitope and the T-helper epitope. Thus, a synthetic peptide vaccine should contain at least two antigenic epitopes, a T-helper epitope and the B-cell or the T-cell epitope [6].

This chapter focuses on the design of B-cell epitope peptide vaccines, discusses their ability to elicit antipeptide antibodies that will neutralize Enteroviruses, and confers immune protection against invading viral pathogens. Peptides chosen as immunogens should contain at least eight and not more than 20 amino acids. Peptides smaller than eight amino acids may elicit antibodies that do not recognize the native protein and peptides longer than 20 amino acids may fold incorrectly and do not represent the conformation of the native protein. The preferred length of synthetic peptides as immunogens should be ranging from 15 to 20 amino acids, targeting the externally exposed regions such as turns, connecting regions, or loops of the native protein [7]. These regions are often enriched with charged and polar amino acids [8]. Information of the native protein structure can be accessed by searching the protein NCBI database (www.ncbi.nlm.nih.gov) Uni-Prot (www.uniprot.org) and ExPaSy (us.expasy.org/tools). B-cell epitope identification can be achieved through experimental methods which are divided into either structural or functional. Structural methods include X-ray crystallography, nucleic magnetic resonance (NMR), and electron microscopy (EM) of the antigen-antibody complexes. Functional methods utilize methods such as surface plasmon resonance, mass spectrometry, as well as immunoassays [9, 10]. B-cell epitopes can be discovered through analysis of the antigen-binding domains of an antibody that are resistant to proteolysis. Thus, an epitope can be generated whilst bound to the antibody. The epitope can be eluted from the bound antibody under non-denaturing conditions. This allows identification of linear as well as conformational epitopes [11, 12]. The antibodies used for identification of the B-cell epitope can be derived from the sera of infected patients [13].

1.2 PEPSCAN for Epitope Discovery

The "PEPSCAN" method is a fast and systematic approach for assessing whether an amino acid sequence will bind to an existing antibody. Synthetic peptides are synthesized on a solid support, usually as overlapping peptides of 9–12 mers covering the whole sequence of a given protein. The peptides are covalently bound to the solid support and the antibody binding properties are identified using antibody marked with peroxidase in an ELISA. Immunization using a recombinant VP1 protein of EV71 was shown to confer protection against lethal EV71 infection in newborn mice, indicating that VP1 contains important antigenic sites or B-cell epitopes that could elicit production of neutralizing antibodies against the virus [14]. Neutralizing synthetic peptides can be identified using either polyclonal or monoclonal antibodies. Polyclonal antisera raised against the foot and mouth disease virus (FMDV) were the first sera used to identify peptides present in the VP1 to bind to anti-peptide antibodies in the PEPSCAN analysis [15]. Short peptides are poor immunogens and need to be conjugated to carrier proteins such as keyhole limpet hemocyanin (KLH) or tetanus toxoid which provide a source of T_H epitopes. Foo et al. [16] used a PEPSCAN strategy in which 95 overlapping synthetic peptides were designed according to the primary sequence of the VP1 capsid protein of EV71. The diphtheria toxoid-conjugated synthetic peptides were injected into mice and the neutralizing activity of the anti-peptide antibodies were determined. One of the synthetic peptides, SP70, was able to elicits neutralizing titer which was only twofold lower than that elicited by a heat-inactivated whole virus [16]. When the antiserum raised against the SP70 peptide was passively administered to newborn mice, the anti-peptide antibodies were able to neutralize viruses injected into mice at a challenge dose of 1000 $TCID_{50}$ and conferred 80 % in vivo protection of EV71 sub-genogroups B2, B4, and B5. Passive protection of heterologous sub-genogroups belonging to C2 and C4 was lower at 70 % [17].

1.3 B-Cell Epitope Prediction Methods

In the absence of a native protein, computational design utilizing reliable in silico bioinformatics tools have led to B-cell epitope predictions. Most of the existing methods of computational B-cell epitope prediction does not consider the conformational structure but depend on a given protein sequence as a continuous amino acid stretch or a linear sequence. Hopp and Woods introduced the first propensity scale method (or amino acid scale based) for predicting linear B-cell epitopes which was dependent on the chemical and physical properties of amino acids [18]. Improved propensity methods such as BepiPred which combines two propensity-scale methods with a hidden Markov model (HMM) was shown to have a statistically significant improvement in performance [19]. Machine learning methods such as ABCPred which uses recurrent artificial neural networks combined with flexible length classifiers for predicting linear B-cell epitopes were proposed by Saha and Raghava [20]. Their program was shown to achieve 66 % accuracy using a window size of 16 amino acids. In an attempt to further improve the accuracy of prediction, a support vector machine (SVM) algorithm was developed and combined with the propensity scale method. This approach was able to attain an accuracy of 72 % on a dataset of 1211 B-cell epitopes [21]. Lin et al. (2013) described the BEEPro (B-cell epitope prediction using Evolutionary information and Propensity scales), a SVM-based learning machine which uses 16 properties to predict both linear and conformational B-cell epitopes. BEEPro achieves an accuracy of 99.29 % with a sensitivity of 0.9604, a specificity of 0.9946, and a correlation coefficient of 0.9281 [22].

In silico approaches for predicting conformational epitopes can be based on the sequence or the structure or both. There are now a few structure-based epitope prediction servers that are widely used and have been validated experimentally to varying degrees. Commonly used servers include the conformational epitope predictor (CEP) which predicts surface-accessible epitopes based on the atomic positional distance between amino acids [23] and the DiscoTope developed by Andersen et al. [24] which uses a combination of amino acid statistics, spatial arrangement and surface accessibility to predict conformational B-cell epitopes. DiscoTope can be used to predict both linear and conformational epitopes. Newer algorithms that try to improve analysis and broaden targets using linear sequences when structures are unavailable include the ElliPro which can align unknown sequences in BLAST and then model the structure with MODELLER [25]. Rubinstein et al. [26] introduced Epitopia for predicting B-cell epitopes in either a three-dimensional structure or a linear sequence which are immunogenic. Epitopia server predicts epitopes based on the physic-chemical and structural geometrical features. The immunogenicity and corresponding probability scores are computed for every amino acid for a 3D structure input or for every amino acid for a sequence input. Performance of the Epitopia as a B-cell prediction tool is judged higher than the CEP, DiscoTope and ElliPro using the same data and the same assessment [26].

1.4 Immunogenicity of Vaccine Constructs

Several excellent reviews on peptide synthesis and conjugation methods, production and characterization of antipeptide antibodies are available for referencing [27–29]. This chapter describes a multi-step algorithmic approach that utilizes both a sequence (ABCPred, BepiPred) and a structure-based Epitopia server to identify a functional peptide epitope for developing the synthetic peptide vaccine. An in silico computational analysis of the sequence and structure models was used to identify highly conserved B-cell epitopes that will elicit cross protection against all EV71 strains and other Enterovirus family members. The combined in silico approach identified three highly conserved 15-mer epitopes across multiple EV subtypes. Epitopes were discounted if their normal position lay buried within the viral coat rather than being presented on the surface. The K1 epitope was structurally more conserved with no more than 6 amino acid substitutions when compared to the VP1 sequence of other virus strains. K2 and K3 (from the VP3 protein) had less sequence conservation with other viruses and represented less exposed, more buried regions. A neutralizing peptide epitope, designated as D, was included for comparison. Other epitopes were identified from VP2 and VP4 but were not as promising as K1, K2, and K3 in terms of their conservation and antigenicity; hence, they were not evaluated as potential vaccine targets.

Following the in silico identification of several potential peptide vaccine candidates, their ability to induce production of neutralizing anti-peptide antibodies in mice was assessed. Each peptide was synthesized as a linear construct containing a universal T helper epitope from Poliovirus by standard F-moc chemistry. These peptide constructs elicited low levels of anti-peptide antibody response after the first vaccination but 6 weeks after the second dose, antipeptide antibodies against D1 and K3 were higher than those elicited by K1 and K2 peptides [30].

2 Materials

2.1 Cell Growth and Maintenance

Dulbecco's minimal essential medium containing glutamine (DMEM) supplemented with 10 % fetal calf serum and 1 % of penicillin/streptomycin (50 μg/ml).

Vero cells (African green monkey kidney cells, ATCC: CCL-81).

2.2 Synthetic Peptides

Four peptides to be evaluated as B-cell epitopes were commercially synthesized using Fmoc-solid phase peptide synthesis. Each B-cell epitope was synthesized as a linear construct containing a universal T helper epitope from Polio virus (KLFAVWKITYKDT).

2.3 Viruses

Human Enterovirus 71(EV71 isolate number CAIG 99018233), Coxsackie A16 (CVA16: isolate number CAIG 9902-2745

-4PMEK9.4.00), Coxsackie B4 (CVB4: isolate number 99039838), Coxsackie B6 (CVB6: isolate number 273370/PBO (14.09.1989), and Echovirus 13(Echo13: isolate number 28-606PMEK (07.06.1990).

2.4 Mice Immunization

BALB/c mice (6–8 weeks old).

Complete Freund's adjuvant (CFA).

Incomplete Freund's adjuvant (IFA).

2.5 Determination of Anti-peptide Antibody Titer by ELISA

1. Sodium carbonate buffer (50 mM, pH 9.6).
2. Synthetic peptide (10 μg/ml of unconjugated synthetic peptide in carbonate buffer).
3. Phosphate-buffered saline (PBS) (20 mM Na_2HPO_4 in 0.15 M NaCl, pH 7.3).
4. Phosphate-buffered saline (PBS) containing 0.05%Tween-20 and 0.5 % BSA (PBST).
5. Blocking solution: 10 mg/ml BSA in PBS.
6. Secondary antibody: Rabbit anti-mouse IgG conjugated to horseradish peroxidase.
7. Enzyme substrate: 3,3′,5,5′-Tetramethylbenzidine (TMB).
8. i-MarkMicrotiter plate reader.

2.6 Purification of Anti-peptide Antibody

Affinity chromatography columns used were Protein A-Sepharose Fast Flow columns (10 mm × 85 mm).

2.7 Neutralization Assay

96-Well plates.

Diluted anti-peptide antibodies.

3-(4, 5-Dimethylthiazol-2-yl)-5-(3-carboxymethoxyphenyl)-2-(4-sulfophenyl)-2H-tetrazolium (MTS).

3 Methods

3.1 Bioinformatics Sequence Approach to Identify B-Cell Epitopes

The Picornavirus capsid protein (VP1-VP4) region was computationally analyzed for hydrophobicity, solvent accessibility, surface accessibility of residues, polarity, and spatial distance orientation relationships. The sequences were obtained from the NCBI Genbank and scored for aforementioned key antigenic attributes against the BLAST query algorithm [31]. Alignment of protein regions was compiled against multiple publicly available database sets and sorted via the clustalW alignment program [32]. Conserved sequences demonstrating homology within the protein data bank listings PDB ID: 3VBS Human Enterovirus 71; PDB ID: 1 BEV Bovine Enterovirus and PDB ID: 1HXS Mahoney Poliovirus were used to construct and verify the model.

3.1.1 Bioinformatics Structure Approach

Sequence alignment models demonstrating >40 % structural conservation with the PDB were used to generate a three-dimensional structural model for the HEV-71 (VP1 and VP3) assemblies, using the Chimera [33] interface to MODELLER [34]. The crystallographic atomic coordinates were reconstructed and uploaded to the epitopia server to estimate the rate of amino acid substitutions at each position in the alignment of homologous proteins [28].

3.2 Assessment of Immunogenicity of Peptide Vaccine Candidates

Four peptide vaccine constructs containing the universal T helper epitope from Polio virus were assessed for their ability to induce anti-peptide antibodies in mice. The vaccine constructs were designated as shown below:

K1 (KLFAVWKITYKDTLMRMKHVRAWIPRPMR).

K2 (KLFAVWKITYKDTLFHPTPCIHIPGEVRN).

K3 (KLFAVWKITYKDTLGFPTELKPGTNQFLT).

D1 (KLFAVWKITYKDTLYPTFGEHKQEKDLEYC).

Groups of six male BALB/c (6–8 weeks old) were inoculated subcutaneously at the base of the tail with each of the four peptides. Animals received 20 μg of peptide at days 0 and 28. The primary inoculation was administered with complete Freund's Adjuvant (CFA) and the subsequent inoculation with the incomplete Freund's Adjuvant (IFA). Sera were obtained from the animals 10 days after each vaccination.

3.3 Enzyme-Linked Immunosorbent Assay

The presence of anti-peptide antibody in mouse sera was detected by enzyme-linked immunosorbent assay (ELISA). Briefly, flat-bottomed polyvinyl microtiter 96-well plates were coated overnight with 10 μg/ml of B-cell peptide construct in 50 μl 50 mM sodium carbonate buffer at pH 9.6 and incubated at room temperature overnight. Unbound antigen was removed and unoccupied plastic surface blocked with 10 mg/ml bovine serum albumin (BSA) in phosphate-buffered saline (PBS) (20 mM Na_2HPO_4 in 0.15 M NaCl, pH 7.3) at 37 °C for 1 h. After blocking, plates were incubated with serially diluted serum from individual mice starting at a 1/100 dilution in PBS containing 0.05 % Tween-20 and 0.5 % BSA at 37 °C for 1 h, and then washed two times with PBS containing 0.05 % Tween 20. One hundred microliters of a 1/2000 dilution of horseradish peroxidase-conjugated rabbit anti-mouse antibody was added to each well and allowed to incubate at room temperature for 1 h. Unbound antibody was removed, and wells washed twice with PBST and rinsed with PBS. Bound antibody was detected by the addition of 100 μl of 3,3′,5,5′-tetramethylbenzidine (TMB) substrate solution. The reaction was stopped after 20 min by the addition of a 1 M sulfuric acid. Color change was detected by an iMark Microplate Reader at a wavelength of 450 nm. The assays were performed in triplicate. Antibody titers were expressed as the reciprocal of the highest analyte dilution that gives a reading above endpoint cutoff [35].

3.4 Neutralization Assay

Neutralizing activity of IgG purified from pooled sera from the immunized mice was measured using an in vitro microneutralization assay in 96-well plates (Imunoblot HB, USA). Purified IgG, (10 μg/ml) at six twofold serial dilutions was pre-incubated with an equal volume of $200TCID_{50}$ of each virus (HEV-71, CVB4, CVB6, CA16, and Echo13) and then used to infect 80–100 % confluent Vero (CCL-81) cells. The cells were washed and then incubated at 37 °C with 5 % CO_2 until the first sign of complete CPE was observed in the virus control wells. At this point, 25 ul of 3-(4,5-dimethylthiazol2-yl)-5-(3-carboxymethoxyphenyl)-2-(4-sulfophenyl)-2H-tetrazolium (MTS) one-shot solution was added to all wells and an absorbance was recorded at 490 nm. The monoclonal antibodies against EV71, CA16, and CVB4 were included as positive controls. Results were read as positive if there was more than a 50 % CPE reduction as compared to naïve controls.

3.5 Statistical Analyses

Confidence interval. Statistical analyses were performed using a one-way parametric ANOVA test with 95 % confidence interval.

References

1. http://www.wpro.who.int/en/.
2. Li YP, Liang ZL, Xia JL, Wu JY, Wang L, Song LF, Mao QY, Wen SQ, Huang RG, Hu YS, Yao X, Miao X, Wu X, Li RC, Wang JZ, Yin WD (2014) A booster dose of an inactivated enterovirus 71 vaccine in Chinese young children: a randomized, double-blind, placebo-controlled clinical trial. J Infect Dis 209:46–55
3. http://www.who.int/biologicals/publications/trs/areas/vaccines/peptide/WHO_TRS_889_A1.pdf.
4. Hijnen M, van Zoelen DJ, van Chamorro C, Gageldonk P, Mooi FR, Berbers G, Liskamp RMJ (2007) A novel strategy to mimic protective epitopes using a synthetic scaffold. Vaccine 25:6807–6817
5. Li W, Joshi MD, Singhania S, Ramsey KH, Murthy AK (2014) Peptide vaccine: progress and challenge. Vaccine 2:515–536
6. Purcell AW, McCluskey J, Rossjohn J (2007) More than one reason to rethink the use of peptides in vaccine design. Nat Rev Drug Discov 6:404–414
7. Thornton JM, Edwards MS, Taylor WR, Barlow DJ (1986) Location of "continuous" antigenic determinants in the protruding regions of proteins. EMBO J 5:409–413
8. Shinnick TM, Sutcliffe JG, Green N, Lerner RA (1983) Synthetic peptide immunogens as vaccines. Ann Rev Microbiol 37:425–446
9. Corti D, Voss J, Gamblin SJ, Codoni G, Macagno A, Jarrossay D, Vachieri SG, Pinna D, Minola A, Vanzetta F et al (2011) A neutralizing antibody selected from plasma cells that binds to group 1 and group2 influenza A hemagglutinins. Science 333:850–856
10. Hager-Braun C, Tomer KB (2005) Determination of protein-derived epitopes by mass spectrometry. Expert Rev Proteomics 2:745–756
11. Suckau D, Kohl J, Karwath G, Schneider K, Casaretto M, Bitter-Suermann D, Przybylski M (1990) Molecular epitope identification by limited proteolysis of an immobilized antigen-antibody complex and mass spectrometric peptide mapping. Proc Natl Acad Sci 87:9848–9852
12. Zhao Y, Muir TW, Kent SB, Tischer E, Scardina JM, Chait BT (1996) Mapping protein-protein interactions by affinity-directed mass spectrometry. Proc Natl Acad Sci 93:4020–4024
13. Olenina LV, Nikolaeva LI, Sobolev BN, Blokhina NP, Archakov AL, Kolesanova EF (2002) Mapping and characterization of B cell

linear epitopes in the conservative regions of hepatitis C virus envelope glycoproteins. J Viral Hepat 9:174–182

14. Wu CN, Lin YC, Fann C, Liao NS, Shih SR, Ho MS (2001) Protection against enterovirus 71 infection in newborn mice by passive immunization with subunit VP1 vaccines and inactivated virus. Vaccine 20:895–904
15. Meloen RH, Barteling SJ (1986) Epitope mapping of the outer structural protein VP1 of three different serotypes of FMDV. Virology 149:55–63
16. Foo DGW, Alonso S, Phoon MC, Rama chandran NP, Chow VTK, Poh CL (2007) Identification of neutralizing linear epitopes from the VP1 capsid protein of enterovirus 71 using synthetic peptides. Virus Res 125:61–68
17. Foo DGW, Alonso S, Chow VTK, Poh CL (2007) Passive protection against lethal enterovirus 71 infection in newborn mice by neutralizing antibodies elicited by a synthetic peptide. Microbes Infect 9:1299–1306
18. Hopp TP, Woods KR (1981) Prediction of protein antigenic determinants from amino acid sequences. Proc Natl Acad Sci 78:3824–3828
19. Larsen JE, Lund O, Nielsen M (2006) Improved method for predicting linear B-cell epitopes. Immunome Res 2:2–9
20. Saha S, Raghava G (2006) Prediction of continuous B-cell epitopes in an antigen using recurrent neural network. Proteins 65:40–48
21. Söllner J, Meyer B (2006) Machine learning approaches for prediction of linear B-cell epitopes on proteins. J Mol Recognit 19:200–208
22. Lin SYH, Cheng CW, Su ECY (2013) Prediction of B-cell epitopes using evolutionary information and propensity scales. BMC Bioinformatics 14(Suppl 2):S10
23. Kulkarni-Kale U, Bhosle S, Kolaskar AS (2005) CEP: a conformational epitope prediction server. Nucleic Acids Res 33:W168–W171
24. Andersen PH, Nielsen M, Lund O (2006) Prediction of residues in discontinuous B-cell epitopes using protein 3D structures. Protein Sci 15:2558–2567
25. Ponomarenko J, Bui HH, Li W, Fusseder N, Bourne PE, Sette A, Peters B (2008) ElliPro: a new structure-based tool for the prediction of antibody epitopes. BMC Bioinformatics 9: 514–521
26. Rubinstein ND, Mayrose I, Martz E, Pupko T (2009) Epitopia: a web-server for predicting B-cell epitopes. BMC Bioinformatics 10: 287–292
27. Hancock DC, O'Reilly NJ (2005) Synthetic peptides as antigens for antibody production. Methods Mol Biol 295. In: Burns R (ed) Immunochemical protocols, 3rd edn. Humana Press Inc., Totowa, NJ, USA, pp 13–25
28. Lee BS, Huang JS, Jayathilaka GDLP, Lateef SS and Gupta S (2010) Production of antipeptide antibodies. Methods Mol Biol 657. In: Schwartzbach SD, Osafune T (eds) Immuno electron microscopy, Humana Press Inc., Totowa, NJ, USA, pp 93–108
29. Trier NH, Hansen PR, Houen G (2012) Production and characterization of peptide antibodies. Methods 56:136–144
30. Kirk K, Poh CL, Fecondo J, Pourianfer H, Shaw J, Grollo L (2012) Cross-reactive neutralizing antibody epitopes against EV71 identified by an in silico approach. Vaccine 30: 7105–7110
31. Altschul SF, Madden TL, Schäffer AA, Zhang ZH, Zhang Z, Miller W, Lipman DJ (1997) Gapped Blast and PS-Blast: a new generation of protein database searchprograms. Nucleic Acids Res 25:3389–3402
32. Chenna R, Sugawara H, Koike T, Lopez R, Gibson TJ, Higgins DG, Thompson JD (2003) Multiple sequence alignment with the clustal series of programs. Nucleic Acids Res 31: 3497–3500
33. Pettersen EF, Goddard TD, Huang CC, Couch GS, Greenblatt DM, Meng EC, Ferrin TE (2004) UCSF Chimera – a visualization system for exploratory research and analysis. J Comput Chem 25:1605–1612
34. Eswar N, Webb B, Marti-Renom MA, Madhusudhan MS, Eramian D, Shen MY, Pieper U, Sali A (2007) Comparative protein structure modelling using MODELLER. Curr Protoc Protein Sci Chapter 2:Unit 2.9
35. Voller A, Bartlett A, Bidwell DE, Clark MF, Adams AN (1976) The detection of viruses by enzyme-linked immunosorbent assay (ELISA). J Gen Virol 33:165–167

Chapter 30

Therapeutic HIV Peptide Vaccine

Anders Fomsgaard

Abstract

Therapeutic vaccines aim to control chronic HIV infection and eliminate the need for lifelong antiretroviral therapy (ART). Therapeutic HIV vaccine is being pursued as part of a functional cure for HIV/AIDS. We have outlined a basic protocol for inducing new T cell immunity during chronic HIV-1 infection directed to subdominant conserved HIV-1 epitopes restricted to frequent HLA supertypes. The rationale for selecting HIV peptides and adjuvants are provided. Peptide subunit vaccines are regarded as safe due to the simplicity, quality, purity, and low toxicity. The caveat is reduced immunogenicity and hence adjuvants are included to enhance and direct the immune response. Although the vaccine has been tested in ART naïve individuals, we recommend future testing of the vaccine during (early started) ART that improves immune function and to select individuals likely to benefit. Peptides representing other epitopes may be used.

Key words HIV-1, Therapeutic vaccine, Functional cure, Peptide, Cytotoxic T lymphocytes, Adjuvant, CD8 T cells, Human trial

1 Introduction

Today about 33 million people are living with HIV-1. Untreated the infection leads to breakdown of the immune system, opportunistic infections and death. The purpose of a therapeutic HIV vaccine or immunotherapy is to eliminate already infected cells including the reservoirs (sterile cure) or enough cells to obtain a new balance between a limited virus replication and a new immunity. This should prevent disease, spreading of HIV, and eliminate the need for lifelong anti-retroviral therapy (ART) as part of a "functional cure" [1, 2]. Although ART may control most HIV replication some replication and chronic inflammation do persist, leading to chronic inflammation, and it is not a cure. However, ART lowers the circulating viral antigen-burden and its immune interference. Although the specific immunity decline during ART, the number of uninfected $CD4^+$ T cells increases and the immune functions improves after a period [3]. Thus, ART may be necessary for obtaining sufficient functional immune induction by

Gunnar Houen (ed.), *Peptide Antibodies: Methods and Protocols*, Methods in Molecular Biology, vol. 1348,
DOI 10.1007/978-1-4939-2999-3_30,

a therapeutic vaccination. Whereas broadly neutralizing antibodies should limit the numbers of infecting virus, the $CD8^+$ cytotoxic T-lymphocytes (CTL) and antibody-dependent cell cytotoxicity (ADCC) should eliminate already infected cells and prevent disease. The natural immunity rarely controls HIV replication. This is in part due to the high viral diversity and the immune focus to dominant sites that can mutate (escape mutations). Thus, a therapeutic vaccine should target multiple novel sites not targeted during the chronic infection that are also conserved. We envision identification and targeting of immune subdominant CTL epitopes that tend to be conserved [4, 5]. This can be done by immunization with multiple rationally selected minimal epitope peptides (8–9 amino acids in length) [4]. Such subunit vaccines are regarded as safe due to the simplicity, quality, purity, and low toxicity. The caveat is reduced immunogenicity and hence adjuvants are included to enhance and direct the immune response [6].

2 Materials

2.1 Selection of Peptides

1. To meet the high diversity of HIV-1 and the HLA tissue types and to focus a cellular immunity to multiple conserved CTL epitopes not (or infrequently) targeted during the HIV infection we have used artificial neural network computers to predict epitopes binding to the most common HLA supertypes [4, 6–8]. Alternatively one can predict epitopes restricted to most common HLA types in a geographic area, e.g., Denmark [9]; however, the selected HLA supertypes were already frequent within Denmark and Guinea-Bissau [4]. Peptide binding to the most common HLA supertypes HLA-A01, A02, A03, B07, and B44 was experimentally confirmed [4]. In fact, an even more promiscuous binding to several other HLA alleles was observed. Infrequent targeting (subdominance) during HIV infection and in vivo processing was confirmed [4]. Selection criteria include then subdominant epitopes defined as binding to MHC-I at 100 nM < Kd < 500 nM and infrequently targeted (e.g., <3/33) by HIV-1 infected ART-naïve individuals plus high conservation (>80 %) among the HIV-1 subtypes and strains. We preferentially selected epitopes within the most target-relevant HIV-1 proteins Gag, Pol, and Nef [4]. These epitopes together will cover 90–100 % of any population. This was confirmed by testing in Denmark and Guinea-Bissau justifying testing in very different populations [4].
2. 15 HIV-1 CD8 minimal CTL epitope peptides were selected (Table 1). The median number of these 15 epitopes simultaneously present per full-length HIV-1 genome is similar in West African and Danish isolates, 10.0 and 11.5, respectively

Table 1
CD8 and CD4 T cell epitopes targeted by the AFO-18 therapeutic HIV vaccine

Peptide (name)	Epitope (name)	HIV target epitope (amino acids)	MHC-anchor-optimized vaccine epitope	HLA-supertype restriction
Pol-934	KY9	KIQNFRVYY	–	A01
Gag-150	RV9	RTLNAWVKV	RLLNAWVKV	A02
Gag-433	FS8	FLGKIWPS	–	A02
Pol-606	KT9	KLGKAGYVT	KLGKAGYVV	A02
Vpu-66	AA9	ALVEMGHHA	ALVEMGHHV	A02
Env-67	NV9	NVWATHACV	NIWATHACV	A02
Vif-23	SI9	SLVKHHMYI	SLVKHHMYV	A02
Vif-101	GM9	GLADQLIHM	GLADQLIHL	A02
Pol-973	KK9	KVVPRRKAK	–	A03
Pol-313	AK9	AIFQSSMTK	–	A03
Nef-73	QY9	QVPLRPMTY	–	B07
Gag-148	AV9	SPRTLNAWV	–	B07
Pol-311	SM9	SPAIFQSSM	–	B07
Pol-592	AA9	AETFYVDGA	–	B44
Nef-107	QY9	QEILDLWVY	–	B44
Gag-298	KY15	KRWIILGLNKIVRMY	–	–
Env-570	VD20	VWGIKQLQARVLAVERYLKD	–	–
PADRE	–	aKXVAAWTLKAAa (X = cyclohexylalanine, a = d-alanine)	–	–

[10]. The vaccine can simultaneously target multiple sites and thereby make viral escape difficult or with the consequence of lower viral fitness. Six of the 7 HLA-A2 epitopes with lower immunogenicity were improved by anchor-amino-acid substitutions to stronger MHC-I-binding amino acids (position 2 and C-terminal amino-acids) (Table 1) [7] which induced enhanced immunity in a clinical trial [11]. To further facilitate induction of new CD8+ T cell immunity three promiscuous CD4+ T-helper cell epitopes were included [7, 11] (Table 1).

3. Several of the chosen epitopes are likely to impact on the viral fitness. Several of these epitopes were included in a study by Matthews et al. [12] where the HLA-associated escape was evaluated in a cohort of 710 individuals infected with C-clade HIV-1. The Pol-311 (SM9) epitope restricted to HLA-B*07:02

and the 8-mer VLPRPMTY restricted to HLA-B*35:01 and contained within Nef-73 (QY9) was associated with viral escape. The latter were described as reverting; that is, the wild type is favored when the HLA selection pressure is removed, suggesting that the mutation has a negative impact on infectivity and/or replicative capacity. In addition, the HLA-supertype epitopes A01-Pol-934 (KY9), A03-Pol-973 (KK9), and A03-Pol-313 (AK9) restricted by HLA-A*30:02, HLA-A*30:01, and HLA-A*30:01, respectively, are associated with HIV-1 escape in the same cohort [4, 12].

4. The peptides are GMP manufactured in sufficient amounts (e.g., >500 mg) and >95 % purity (by HPLC), QA by mass spectrometry analysis, amino acid analysis, tested for peptide and water content, residual solvents (TFA, DMF, MeCN) after freeze-drying, and with satisfactory low endotoxin and bioburden [6].
5. Vaccine preclinical testing according to EMA guidelines for repeated dose toxicity, preclinical pharmacological and toxicology testing, e.g., on Göttingen mini-pigs, including short (<24 h) and long-term stability testing of all individual peptides in the mix, and of the CAF adjuvant, and of the final mix AFO-18 [6].

2.2 Selection of Adjuvants

1. The CAF01 adjuvant is a synthetic two-component liposomic adjuvant comprising the quaternary ammonium dimethyl-dioctadecyl-ammonium (DDA) and the C-type lectin-receptor ligand trehalose 6,6′-dibehenate (TDB), an analogue of mycobacterial cord factor trehalose 6,6′-dimycolate. CAF01 induces a balanced Th1/antibody/CTL response in several animal models and has an acceptable safety profile in humans. The components DDA and TDB were synthetically manufactured and the CAF01 adjuvant was manufactured as described [6].
2. Alternatively, the more CD8 T cell-inducing CAF09 can be used consisting of the same DDA liposomes but stabilized with monomycoloyl glycerol (MMG)-1 and combined with the TLR3 ligand, poly(I:C) (a synthetic dsRNA) [13].

2.3 Selection of Vaccine Recipients

1. Criteria for receiving the vaccine in a clinical trial can be as follows (*see* **Note 1**):

   Male or female, age 18–40 years; ART for more than 6 months with viral load <100 copies/ml and CD4 T cell count >400 cells/μl blood; blood biochemistry and hematology values within normal range; not receiving other experimental or immune-modulating drugs within the last 3 months; no autoimmune or other active chronic diseases.
2. To identify persons more likely to benefit from the vaccination one can select individuals likely to have lower HIV reservoir, e.g. early or acute started ART, low inflammation markers, and low inhibitory markers, e.g., low Treg and/or PD-1.

3. It is important to select enough participants to demonstrate the effect in clinical trials; for example obtain a 90 % likeliness to detect a 50 % reduction in viral reservoirs due to the vaccination and to make frequent follow-ups for years.
4. To facilitate evaluation of effect parameters the nadir CD4, HLA type with four digits (to rule out protective alleles or monitor HLA-specific vaccine responses), HIV RNA sequences (to look for the presence of vaccine target sequences), and viral set point before ART should be known before vaccination.
5. It is important to inform the participants also of terms that appear in the study and explain the risk/benefit, e.g., of analytical treatment interruption, ATI, in details *See* also **Note 2**.

2.4 Selection of Evaluation Parameters

1. To monitor the effect one can apply a supersensitive RT-Q-PCR [14] to measure lowering of residual viral load in the range of 1–100 copies/ml blood. No standardized tests are available for precise measuring the viral-DNA reservoir [15], e.g., in tissue, but as surrogate one could measure cell associated HIV RNA transcripts, HIV-1 DNA copies per 10e5 PBMC or $CD4^+$ resting memory T cells, integrated versus non-integrated HIV DNA, and HIV DNA in gut biopsies (gut reservoir).
2. IFNγ ELISPOT and multicolor Flow-cytometry measuring of the vaccine-induced HIV-specific CD8 and CD4 T cell immunity (e.g., CD107a expression (marker of cytotoxicity) and intracellular IFNγ, TNFα, IL-2, MIP1β cytokine profile after peptide stimulation) [16].
3. It is recommended to include also functional tests such as viral inhibition assay (VIA) of CD8 CTL [17] and $CD8^+$ T cell proliferation. An elimination of residual HIV replication by the vaccine would also lower the degree of any systemic inflammation as measured by decrease in inflammatory markers (e.g., sC14, IL-13, IL-21, IL-2, IFNγ).

3 Methods

1. The lyophilized mix of 18 peptides is stored at −20 °C. Just before use the vial is thawed. CAF adjuvant solution is added as solvents. The adjuvanted ready-to-use vaccine is named AFO18.
2. 1.25 ml AFO18 containing 0.25 mg of each peptide in CAF01 corresponding to 625/125 μg DDA/TDB per dose is injected i.m. on weeks 0, 2, 4, and 8, with the possibility of further boostings if necessary. A placebo group may be injected with saline only as it may be unethical to use adjuvant alone that has no potential benefit but only possible adverse reactions.

3. Ultimate test of antiviral effect and benefit will be analytical treatment interruption (ATI) if functional antiviral immune tests indicate antiviral effect. Although ATI is risky, a short (e.g., 16 weeks) ATI in patients with stable suppressive ART and CD4 > 400, a CD4 nadir of >200, and without concomitant disease is considered safe [1, 2]. Such ATI is acceptable in the context of a well-monitored therapeutic vaccine trial with timely resumption of ART in case of defined rebound and/or CD4 drop.

4 Notes

1. This method has been used in HIV-1 infected individuals not in ART from Denmark [16] and West Africa [18] to demonstrate safety and the possibility to induce a new immunity even in untreated individuals with already high viral load. Due to already impairment of immune capability in untreated HIV-1-positive individuals, even with acceptable CD4 count >400, this protocol should be tested during (early started) ART as described here).
2. Probably combination therapies will be considered to further optimize an antiviral effect. This could include pharmaceutical induced reactivation of latent HIV reservoirs, e.g., by HDAC inhibitors, use of anti-PD-1 antibodies at vaccinations [1], and/or combination with another vaccine component inducing broadly neutralizing antibodies and broad ADCC.

Acknowledgements

This work was supported by grants from DANIDA, EDCTP, AIDS Fondet, Danish Medical Research council. We thank all the scientists and technicians from the Virus R&D Laboratory, Statens Serum Institut, Copenhagen, Denmark; the medical staff and participants at the university hospitals of Copenhagen, Hvidovre, Odense; and Hospital Nacional Simão Mendes and The Bandim Health Project, Bissau, Guinea-Bissau.

References

1. Vanham G, van Gulck E (2012) Can immunotherapy be useful as a "functional cure" for infection with human immunodeficiency virus-1? Retrovirol 9:72–93
2. Pantaleo G, Levy Y (2013) Vaccine and immunotherapeutic interventions. Curr Opin HIV AIDS 8:236–242
3. Jensen SS, Hartling HJ, Tingstedt JL et al (2014) HIV-specific ADCC improves after antiretroviral therapy and correlates with normalization of the NK cell phenotype. J AIDS. doi:10.1097/QAI.429
4. Karlsson I, Kløverpris H, Jensen KJ et al (2012) Identification of conserved subdominant HIV

type 1 CD8+ T cell epitopes restricted within common HLA supertypes for therapeutic HIV type 1 vaccines. AIDS Res Hum Retroviruses 11:1434–1443

5. Frahm N, Kiepiela P, Adams S et al (2006) Control of human immunodeficiency virus replication by cytotoxic T lymphocytes targeting subdominant epitopes. Nat Immunol 7:173–178
6. Fomsgaard A, Karlsson I, Gram G et al (2011) Development and preclinical safety evaluation of a new therapeutic HIV-1 vaccine based on 18T-cell minimal epitope peptides applying a novel cationic adjuvant CAF01. Vaccine 29:7067–7074
7. Corbet S, Nielsen HV, Vinner L et al (2003) Optimization and immune recognition of multiple novel conserved HLA-A2, human immunodeficiency virus type 1-specific CTL epitopes. J Gen Virol 84:2409–2421
8. Sidney J, Peters B, Frahm N et al (2008) HLA class I supertypes: a revised and updated classification. BMC Immunol 9, doi: 10.1186/1471-2172-9-1
9. Schubert B, Lund O, Nielsen M (2014) Evaluation of peptide selection approaches for epitope-based vaccine design. Tissue Antigens 82:243–251
10. Vinner L, Holmgren B, Jensen KJ et al (2011) Sequence analysis of HIV-1 isolates from Guinea-Bissau: selection of vaccine epitopes relevant in both West African and European countries. APMIS 119:487–497
11. Kloverpris H, Karlsson I, Bonde J et al (2009) Induction of novel CD8+ T-cell responses during chronic untreated HIV-1 infection by immunization with subdominant cytotoxic T-lymphocyte epitopes. AIDS 23:1329–1340
12. Matthews PC, Prendergast A, Leslie A et al (2008) Central role of reverting mutations in HLA associations with human immunodeficiency virus set point. J Virol 82:8548–8559
13. Korsholm KS, Hansen J, Karlsen K et al (2014) Induction of CD8+ T-cell response against subunit antigens by the novel cationic liposomal CAF09 adjuvant. Vaccine 32:3927–3935
14. Mens H, Kearney M, Wiegand A et al (2011) Amplifying and quantifying HIV-1 RNA in HIV infected individuals with viral loads below the limit of detection by standard clinical assays. J Vis Exp 55:e2960
15. Ho Y-C, Shan L, Hosmane NN et al (2013) Replication-competent non-induced proviruses in the latent reservoir increase barrier to HIV-1 cure. Cell 155:540–551
16. Karlsson I, Brandt L, Vinner L et al (2013) Adjuvanted HLA-supertype restricted subdominant peptides induce new T-cell immunity during untreated HIV-1-infection. Clin Immunol 146:120–130
17. Saez-Cirion A, Lacabaratz C, Lambotte P et al (2007) HIV controllers exhibit potent CD8 T cell capacity to suppress HIV infection ex vivo and peculiar cytotoxic T lymphocyte activation phenotype. Proc Natl Acad Sci U S A 104:6776–6781
18. Roman VRG, Jensen KJ, Jensen SS et al (2013) Therapeutic vaccination using cationic liposome-adjuvanted HIV type 1 peptides representing HLA-supertype-restricted subdominant T cell epitopes: safety, immunogenicity, and feasibility in Guinea-Bissau. AIDS Res Hum Retroviruses 29:1504–1512

INDEX

Gunnar Houen (ed.), *Peptide Antibodies: Methods and Protocols*, Methods in Molecular Biology, vol. 1348,
DOI 10.1007/978-1-4939-2999-3, © Springer Science+Business Media New York 2015

E

F

G

H

P

R

S

T

If you have any concerns about our products,
you can contact us on
ProductSafety@springernature.com

In case Publisher is established outside the EU,
the EU authorized representative is:
Springer Nature Customer Service Center GmbH
Europaplatz 3, 69115 Heidelberg, Germany

Printed by Libri Plureos GmbH
in Hamburg, Germany